The Soil

An introduction to soil study in Britain

The Soil

An introduction to soil study in Britain

F M Courtney M.Sc., Cert. Ed.

Lecturer in Geography, Manchester Polytechnic

S T Trudgill Ph.D.

Lecturer in Geography, University of Sheffield

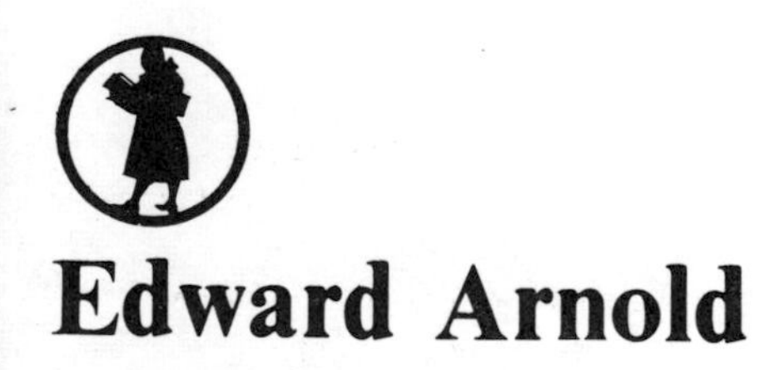

Edward Arnold

First published 1976
by Edward Arnold (Publishers) Ltd
25 Hill Street, London W1X 8LL

ISBN: 0 7131 1987 X

Printed in Great Britain by Butler & Tanner Ltd, Frome and London

Contents

Preface

With the large number of soil books that there are on the market some justification for the present volume seems necessary. After having taught various courses at school, polytechnic and university levels, the present writers concluded that there was no book which, whilst making no assumptions about previous knowledge, nevertheless managed to make soil study interesting and stimulating. (Whether we have succeeded in the latter only time, our readers and our reviewers will tell!)

This book is therefore intended for sixth-formers, students in colleges of education and first-year undergraduates in polytechnics and universities. It is also intended for teachers who may be specialists in other areas but relative newcomers to soil study. The book is written by two geographers and is therefore, to some extent, written to geographers. Nevertheless basic soil study is transdisciplinary and the book should also be useful to those studying agriculture, biology, ecology, environmental studies and geology.

The book concentrates on, and takes its examples from, British soils. The writers feel that for far too long there has been an undue emphasis on foreign soils in British teaching and examination syllabuses. While there are sound arguments to support studies of world soil problems this has often been at the expense of any serious understanding of local soils. It seems likely that students benefit from learning about soils which they themselves can study at first hand. So this book has no reference to chernozems and krasnozems – other than that which you have just read.

Although written in a sequential manner it is hoped that the use of cross-references will enable the book to be used as a basic reference text. Chapter one deals with the fundamental reactions involved in the development of soil. Chapter two examines the components of soil, their properties and the way in which these may combine to influence the overall soil characteristics. Then, in Chapter three, this knowledge is applied to a deeper understanding of individual soil types. Chapter four locates soil as a component in the ecosystem and deals with the circulation and flows of nutrients, energy and water. Chapter five applies the knowledge of soils so far discussed in a context of soil management, not only in an agricultural sense but also touching on the wider theme of soil as a resource. Chapter six takes a practical view of soil study, especially of the description and mapping of soils in the field. Some problems and schemes of soil classification and of soil research are discussed.

Soils are complex phenomena and it is therefore extremely difficult to treat them in a simple manner: some discussion of chemistry, geology, physics and biology has been necessary in order to be able to approach an understanding of many soil processes. Moreover, as is explained in Chapter six, soils vary both in their nature and in their scale of variation from place to place: this poses a serious problem in soil field studies. Because of this field work has often been either very complicated or ultra-conservative and merely descriptive. Classifications can be both unwieldy and difficult to grasp at first sight. Some attempt is made to attack these problems head-on in this book. This means that the authors have inevitably laid themselves open to criticisms of over-simplification. Although this is a risk, the authors have tried to give a sound background of basic principles and an awareness of some of the problems of soil study which will prepare the reader for more advanced soil texts. Simply then, we have tried to impart the basic ideas of components, interactions and processes in the soil.

The authors are indebted to a number of individuals. Special thanks are extended to Dr Leonard Curtis of the Department of Geography, University of Bristol, who first promoted our interest in soils and has subsequently provided friendly encouragement and help throughout our undergraduate and postgraduate careers. Other staff at Bristol also deserve special thanks, notably Professors Peel and Haggett for providing such an excellent atmosphere in which to learn and work and Dingle Smith whose enthusiasm for field work spurred us into rewarding activities when we might otherwise have stayed at home.

Both authors are grateful to students and staff on Field Studies Council courses at Slapton Ley Field

Centre (especially Bob Troake, Maggie Calloway and Tony Thomas) for helpful and lively criticisms. Also thanks to Dr Steve Reynolds of the South Pacific Regional College of Tropical Agriculture, Samoa, for helpful teaching ideas. Credit is due to the cartography staff of the Department of Geography, University of Strathclyde, who have painstakingly drawn the diagrams and to the typing staff for tackling the manuscript in such a competent manner.

FMC would like to thank his former colleagues in the Soil Survey of England and Wales (especially Ben Clayden, David Cope and Derek Findlay), pupils at Filton High School and students at Manchester Polytechnic, the latter for making appropriate remarks on earlier drafts. Dr Don Bayliss of Manchester Polytechnic is thanked for sustained encouragement.

STT thanks students and staff in the Department of Geography, University of Strathclyde (especially Professor Howe) for continued encouragement and also Dr Valerie Haynes, Dr Gareth Jones and Dr Keith Smith for fruitful discussion.

Finally we would both like to thank those friends without whom we would have given up long ago – Dr David Briggs, Dr Stephen Nortcliff and Darrel and Valerie Weyman. Not least, thanks to Catherine Courtney and to our parents for so much help.

Frank Courtney
Stephen Trudgill

1

Soil development

1.1 Rock and soil

Rock weathering

Soil develops when rock at the surface of the earth is changed by a series of processes, collectively known by the term *weathering*. The rock is weathered and broken down by the combined action of water, gases and living matter. The formation of soil is not just a matter of the disintegration of rock; while the rock is disintegrating it is exchanging material with its immediate environment. A true soil is therefore a rock which has exchanged some material with its environment and the soil now incorporates not only rock but also water, gases and both living and dead organic matter.

Rocks and equilibrium

When rock breaks down to form soil it is tending to come into *a state of equilibrium* with its environment. By using the phrase 'a state of equilibrium' it is meant that an object is adjusted to the external forces acting upon it.

For example, if you were to place a beaker of hot water in a cold room then the water would cool until it was the same temperature as the room. The water would at that stage be adjusted to its external environment and would be in a state of equilibrium with the temperature in the room. Similarly one litre of hot water mixed with one litre of cold water would mix to produce two litres of warm water. In this case both items have adjusted to each other, resulting in a compromise.

The principle of equilibrium can be stated in a more general way:

matter tends to change by the loss or gain of energy into a form where energy differences between the matter and its environment (or between two sets of matter) are minimized.

A piece of rock several thousand feet down in the earth's crust will be subject to great pressure from the weight of the overlying rock. Because of this pressure the rock is molten. This is an equilibrium adjustment to external forces. If the rock was nearer the surface of the earth it would have less pressure on it. It will be cooler and the rock will crystallize. This is also an equilibrium adjustment to external conditions.

Eventually, by earth movements or the erosion of the overlying rock, the piece of rock may find itself at the surface of the earth. The rock will have inherited characteristics from the place of its origin in the earth's crust, but it is now in a new environment at the surface. It is in *disequilibrium* with the new environment. That is, it is not in equilibrium with the conditions at the surface of the earth where cooler (and changeable) temperatures occur and where water, air and organisms are present. The rock has to adjust to the new conditions.

The adjustment can take many forms and may vary in the amount of time it takes, according to the nature of the surface conditions. In desert conditions the simple disintegration of rock occurs. The production of sand can be viewed as an equilibrium adjustment to harsh conditions. The rock cannot withstand the expansion and contraction caused by temperature changes, but the sand can expand and contract freely. Soil is not formed, however, because there is insufficient water and life to be incorporated into the disintegrated rock.

The material that disintegrates to produce the soil is called the *parent material*. It may be *igneous* rock, which had its source deep in the earth's crust as described above. It may be a *sedimentary* rock which has been formed from the deposition of previously weathered rock. Although these rocks have not been as deep in the earth's crust as igneous rocks and so have not become molten, they still inherit characteristics from the place where they were first deposited, for example in a fresh water or marine basin. As soon as they reach the surface of the earth they will start to alter, in response to the new conditions they have met. The third group of rocks are the *metamorphic* rocks, which are sedimentary rocks altered by heat or pressure. Also many soils in Britain are developed on *unconsolidated deposits* such as river alluvium or glacial drift. Again these deposits inherit characteristics from having been laid down under water or ice and when exposed at the surface begin to adjust to surface conditions and incorporate rain water, gases, organisms and organic matter to form soil.

The soil system

The approach adopted by systems analysis is extremely useful in the study of soil development. The object under study is termed the *system* and the workings of the system are divided into *inputs, outputs* and *internal processes.* Using this approach a soil can be studied in a similar way to a processing factory with raw material input, internal manufacturing processes and the output from the system (Figure 1).

In the case of the soil system, the system under study is the soil between the living plant above and the unaltered parent material below, whether this be igneous or sedimentary rock or an unconsolidated deposit. The inputs and outputs of a small area of soil (Figure 2) can be listed:

Inputs

1 Nutrients from decaying rock. (Nutrients are the chemicals used as plant food.)
2 Water from the atmosphere.
3 Gases from the atmosphere and the respiration of soil animals.
4 Organic matter from decaying vegetation and animals.
5 Excretions from plant roots.

Outputs

1 Nutrients taken up by plants.
2 Nutrient losses into water passing through the soil.
3 Losses of soil material by soil creep downslope.
4 Evaporation.

However, the soil system is not a simple matter of input and output, and under natural conditions *recycling* will occur. For example, the nutrients lost as an output (output 1) may well return in leaf litter the following autumn (input 4). In this way many nutrients are recycled through the system. Obviously if the vegetation is removed by a crop the nutrient store of the soil will gradually be depleted until fertilizers will have to be added to the soil to replace them.

Movement in the system

The movement of *water* in the soil governs most of the processes in the soil. It governs the removal of nutrients in drainage waters and the biological processes within the soil. It influences most of the *internal processes* of the soil, whether they be chemical, physical or biological.

Within soil, nutrients can be moved from their original positions, transported through the soil and deposited higher or lower in the profile (or vertical soil section). If the dominant water movement in the soil is downwards, as in regions of high rainfall and where the soil is very porous, the nutrients will be transported downwards in the profile. The loss of nutrients from the upper part of the soil downwards is referred to as *leaching*. The transport and deposition of various soil constituents to different depths in the soil leads to the formation of horizontal layers within the soil. These are termed soil *horizons* and can often be distinguished as differently coloured layers in a section dug through soil in a pit or roadside cutting. Further details of soil horizon distinctions are given in Chapter six and section 3.1.

1.2 Agents and processes of rock weathering and soil development

We have seen that soil is formed by the interaction of the soil parent material with its environment. But how does a parent material incorporate water, gases and organic material from the environment to form soil? This section looks at the processes which enable the equilibrium adjustments of parent material to take place. The soil-forming processes that result from the exposure of the material to water, air and life are considered in turn.

Processes associated with exposure to water

The most important process is that of dissolving. Associated processes are those of *hydrolysis* (the break down of minerals by water) and *hydration* (the incorporation of water into the mineral structure).

DISSOLVING

Materials *dissolve* into water to form a *solution* of the material. Dissolving (the verb to dissolve) is the process, and the solution (the noun) is the resultant product.

Before we can understand how solution processes work in the soil it is necessary to refer to a knowledge of basic chemistry. When a soluble material (the *solute*) comes into contact with water (the *solvent*) small particles of the solute move out from the solid into the water. This is the process

INTERNAL

PROCESSES

Figure 1 The basic systems approach

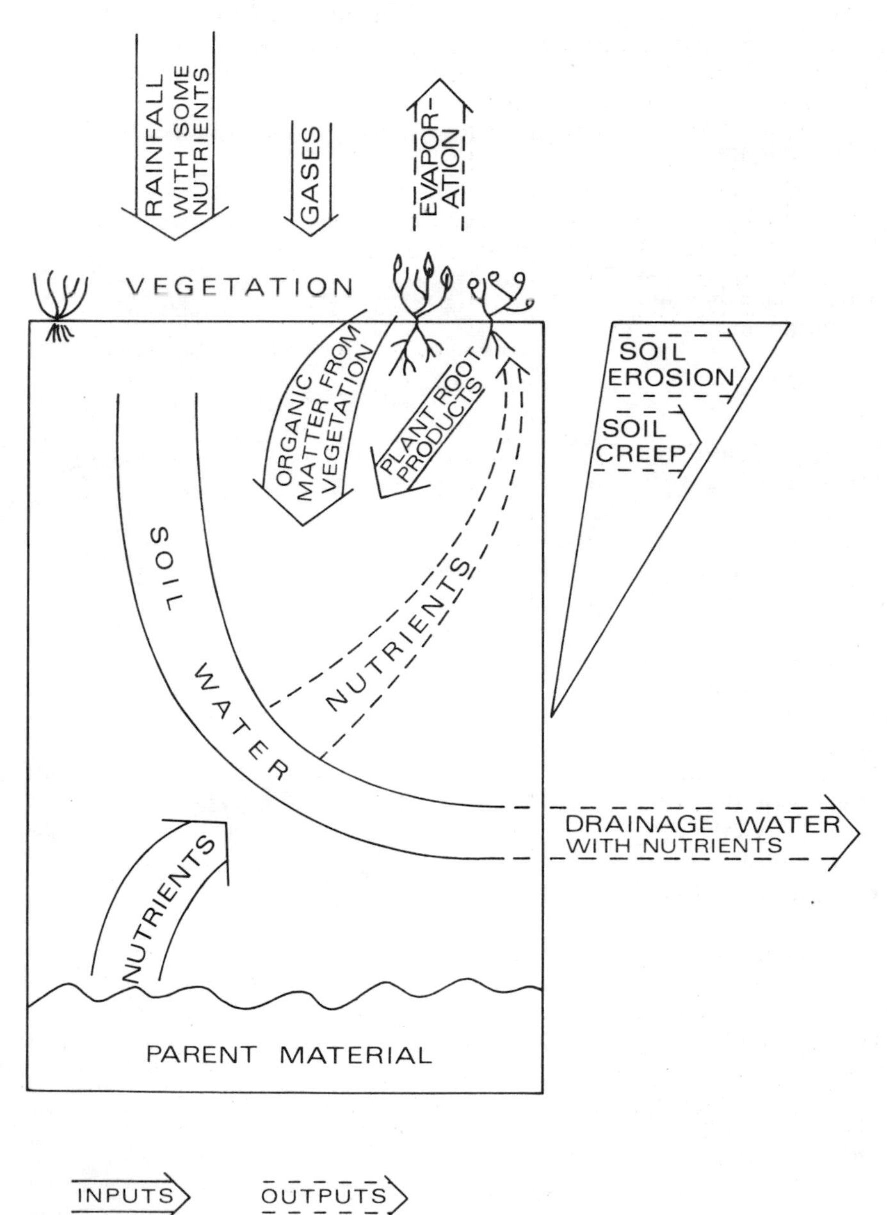

Figure 2 The major soil inputs and outputs

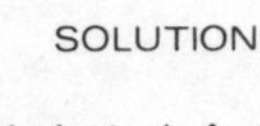
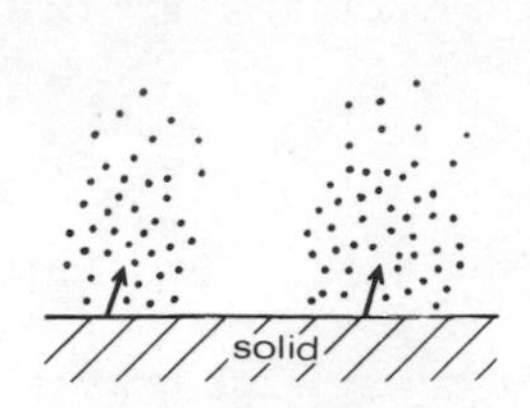
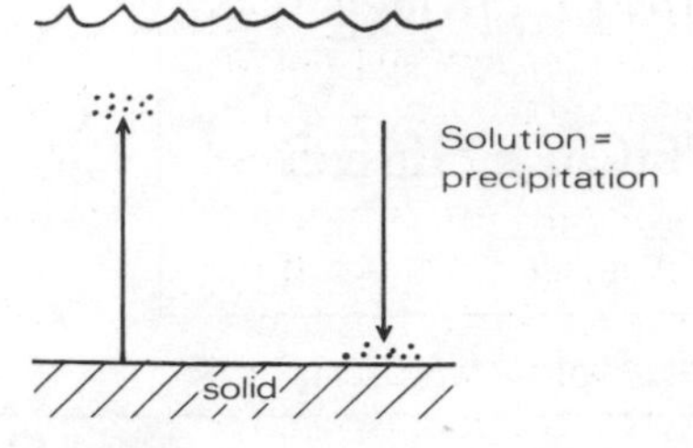

Figure 3 The process of dissolving

of simple *dissolving* (Figure 3). When many solutes dissolve they split up (or *dissociate*) into their constituent parts (though not all substances dissociate when dissolved and can exist as compound molecules in solution). These constituents are electrically charged and are referred to as *ions*. Common salt, for instance (sodium chloride), dissociates in water to yield separate sodium and chloride ions:

$$\underset{}{NaCl} \longrightarrow \underset{\text{cation}}{Na^+} \text{ and } \underset{\text{anion}}{Cl^-}.$$

Sodium chloride in bulk is electrically neutral but dissociates to give *positively charged* sodium *cations* and *negatively charged* chloride *anions.*

Solute ions can also move back from the water to the solid and this is the process of chemical *precipitation.* Obviously, while the solid is dissolving the net movement of ions will be from the solid to the water. But at a certain stage it will be found that as many ions are moving from the water to the solid as are moving in the opposite direction. In other words precipitation equals solution. A state of equilibrium has now been reached; this is called *saturation.* The concentration of solute ions in the water at the equilibrium saturation state defines the *solubility* of the solute. Figure 4 demonstrates the amounts of various chemical elements found in soil water.

This basic chemical knowledge of the solution process can now be applied to the soil-forming situation. The constituents of the soil parent material will each possess a different solubility. The more soluble constituents will be easily washed away by rainwater and the least soluble will remain as the framework or skeleton of the soil. The formation of soil horizons, mentioned in section 1.1, is influenced largely by the solubility of soil materials. The more soluble chemicals are carried further down the soil profile while the least soluble chemicals remain undissolved in the upper layers of the

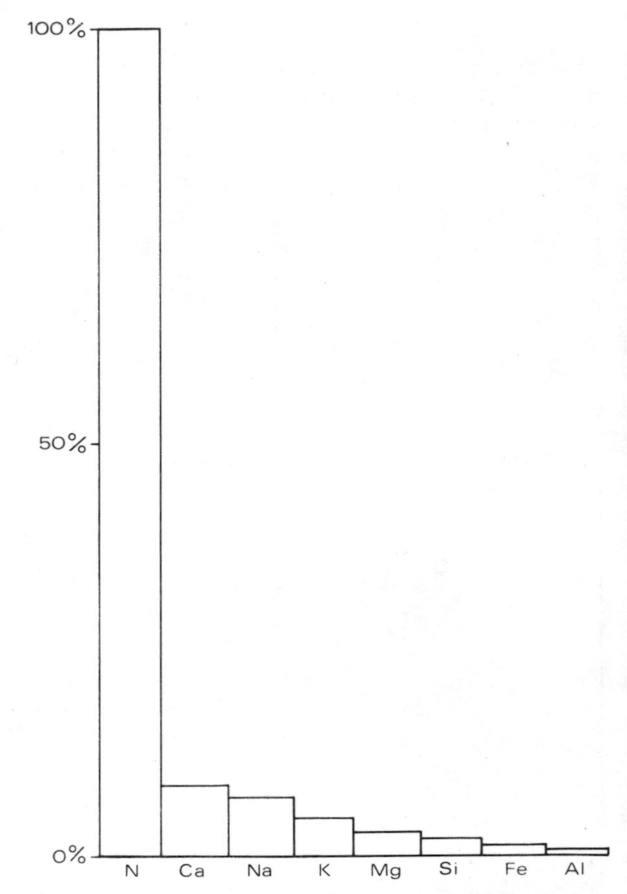

Figure 4 Relative amounts of chemical elements found dissolved in water from a mineral soil (expressed as a percentage of the amount of N). *Note:* Percentages are only approximate

soil. Rain water percolates through the upper layers of the soil, dissolving material slowly. Eventually the state of saturation will be reached and no further material can be dissolved. The materials are then redeposited, or precipitated, lower down the soil profile (Figure 5).

It can be seen that the process of solution is an important way in which the minerals of the parent material of the soil react to the presence of water. Therefore the solubility of the soil parent material and the amount of rainfall (solvent) are important factors in the formation of soil, and an insoluble material will not break down easily to form soil, however much rain falls on it. Soluble material on the other hand will be readily moved in wetter areas, the degree of movement down the soil profile depending upon the amount of rainfall input to the soil.

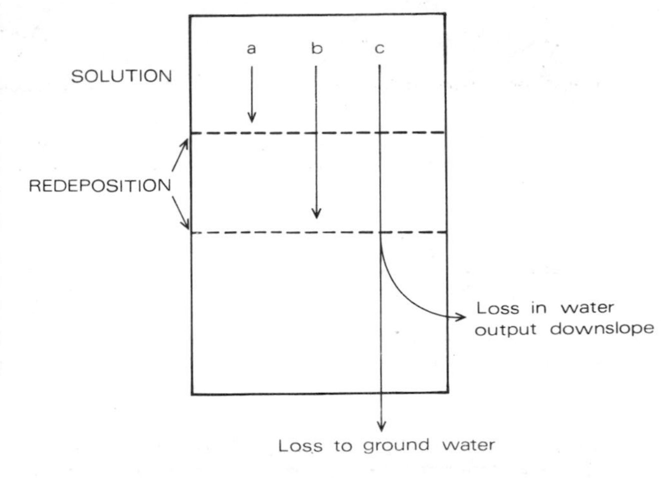

a. Relatively insoluble elements

b. Moderately soluble elements

c. Very soluble elements

Figure 5 Solution and redeposition in the soil profile

HYDROLYSIS

Hydrolysis is the breaking down of minerals (in the parent material) by hydrogen ions and hydroxyl ions derived from water. Thus the mineral *combines* with the water rather than simply dissolving in it by dissociation. Some water in the soil exists in a partly dissociated state, that is the H_2O is already partly split up into H^+ and OH^- ions and the hydrogen ion (H^+) is particularly important in the attack of minerals.

In solutions of pure chemicals in the laboratory both H^+ and OH^- ions are involved in hydrolysis. For cations (M^+) the equation is:

$$M^+ + H_2O \rightarrow MOH + H^+.$$

For anions (X^-) the reaction is:

$$X^- + H_2O \rightarrow HX + OH^-.$$

Hydrolysis is thus the reaction of a solid ion with water to form an associated ion species plus H^+ or OH^-.

In natural situations complex minerals exist and hydrolysis is not as simple as in the above equations. Often both cation hydrolysis and anion hydrolysis occur together. Frequently cations in a mineral combine with OH^- ions from water and then the cations are replaced by H^+ ions later. An example of a complex reaction is that of microcline feldspar reacting with water.

$$\underset{\text{microcline feldspar}}{KAlSi_3O_8} + \underset{\text{dissociated water}}{H^+ + OH^-} \rightarrow \underset{\text{hydrolysed mineral}}{HAlSi_3O_8} + \underset{\text{potassium hydroxide}}{KOH}.$$

The hydrolysed mineral containing the H^+ ion is unstable and usually breaks down.

Minerals composed of weakly ionised cations combine with OH^- of water more than with H^+ ions. Minerals composed of weakly ionised anions, such as the silica-rich minerals, take up H^+ ions from the water more than the OH^- ions.

An important factor in natural processes is that pure water rarely exists because it is usually dominated by the hydrogen ion, H^+. These are derived from organic acids from decaying humus and the dissociation of carbonic acid in water:

$$CO_2 + H_2O \rightarrow \underset{\text{carbonic acid}}{H_2CO_3} \rightarrow H^+ + HCO_3^-.$$

Plant roots and exchangeable hydrogen ions on acid clays also supply hydrogen ions. Thus in soil water containing much organic matter and many plant roots and acid clays the hydrogen ion tends to be the most important factor, leading to the dominance of mineral anion hydrolysis.

Many minerals that make up igneous and metamorphic rocks are rich in silica (SiO_2). The atoms in many silicate minerals are arranged in pyramids (called silica *tetrahedra*). These tetrahedra are held together by other atoms, especially calcium (Ca) or magnesium (Mg). The hydroxyl and hydrogen ions from water attack these atoms that link the tetrahedra and replace them with hydrogen ions. The silica tetrahedra which are linked by hydrogen ions are unstable and soon break apart. Thus the silicate minerals break down under attack from water which renders the minerals unstable (Figure 6).

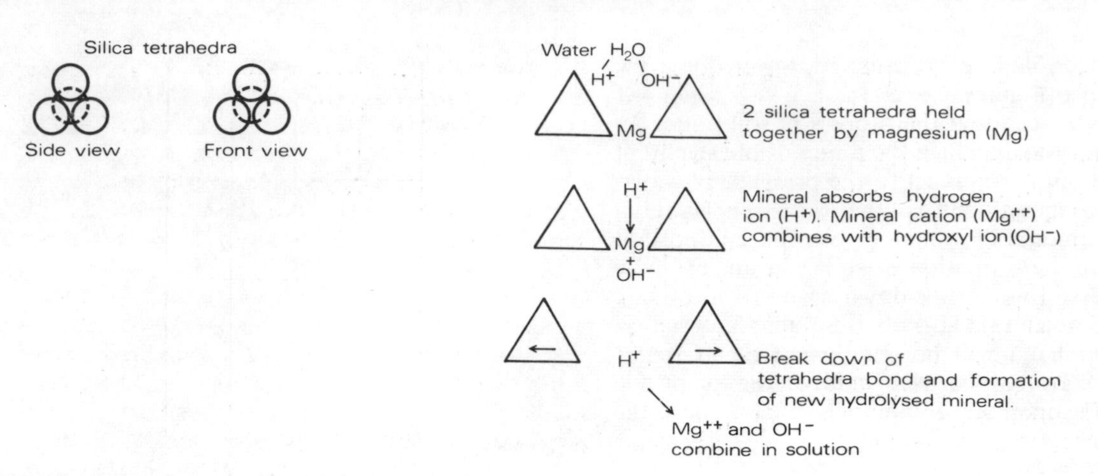

Figure 6 Hydrolysis of silicate minerals

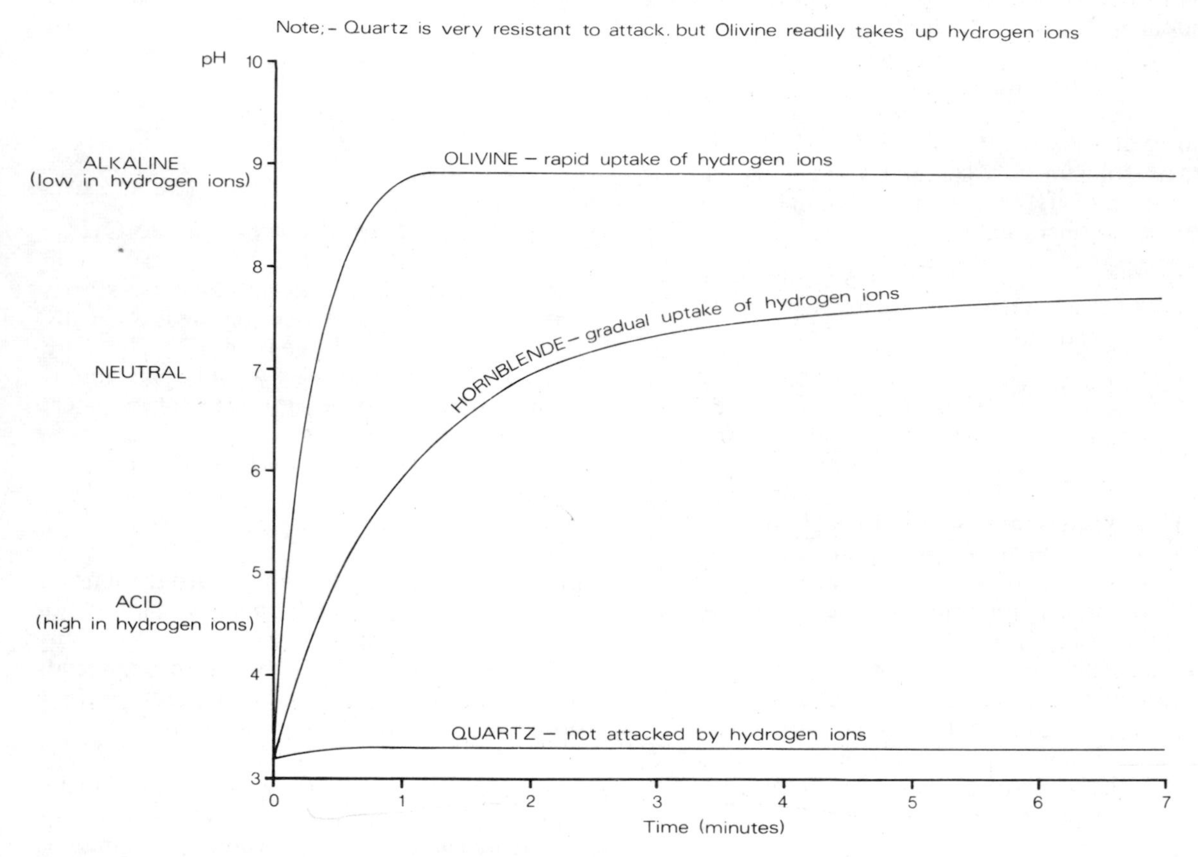

Figure 7 Hydrolysis: the uptake of hydrogen ions by three minerals: olivine, hornblende and quartz

It is possible to watch the process of hydrolysis occurring in a laboratory. If a silicate mineral is ground down to a fine powder and placed in water rich in hydrogen ions it is possible to measure the decrease in the amount of hydrogen ions in the water as they are absorbed into the mineral and cause the silica tetrahedra to break up. The amount of hydrogen ions in water is termed the pH of the water, where pH is the *negative logarithm of the hydrogen ion concentration* (i.e. 0·0000001 gram/litre hydrogen ions in water = pH 7). pH 7 is neutral, pH 2–6 is acid and pH 8–12 is alkaline. Using a pH meter or pH papers it is possible to detect a pH change from acid to alkaline as silicate minerals react with water and take up hydrogen ions (see Figure 7 and section 2.11).

HYDRATION

Some minerals can react to the presence of water by incorporating it directly into their crystal structure. The mineral anhydrite ($CaSO_4$), for instance, although not common in British soils, can occur in tropical soils and may be used to illustrate the process of hydration.

Anhydrite takes up water to form gypsum ($CaSO_4 . H_2O$):

$$\underset{\text{anhydrite}}{CaSO_4} + \underset{\text{water}}{H_2O} \rightarrow \underset{\text{gypsum}}{CaSO_4 . H_2O}.$$

The uptake of water can alter the solubility of the minerals. In this example gypsum dissolves far more rapidly than anhydrite, taking fifteen days to reach saturation, whereas anhydrite takes thirty days to reach saturation.

Climate

While the nature of the parent material will be important in soil formation the climatic conditions prevailing in the environment of the parent material will be equally important. It can be seen from the above description of the processes that the amount of water present will be crucial to the amount of solution that occurs. Furthermore the occurrence of freezing in cold temperate climates will be important in the initial stages of rock weathering that lead to soil formation.

Geomorphologists emphasize the importance of water freezing in cracks in rock, expanding and breaking up the rock. This is an important landforming process and helps to degrade cliffs and other rock faces. The significance of this process for soil study is that it provides a large number of surfaces for solution processes to act upon. A shattered piece of rock can be more easily weathered by solution and hydrolysis than can one single large block of rock.

Processes associated with exposure to air

The most important process is that of oxidation. Minerals in the soil parent material may take up oxygen from the atmosphere. This is a spontaneous reaction as the minerals are unstable in the presence of oxygen. Thus, when exposed to the air they spontaneously oxidize to achieve a chemically more stable form. This is, therefore, a good example of a mineral reacting to its new environment at the surface of the earth. Oxidation occurs as the mineral attempts to come into a state of equilibrium with its environment.

The characteristic brownish or reddish colours of most soils are due to the presence of oxidized iron. Ferrous iron (Iron II compounds) can be oxidized to ferric iron (Iron III compounds) which is red. In waterlogged soils, where air cannot easily penetrate, the soil is bluish or greenish. Here the colour comes from the unoxidized ferrous iron which is characteristically blue-green. Red mottles can be seen in some waterlogged soils where air has been able to penetrate (e.g. down old root channels) and oxidize the iron (see Section 3.1).

Processes associated with exposure to plants and animals

The presence of life is critical to the formation of a true soil. If the environmental conditions are not suitable for plant and animal growth a true soil will not be formed. Plants and animals have two important functions in the soil formation processes:

1 They provide organic matter (through decay) which accumulates on the surface of the soil as a layer of humus. This organic matter has properties which influence the solubility of soil minerals.
2 Soil animals mix soil particles and help to aerate minerals.

It is very difficult to separate processes which are purely chemical from those which are biological in soil development. Usually biological processes have a profound influence on the chemical processes like solution. Thus the word *biochemical* can be applied to many of the soil processes.

The most significant biochemical process of soil formation is *chelation*. The word comes from the Latin word *chele* which means claw. This aptly describes the process by which mineral ions are firmly grasped in the molecular structure of organic compounds. Elements like calcium, magnesium and iron are firmly attached in the molecular structure of organic compounds like humus acids.

If the soil has a thick layer of organic matter (humus) on its surface it means that any water percolating into the mineral material below will be charged with a large supply of organic compounds.

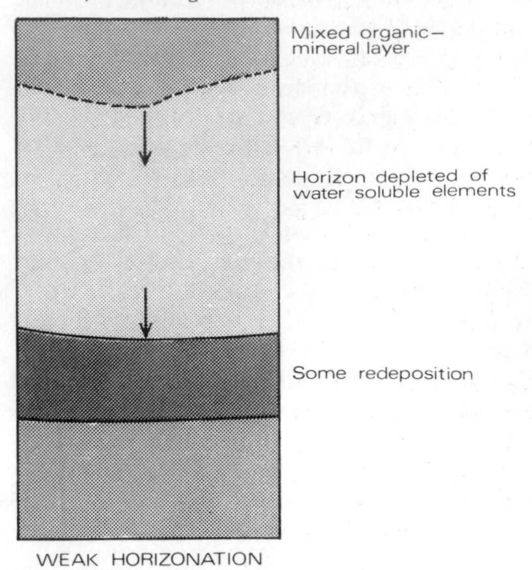

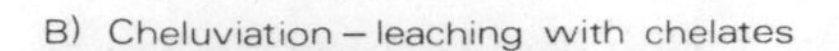

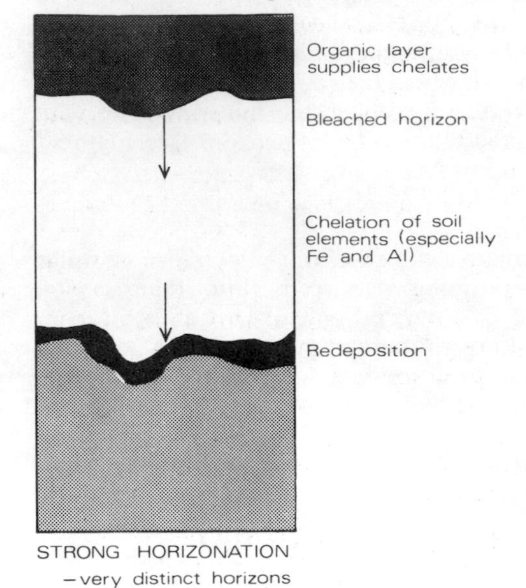

Figure 8 Leaching and cheluviation

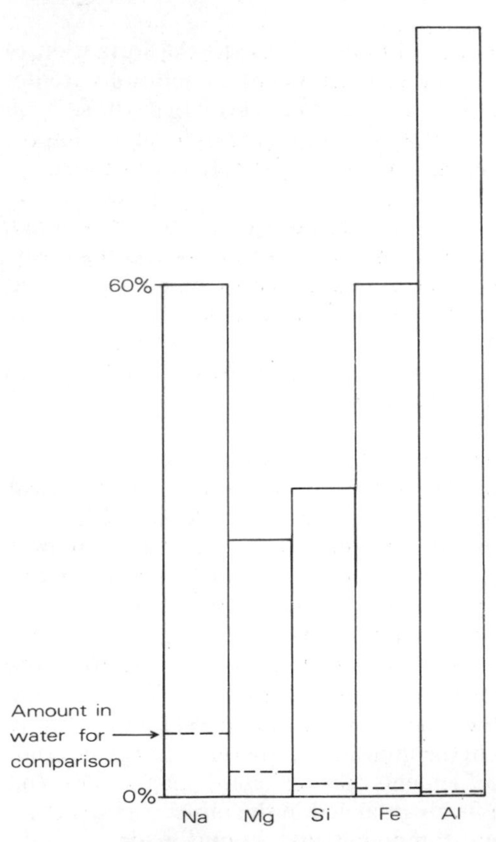

Figure 9 Relative amounts of chemical elements dissolved in water from soil rich in organic acids (vertical axis as in Figure 4). *Note:* Percentages are only approximate

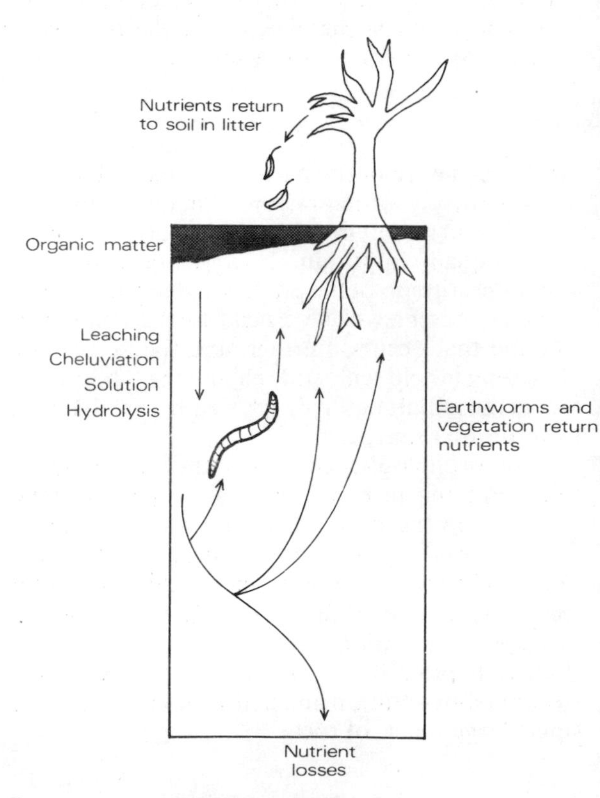

Figure 10 Nutrient movement in soil

Many of these compounds washed from the humus into the mineral soil below are capable of chelation. As the water percolates through the soil the organic compounds will take up mineral ions from the soil solids. In this way elements can be moved from the upper layers of the soil and washed down the soil profile. This washing process is termed *cheluviation.*

Cheluviation is responsible for the white, bleached layers which occur just below a layer of organic matter in some soil profiles (particularly in soils known as podzols). These bleached layers are extremely poor in nutrients as cheluviation is more efficient at removing elements than is the process of simple solution. Darker layers of redeposited organic matter and nutrients can be seen below these bleached horizons. These horizons are usually iron-rich, as iron is very easily moved by cheluviation (Figures 8 and 9).

As well as chelatory powers organic matter has the ability to produce hydrogen ions (which are important in hydrolysis). Thus the presence of organic matter on the surface of the soil greatly increases the movement of elements down through the soil profile.

Biological processes not only contribute to the movement of nutrients away from the surface of the soil but bacteria and fungi decompose the organic matter and release many of the nutrients that would otherwise be locked up in the organic compounds. These nutrients may then be available for use by plants. Also, the burrowing actions of animals like earthworms and ants may bring soil back nearer the surface and will thus bring nutrients which have previously been washed down the soil profile back to the surface (Figure 10).

Finally it should be emphasized that while we began this section with reference to a knowledge of simple chemistry, most of the processes in the soil are very complex. It is impossible to separate biological, chemical, atmospheric and hydrological processes in a soil. Each factor interacts to produce a complex mixture of rock particles, organic matter, water and nutrients that constitute a soil.

Summary

Soil can be viewed as the equilibrium product of the reactions of such materials as rocks, glacial deposits and alluvium with their environmental conditions. Chemical reactions, such as solution, oxidation, hydrolysis and hydration, occur as a result of the presence of water and air in that environment. Moreover, physical actions, such as freezing, may result from cold environmental conditions. The presence of life means that biochemical reactions, such as chelation, can take place and also that organic matter occurs. A soil is formed as a result of all these reactions. The soil becomes organized into horizontal layers, or horizons, as a result of leaching and of organic processes. Vegetation helps to offset leaching by the recycling of nutrients. Soil is thus a product of the soil parent material reacting to its environmental conditions through various chemical and biological processes taking place within the soil profile over time.

2

Soil components and soil properties

2.1 Soil material

What is soil made of? In order to understand how soil develops and to understand the inner workings of soil we must examine the separate components that make up soil. Also, in order to understand soil sufficiently to be able to manage it efficiently for agriculture we must understand how the soil components may react to different agricultural practices. Each component has particular properties and functions which influence how a soil behaves as a whole. We shall consider each component in turn: mineral matter, organic matter, water, air, biota (plants and animals) and nutrients.

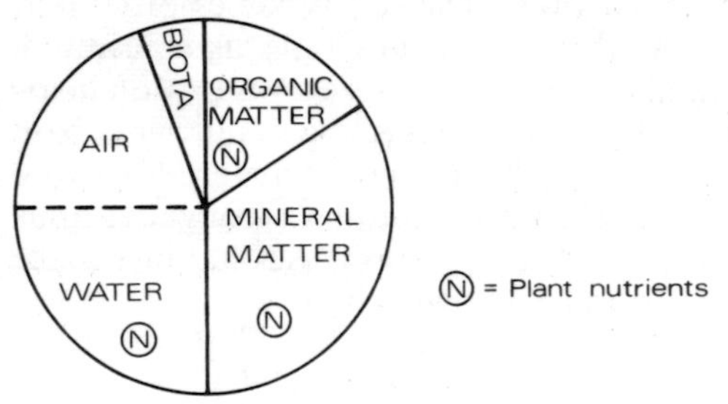

Figure 11 Relative proportions of soil components in an average agricultural loam

In sections 1.1 and 1.2 it was shown how the soil consists of a mineral 'skeleton' which is the weathered parent material. It was also shown that as the weathered material interacted with its environment it incorporated other components to form soil. In an average agricultural soil (a good, fertile loam) *mineral matter* from weathered rock and *organic matter* from plants and animals take up about half the volume of the soil and are thus the main soil components (Figure 11). *Air* and *water* take up the other half of the soil body. Air and water occupy the pore spaces between the mineral particles, and as water increases, with an input of rainfall, the amount of air decreases. As the soil dries (by evaporation and drainage) the amount of air in the pore spaces increases. Thus, in Figure 11 the boundary between air and water is drawn with a dotted line to indicate a fluctuating proportion. The large *soil animals* and *plant roots* take up what would otherwise be air space. The *smaller biota* (bacteria, fungi and actinomycetes) are either present in the soil water or are distributed in the organic and mineral matter. *Nutrients* available to plants are found in the organic matter, in the soil water or in the mineral matter, but a large proportion is found in combination with compounds composed of both organic and mineral matter. These compounds are called the *clay–humus complexes* and they hold the biggest reserve of nutrients.

2.2 Soil mineral matter

Primary and secondary minerals

Soil minerals are derived from the minerals in the parent material by the weathering processes described in section 1.2. As soil formation is a continuous process the mineral matter can be conveniently divided into:

1 *Primary minerals*– those remaining unaltered from the original parent material.
2 *Secondary weathering products*– those produced by the weathering reactions.

While the primary minerals are those which were originally present in the soil parent material the secondary minerals are those which are produced in the soil. Thus the primary minerals remain in the soil during the soil-forming process and are those minerals which are relatively insoluble. They include such minerals as quartz, which are very resistant to weathering (see Figure 7). The secondary decomposition minerals include the products of the equilibrium reactions discussed in Chapter 1 and therefore include oxides and hydroxides of primary minerals which form as a result of exposure to air and water.

Clays

The exact composition of a clay depends upon the mineralogy of the parent material and the weather-

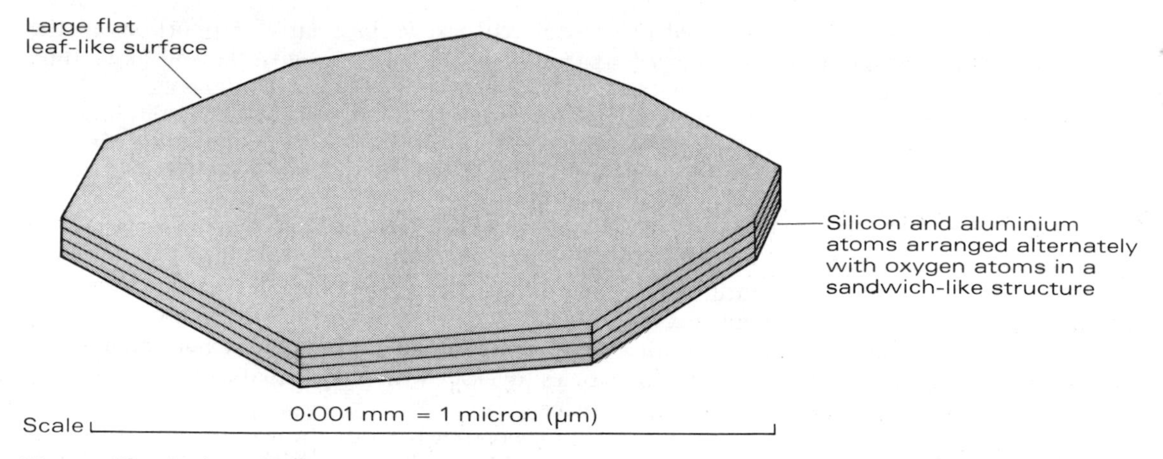

Figure 12 A clay micelle

ing environment. Silicate minerals, as was shown in Figure 6, may be prone to hydrolysis. One of the most important results of this reaction is the production of clays. Clays are minute particles composed of silicon (Si), aluminium (Al) and variable amounts of oxygen (O) and hydrogen (H) which are left after weathering reactions.

There are four main types of clay found in soils. Three, *kaolinite*, *montmorillonite* and *illite*, have a recognizable crystal structure. The fourth type, *allophane*, is non-crystalline and has no recognizable form. *Allophane* may include a number of different chemical types, but these are difficult to investigate because of the lack of structure; the terms the *Allophane Group* or the *Amorphous Clays* are often used to describe these clays. The first three types of clays mentioned have their atoms arranged in layers, like a sandwich. They belong to the group of silicate minerals termed *phyllosilicates* (*phyllo* = leaf-like) where sheets or leaves of atoms are laid down on top of one another. The basic clay structure is the *micelle* composed of several layers. The structure of a micelle is illustrated in Figure 12.

The three clay types are recognized by differences in their crystal structure, and these are shown in Figures 13–15.

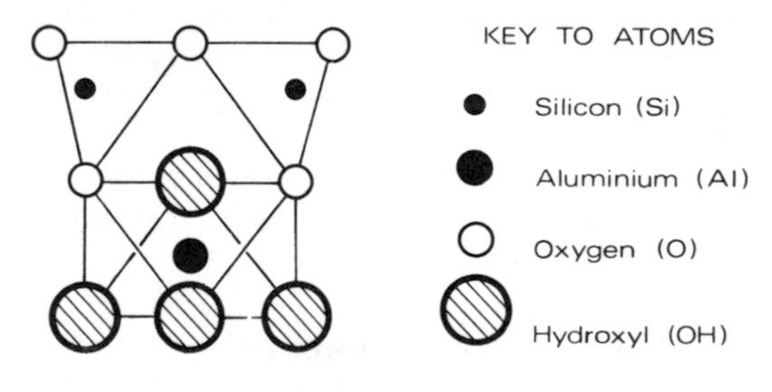

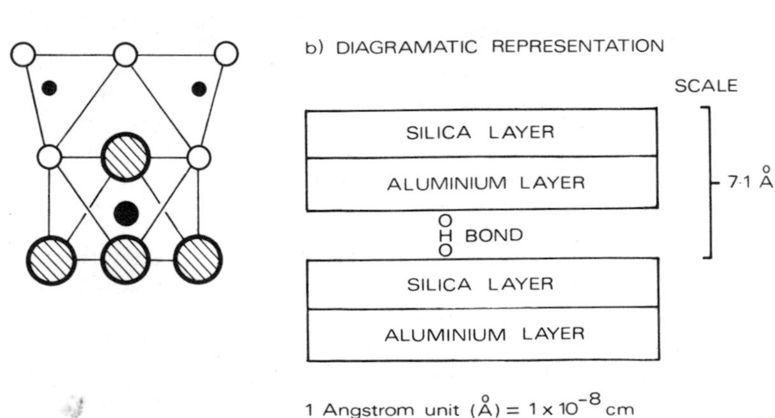

Figure 13 The structure of kaolinite

Kaolinite has one layer of silicon atoms joined to one layer of aluminium atoms by a row of shared oxygen atoms. This structure is then joined to the next silicon–aluminium layer by a bond consisting of hydrogen and oxygen (Figure 13).

Montmorillonite is similar except that it has an extra silicon layer in each unit. Thus the unit is a silicon–aluminium–silicon one. Moreover the units are joined by a weak water bond and not an O–H–O bond as in kaolinite (Figure 14).

Illite is almost identical to montmorillonite except that it has a very strong bond between the silicon–aluminium–silicon structure and this bond is made of potassium (K) (Figure 15).

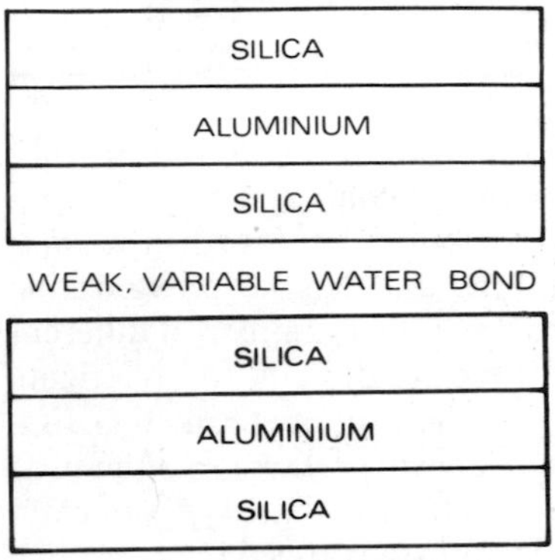

SILICA
ALUMINIUM
SILICA
WEAK, VARIABLE WATER BOND
SILICA
ALUMINIUM
SILICA

Figure 14 The structure of montmorillonite

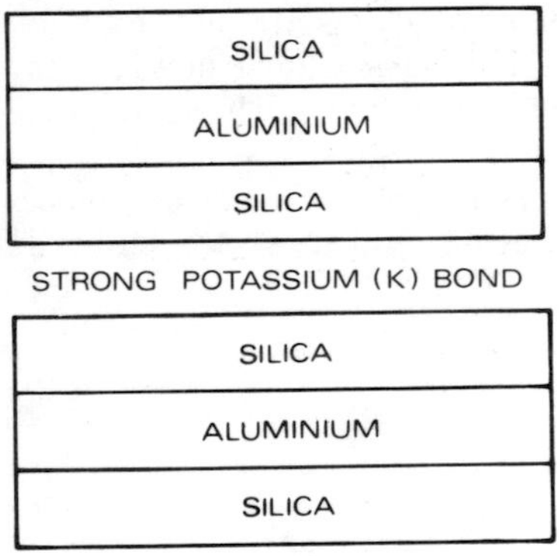

SILICA
ALUMINIUM
SILICA
STRONG POTASSIUM (K) BOND
SILICA
ALUMINIUM
SILICA

Figure 15 The structure of illite

2.3 Soil texture

Soil texture refers to the degree of coarseness or fineness of the mineral particles of the soil. These range in size from the large fragments of weathering rock down to the minute particles of about 1–2 microns in diameter. As the soil parent material may have been derived in a number of different ways and as weathering tends to break up the larger particles, a soil usually has a mixture of particles of different sizes. The relative proportions of the particles of differing sizes give the soil its texture.

The understanding of soil texture is a crucial step in the investigation of soil characteristics and soil behaviour as it influences many other factors, especially structure and the availability of water and nutrients.

A *coarse textured soil* has a high percentage of large particles of sand size and a *fine textured soil* has a high percentage of small particles, especially of clay. A *medium textured soil*, usually referred to as a *loam*, contains a mixture of coarse and fine particles together with intermediate, or silt-sized, particles.

It is possible to study the texture, or particle size, of a soil by rubbing a small, moist sample of soil between the fingers. Coarse particles of *sand* can be felt easily. *Silt* can be recognized by its soapy feel and a *clay* soil is pliable and can be moulded into various shapes. A clay soil can be rolled into long threads which do not break and also it stains the hands (often a brown or reddish-brown colour). A loamy soil can be moulded to some degree but the shapes easily break up. (The description of soil texture in the field is dealt with in detail in section 6.5.)

While the above 'Field Method' of assessing texture is an extremely useful guide to soil type and behaviour, more precise information can be gained by the analysis of particle size in a laboratory. The particles are classified into groups of sizes (sand, silt and clay) and each class is separated by measurements of the diameter of the particles. Two classification schemes are often used, a so-called International scale and an 'American' scale, the latter derived in the United States.

Classification of the sizes of soil particles

Texture class	*International scale* (mm)	*American scale* (mm)
Coarse sand	2·0–0·2	2·0–0·25
Fine sand	0·2–0·02	0·25–0·05
Silt	0·02–0·002	0·05–0·002
Clay	less than 0·002	less than 0·002

As these scales use slightly different definitions of fine sand and silt it is important to state which scale is being used when texture results are being reported.

The amounts of soil in each size class are commonly determined by two laboratory methods: sieving and sedimentation. These techniques are described in detail in Appendix 1.

Sieving

In this technique sieves are used with meshes of varying sizes. Thus a sieve of mesh size 0·2 mm would allow fine sand, silt and clay particles to pass through, but would retain the coarse sand particles. A stack of several sieves is normally used, with the larger mesh sieves placed above the smaller meshes (Figure 16). The dried soil is poured into the top sieve and the whole stack is shaken on a mechanical shaker. The particles larger than any particular sieve mesh are retained by that sieve.

After sieving, the soil on each sieve is turned out on to a large piece of paper and the back of the sieve is brushed over the paper. A wire brush is used for the coarser wire sieves and a nylon brush for the finer nylon sieves. The soil on each piece of paper is carefully transferred to a labelled beaker and weighed. The result for the weight retained on each sieve is expressed as a percentage of the whole soil; for example, the table below shows the results for a 150 g soil sample.

Sieve mesh (mm)	*Texture class* (*International scale*)		*Weight retained on sieve* (g)	*Percentage*
0·2	Coarse sand		80	53·3
0·02	Fine sand		40	26·6
0·002	Silt		10	6·6
Pan	Clay		20	13·3
		total	150	99·8

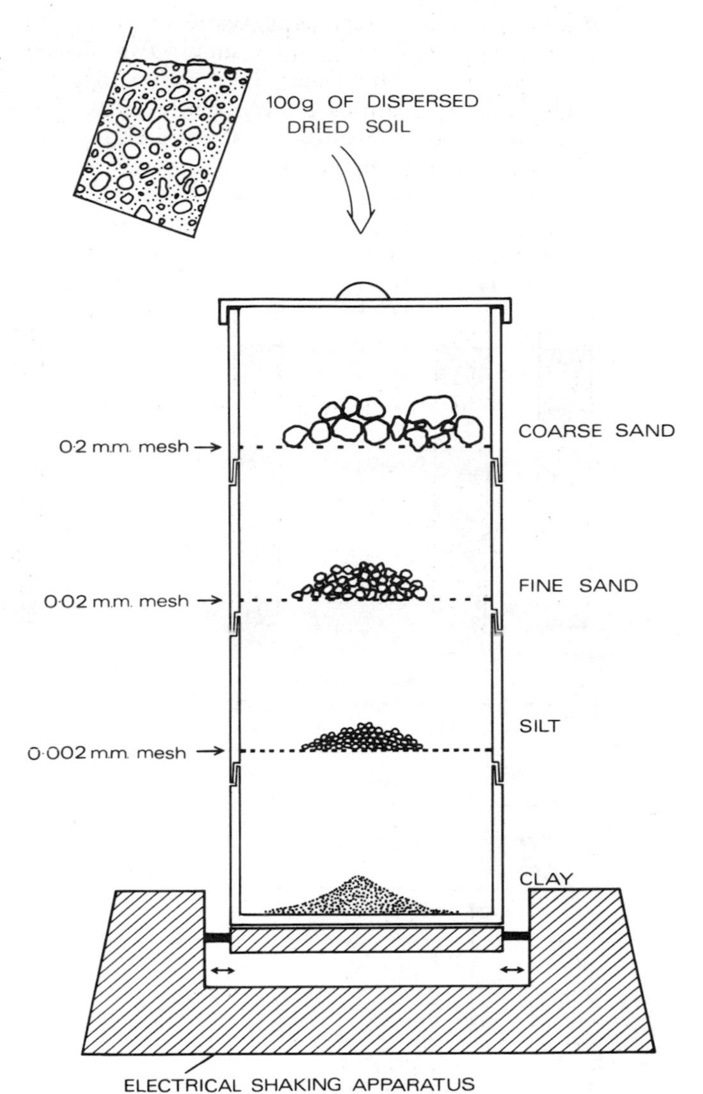

Figure 16 The principle of sieving

Sedimentation

Whereas sieving is least accurate at the fine end of the texture scale sedimentation is most accurate in the measurement of small particles. Therefore if both methods are used together a more complete soil analysis can be made than if one method is used alone. The Soil Survey of Great Britain use Sedimentation as the standard method for assessing texture.

In the sedimentation technique the dispersed, dried and weighed soil sample is poured into a column of water (Figure 17). The larger grains settle out almost immediately, but the smaller particles settle out very slowly. If the settling rate is measured it is possible to work out the sizes of the particles which are settling using *Stoke's Law* (first proposed in 1851), which states that *the settling rate of a particle is proportional to the diameter of the particle.*

All the sand will have settled out from the top of the cylinder after about five minutes and the silt after about eight hours, the precise timing depending upon the temperature. There are two methods of assessing the rate of sedimentation:

1 using a hydrometer;
2 sampling with a pipette.

The *hydrometer* measures the density of the liquid. If there is a large amount of soil suspended in the water the density will be high and the hydrometer will ride high in the water. As particles settle out the hydrometer will sink.

Alternatively a 20 ml sample of the water can be drawn off with a *pipette* from 10 cm below the surface. The sample is carefully evaporated and the residue is weighed. (Accurate measurement is necessary as weight differences will be small.) This weight gives a measure of the solids still in suspension.

The results for texture analysis can be displayed on a pie diagram (Figure 18) or on a triangular graph (Figure 19). If the results given in Figure 18 are plotted on Figure 19 it can be seen how the triangular graph works, confirming that sample 1 is a clay loam, sample 2 loamy sand, 3 a clay and 4 a silty clay loam. Note that sand and silt have to

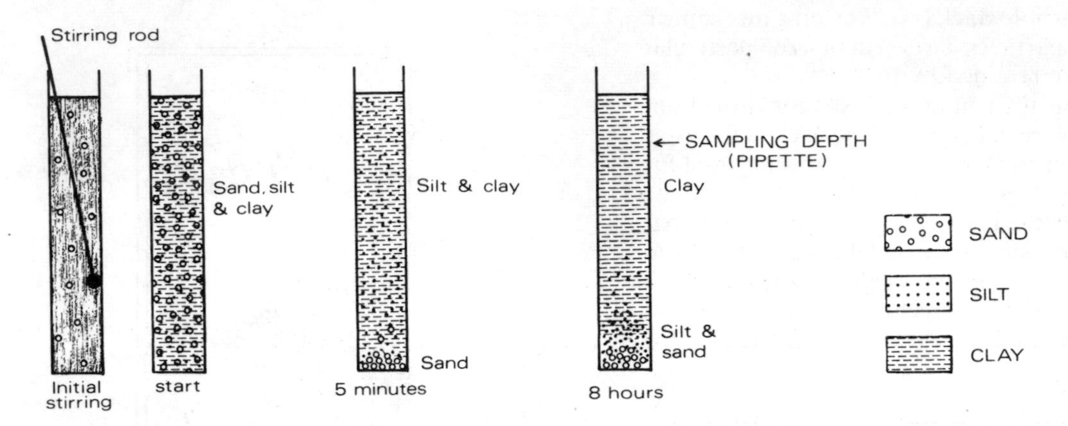

Figure 17 The principle of sedimentation

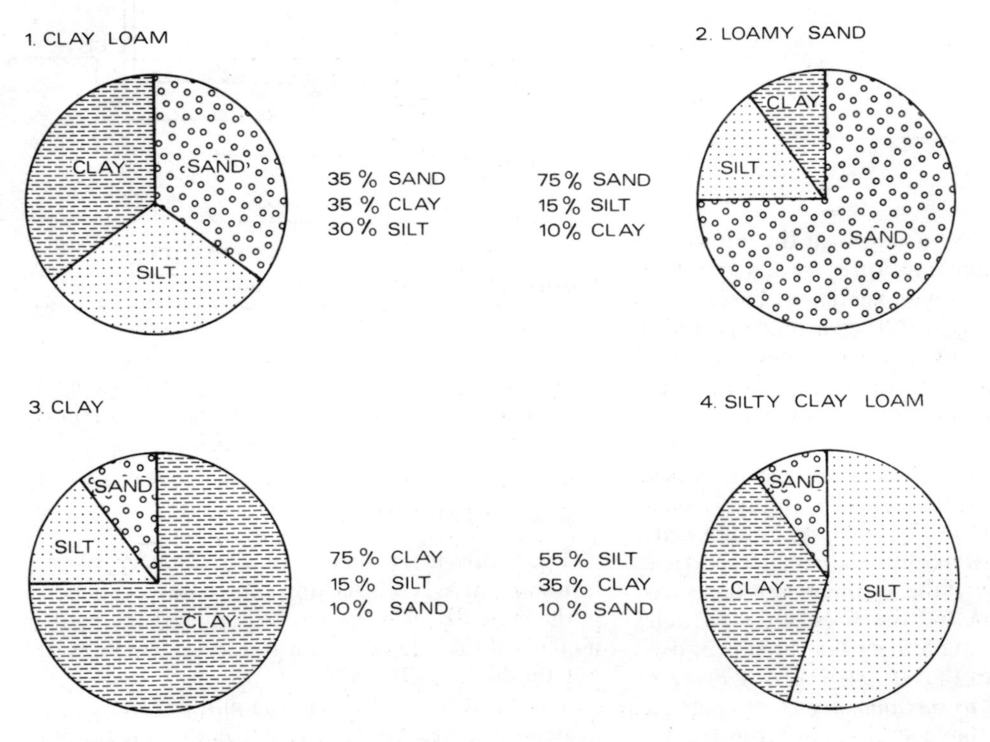

Figure 18 Texture of different soil types

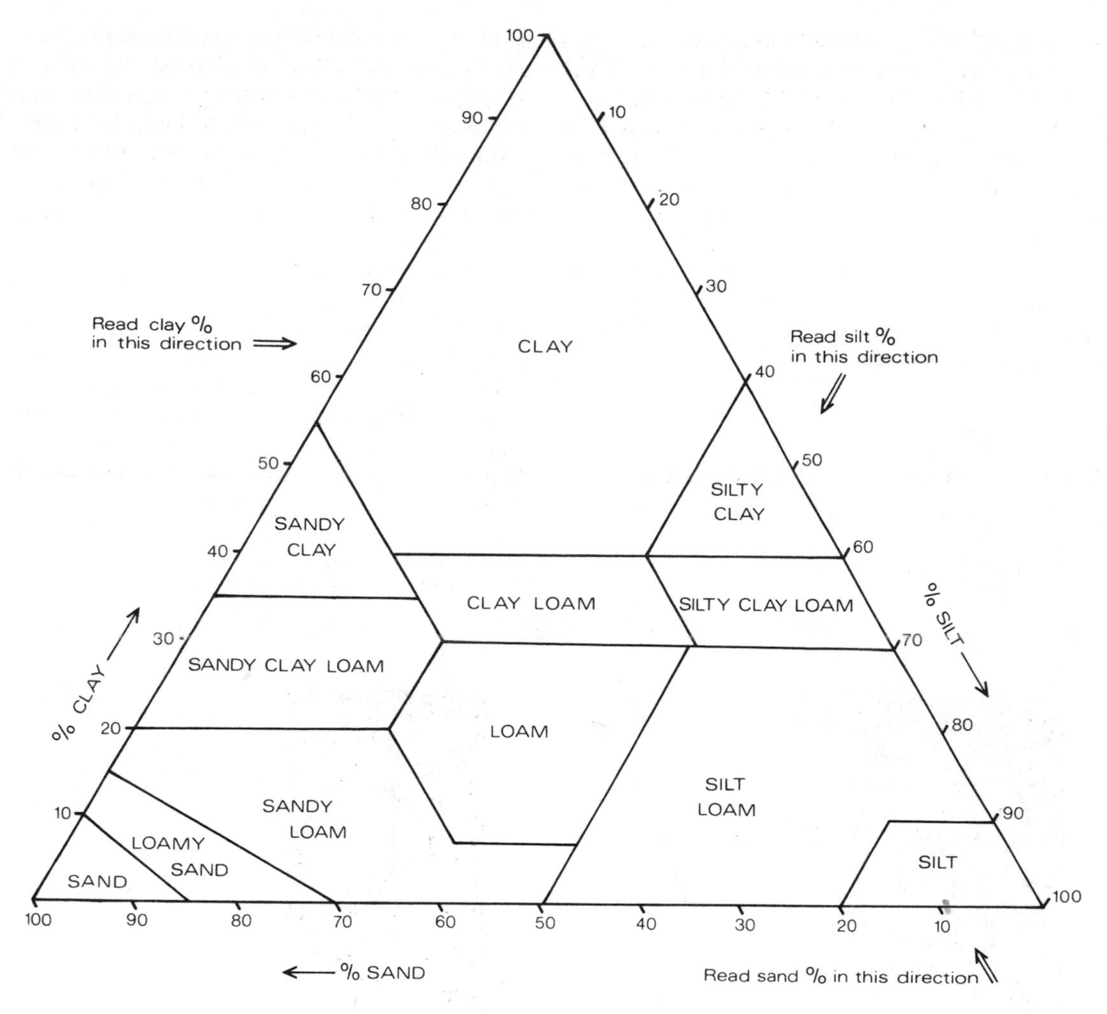

Figure 19 Soil texture

be quite pure in composition to be classed separately (bottom right and left hand corners), but that clay can have quite a large proportion of silt and sand in it. This is because the properties of clay are very dominating and soil reacts in a manner which is a result of the clay fraction even though quite a high proportion of the soil may be silt or sand.

A simple adaptation of the sedimentation technique is to pour soil into a beaker of water and stir it. If it is then left for a day or two, the particles will settle out, sand first, then silt, and finally clay. The depth of each type of particle can then be measured and calculated as a rough percentage to give a rough estimate of soil in each particle-size class.

Once texture has been assessed it is possible to make some statement about the properties of the soil as a whole. *Sand* particles impart to the soil a porous open texture, but contribute little in the way of nutrients. *Silt* particles are usually smaller sand particles (e.g. of quartz), and similarly contribute little in terms of nutrients. Moreover they do not give a porous open texture, but the particles can be packed together, giving a dense soil. Some soils which are a problem to cultivate have a very high silt content. *Clay* soils on the other hand hold rich reserves of nutrients (see under soil nutrients, section 2.10). They are also noted for their properties of shrinking when drying and expanding when wet. The largest clays with the weakest micelle bonds expand and contract more than the smaller clays with stronger bonds. Thus montmorillonite is more expansive than kaolinite. It is the contraction of clays that causes soil to crack when it is dried.

Knowing the properties of each soil-texture component, it can be seen that the loam, with its mixture of sand, silt and clay, is in many ways the most desirable from an agricultural viewpoint. The sand would give it a porous and well-aerated nature, the silt and clay would retain some moisture and the clay would hold nutrients in store. Conversely a soil composed of one dominant component would be difficult to manage. Purer clays tend to be rich in nutrients, but this is of little use to plant growth as the clay holds up water, giving a sticky, intractable, dense soil. The nutrient reserve may only be made use of by artificially draining the soil and also by improving the porosity and aeration of the soil by adding organic matter (or even sand though this is not a common practice).

2.4 Soil structure

Cultivation of many soils would be virtually impossible were it not for the fact that the individual particles making up a soil are usually aggregated into *soil structures.* Groups of clay, silt and sand particles are found stuck together in aggregates or structures (often referred to as *peds*) which have a profound influence on soil properties.

If it were not for soil structures the soil would have few pores through which air, water and plant roots could pass. Thus a compacted, structureless soil is of little value for agriculture as roots cannot penetrate, water is not available and the soil is not aerated. Soils with bad structures are often *layered* or *blocky* while a soil suitable for plant growth has a number of small structures like breadcrumbs in shape and size; this is referred to as a *crumb* structure. Subsoils having tall and thin structure (*prismatic* or *columnar*) may aid the drainage of the profile as water can rapidly percolate down the sides of the structures. The types of soil structures are illustrated in Figure 20.

The type of structure is very much governed by the soil texture. In turn the overall properties of the soil (and therefore its value to agriculture) may depend greatly on the structures that are present.

How are the structures held together? Organic matter, chemical cements and clay play important roles. Of these, that of clay is the most important. Humus may help to stick smaller particles together as may the slimes secreted by earthworms and actinomycetes (see under soil organisms, section 2.9). Some nutrients in the soil can also act as cementing agents, especially calcium carbonate, and consequently a soil rich in calcium may have very well cemented structures (in fact in the tropics this may become too hard and the soil may set rock hard, forming calcrete). The cements used in the building industry are made from chalk and clay as raw materials.

While organic and chemical cements play important roles in forming structures a crucial process is that of *clay flocculation.* Clay particles can exist in

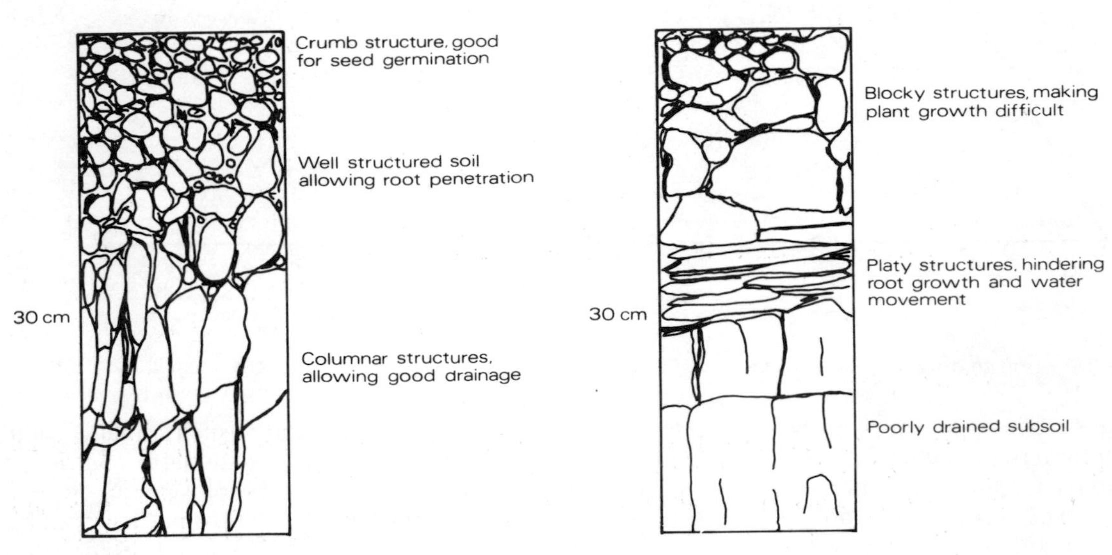

Figure 20 Soil structures

two states: *dispersed* and *flocculated*. Dispersed clay particles are separated from each other in individual 'coats' of cations, usually sodium. Flocculated clay particles are linked together by other cations, especially calcium, into clay aggregates or *floccules*. Dispersed clay soils are very dense and the particles are closely packed. This is why cultivation of saline soils, rich in sodium, is difficult; before cultivation can proceed the soil first has to be reclaimed by washing out of the salts, especially sodium chloride, from the soil. A practical example of this is found in the Netherlands where land reclaimed from the sea has to be leached with fresh water to remove the sodium salt and then treated with gypsum (calcium sulphate). In this case the calcium replaces the sodium coat round the clay and allows the clay particles to flocculate or stick together. When the clay is flocculated it forms larger particles, which give the soil a greater porosity but also retain the water-holding and nutrient-storing capacities of clays. Thus it is important to investigate the texture of the soil and its organic content and also the content of sodium and other nutrients, especially calcium, in order to understand soil structure.

We have already seen that texture may affect structure. Crumb structures are usually associated with loamy textures. Platy and columnar structures are found in clayey soils, although platy structures tend to be associated more with clay or silt soils where compaction has occurred. The columnar structures are formed by the shrinking and expansion of clays due to drying and wetting.

Ploughing can alter structures radically and this is the topic for discussion in section 5.2. Ploughing breaks up large, lumpy soil structures so that the soil is suitable for plant growth. However, this should not be carried out when the soil is very wet, especially on silt or clay soils prone to compaction, otherwise the structures may collapse completely. The weight of the tractor wheels may reduce the soil to a platy, compressed mass which is difficult to break up again later. Ploughing should therefore reduce the soil structures to a suitable size for seedling establishment and plant growth, but it should not damage and compress the structures so that plant growth is hindered.

2.5 Soil fabric

Soil fabric refers to the arrangement of soil particles on a very small scale, i.e. it is the small-scale soil structure. Soil fabric is studied by looking at the organization of a soil under the microscope. It is possible to study the microscopic structures of a soil in thin sections just as rocks are studied by the geologist.

In thin sections under a microscope the soil *plasma* can be recognized. This is the characteristic product of soil development, being a mixture of parent material and organic matter. It is an amorphous combination of humus, clays and chemical compounds (e.g. iron oxide), and is produced by the secondary weathering processes (see section 1.2) and by the incorporation of organic matter. The presence of organic matter distinguishes the soil plasma from the underlying mineral material, and the presence of mineral matter in the plasma distinguishes it from the overlying purely organic horizons. The characteristic of soil plasma is that *mineral and organic matter are virtually inseparable.* This is the result of the equilibrium reactions whereby the weathered rock reacts with its environment in an attempt to achieve a form which is stable in the presence of water, air and organisms.

Obviously it is necessary to solidify the soil in some way before it can be cut into a thin section. This is done by soaking the soil in a resin solution. The solution hardens on drying and forms a solid block. The resin block containing the soil can then be sliced into thin sections. The process of solidification is termed *impregnation* with resin.

Using impregnation methods much useful information can be gained about the minerals present in the soil. The primary minerals can be identified with little difficulty as they are similar to the unaltered rock minerals studied by geologists. The secondary, altered, minerals are usually more difficult – often being opaque and having no distinct outline.

If care is taken to impregnate the soil with resin carefully (while removing the air in the soil pores with a vacuum pump at the same time as the resin is introduced) the microstructures of the soil will be preserved in their natural orientation. In thin section the dark amorphous soil plasma and the pore spaces within and between the soil structures can be seen amongst larger sand grains (Figure 21). The pores of varying sizes have different functions (see also under Soil water, section 2.7).

Macropores – over 75 μm wide (0·075 mm): rapid transport of water and air

Mesopores – 75–30 μm wide: reservoirs of water for plants

Micropores – less than 30 μm wide: here the water is unavailable to plants

2.6 Soil organic matter

This is derived from decaying leaf litter and faecal material from animals. Sometimes the plant

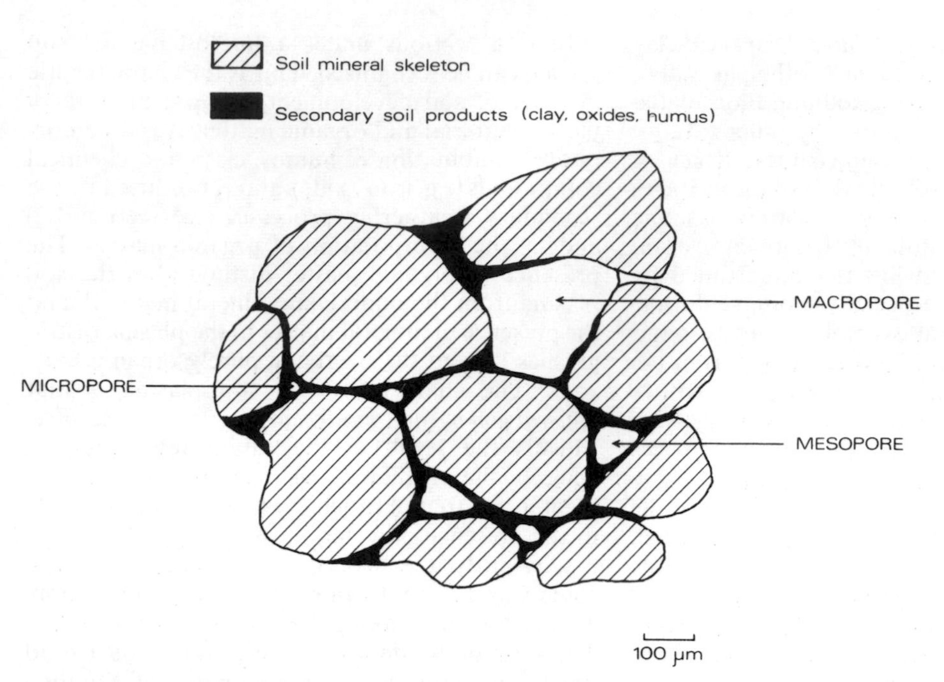

Figure 21 Soil pores

remains are still visible, but as the organic matter decays the *leaf litter* loses its structure and the individual leaves are no longer recognizable: *humus* has been formed. The term *humification* is used to describe this breakdown process from recognizable plant remains to an amorphous black or brown, almost jelly-like substance. Figure 22 illustrates the process of humification.

When the organic matter can be recognized as a distinct layer on top of the soil profile it is termed *discrete humus* because it is separate from the soil mineral matter. However, as humification proceeds the humus becomes incorporated into the soil mineral matter, especially to form *clay–humus complexes* in the soil plasma. It is then referred to as *intimate humus*.

In the soil profile three layers of discrete humus can often be recognized, the L, F and H layers:

- L leaf litter, leaves and other plant remains are recognizable;
- F the fermentation or humification layer, where decay is active;
- H the humus layer; plant remains are unrecognizable.

The H layer merges into the top layers of the soil where mineral and organic matter are mixed (see also section 6.5).

Whilst soil organic matter within *one* soil profile can be separated into discrete and intimate humus, when *different* soil profiles in different areas are compared three distinct types of discrete humus can be recognized. These occur in relation to the nature of the soil-forming environment and are called *mull*, *moder* and *mor*.

Mull is a soft, blackish material, crumbly when dry and rich in nutrients. *Mor* is a raw, fibrous, acid humus, poor in nutrients. *Moder* is an intermediate form.

Mull is produced by the action of fungi, bacteria and earthworms where the soil is not too acidic. It is common in lowland hardwoods (e.g. oakwoods), and in many fertile grasslands. Mor is formed in upland heath and bog environments which are wet and acidic and are not favourable for the activity of bacteria, fungi and earthworms. A mor-like mull, or moder, may be produced as a transitional state.

Since organic matter holds nutrients in reserve and since it can affect structure it is often important to determine the amount of organic matter in a soil in order to understand a soil's behaviour. One simple way of assessing organic matter in a soil is to weigh a small dry sample of soil which contains organic matter and to ignite it in a crucible in a muffle furnace. Details are given of the method in Appendix 2. The organic matter oxidizes and the resulting weight loss is used as a measure of organic matter content.

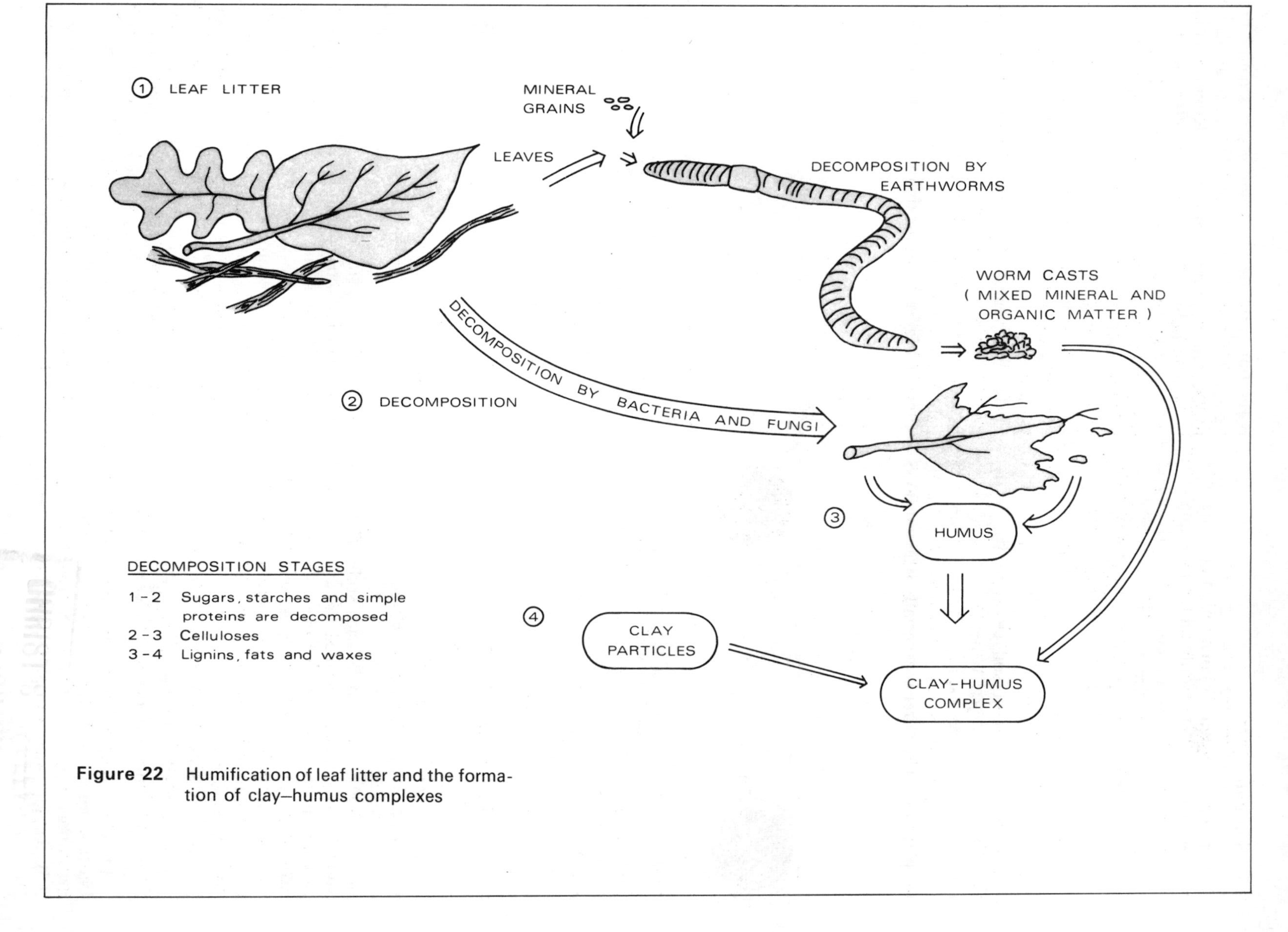

Figure 22 Humification of leaf litter and the formation of clay–humus complexes

2.7 Soil water

The amount of water (W) in a soil at any one time will depend upon the amount added by rainfall (R) minus the amount lost by evaporation (E), plant transportation (T) and drainage (D). This relationship can be expressed in a mathematical way using the sign $\propto$ to indicate 'proportional to' thus:

$$\underset{\text{input}}{W \propto R} - \underset{\text{output}}{(E+T+D)}.$$

The most important soil characteristic is the *water retention capacity*. Water may flow quickly through coarse, porous soils and not be retained or it may be retained for some time and be available for plant use. The latter is especially true if it flows through tightly packed clay-rich soils. The availability for plant use is termed the *available water capacity*. Both the retention and availability of soil water depends upon *pore space*. As seen earlier this will depend upon texture and organic matter as these both influence structure and fabric (Figures 20 and 21).

Two pore space properties are important:

1 total pore space (soil porosity), and
2 the distribution of pores of different sizes.

Of these the second is the most important because pores of differing sizes have differing properties regarding soil water (see Soil fabric, section 2.5). Rapid water transport occurs in the larger pores, but in the smaller pores water is held more tightly. The water may be held in a state available to plants in larger pores, but in smaller pores it may be held so tightly that plants cannot extract it.

The behaviour of water in a soil can be understood in terms of the forces acting upon it. Water is attracted to the solid soil particles by surface tension. Opposing this are the agents which exert forces away from the solid particles. These are the forces of gravity, the action of plant roots and evaporation.

The effectiveness of the force exerted by surface tension depends upon how far the water surface is from a solid particle. A thin film is held extremely tightly, but a thick film is held with less force, as shown in Figure 23. In the figure the water is shown as surrounding a single grain to illustrate the point of how tension varies with film thickness. But in actual soils the grains are arranged to form pore spaces between the grains. It is the size of these pores which is the important factor. In a very small pore all the water will be near a solid particle, the surface tension will be very strong and the water will be tightly held. However, in a large pore the water will be further away from the solid particles, the surface tension effect will be lower and the water will be less tightly held. Indeed if a large pore is completely full of water the water will be held on to the particles only very lightly.

In the large pores where the surface tension effect is small, the forces exerted by gravity and plant

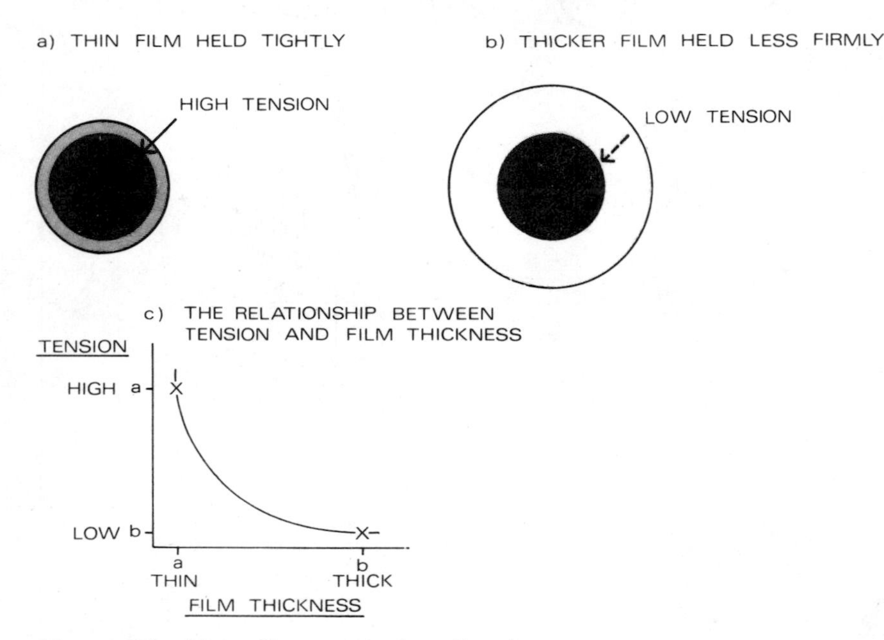

Figure 23 Water films on single soil grains

roots may be greater than the surface tension. Thus water, instead of being held on to the particles, can be moved by the dominant force. This is why in section 2.5 the macropores transport water rapidly and why in the micropores the water is unavailable to plants. In these micropores, water is held so tightly to the solid that plants cannot exert enough suction to extract the water.

Water in a soil can be classified according to the tension at which it is held in the soil. The tension is measured in atmospheres of pressure (atm) and the higher the figure the more firmly the water is held. Water in very small micropores close to solids is held at a tension equal to 10 000 atm. Evaporation can exert a force of up to about 30 atm, plant roots can exert up to about 15 atm and gravity up to $\frac{1}{3}$ atm. The situation can be summarized:

water held at 30–10 000 atm: held tightly to grain surfaces in small pores
water held at 15–30 atm: moved only by evaporation; usually only moved as water vapour
water held at $\frac{1}{3}$–15 atm: water can be extracted by plant roots
water held at less than $\frac{1}{3}$ atm: water can be moved under the influence of gravity

Water in soil is usually divided into three types according to its behaviour. Water held at less than $\frac{1}{3}$ atm and which can be moved by gravity is termed *gravitational water*. Water held fast in the soil, firmly adhering to the solid, is termed *hygroscopic water*. Between these states water is given the general term *capillary* or *matric water*. Matric is the American term and this type of water is water held in the soil matrix, or framework.

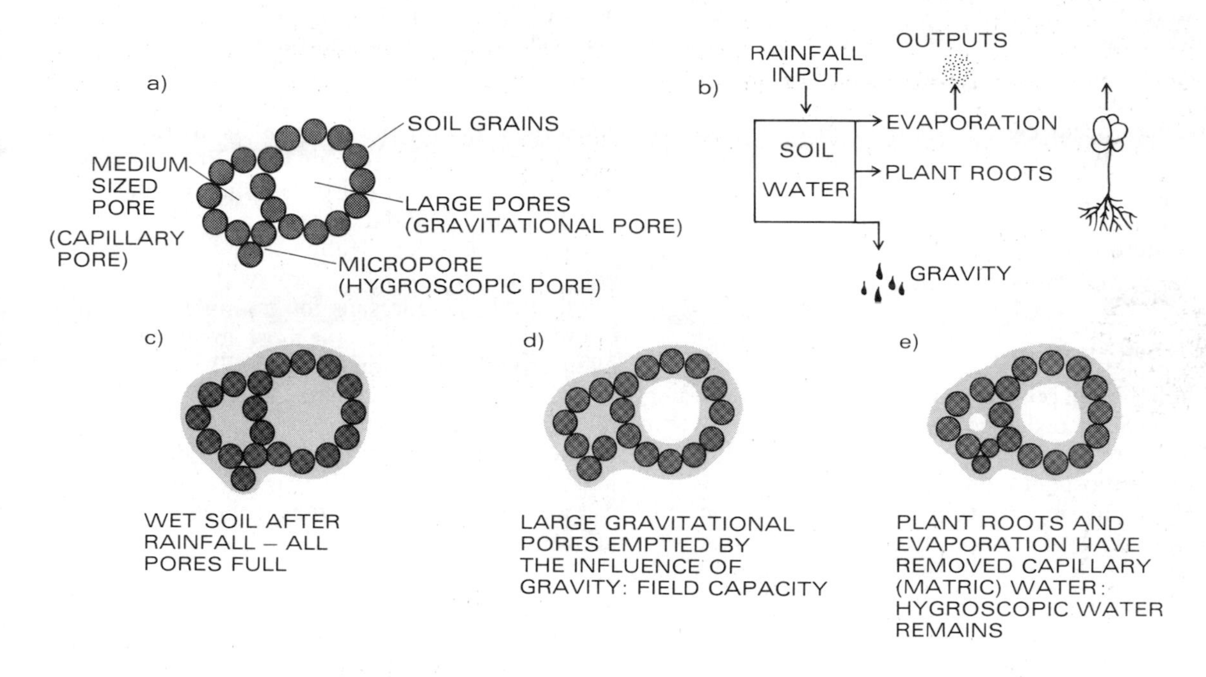

Figure 24 Behaviour of water in soil pores during drying

From this division of water types we can divide soil pores into gravitational, capillary and hygroscopic pores. The classification of soil water and soil pores can be summarized:

1. *Gravitational water*–that water held in the large pores, possessing only a weak attraction to the solid. It is held at less than $\frac{1}{3}$ atm and can be moved under the influence of gravity.
2. *Capillary water* (*or matric water*)–The water held in medium-sized soil pores (in the capillary pores) at between 30 to $\frac{1}{3}$ atm.
3. *Hygroscopic water*–water held in very small pores, adhering firmly to the solid with a tension of up to 10 000 atm. (It may be removed artificially by 'boiling off' by heating the soil to 105°C.)

The types of soil water are illustrated in Figure 24 by considering a wet soil during drainage.

Soil water classification and plants

As plant roots can exert a tension of up to about 15 atm the water in a soil can be classified as unavailable (held above 15 atm) and available (less tightly held, at below 15 atm). Thus when soils dry out so that no water is left which is held at less than 15 atm the *wilting point* is reached. Only if the soil is wetted again will water be available to plants. After a rainstorm the soil may well be saturated with water. Assuming the soil is well drained, the gravitational water will soon drain off and when this has occurred the state of *field capacity* is reached. Water available to plants is that held between field capacity ($\frac{1}{3}$ atm) and wilting point (15 atm).

A method of measuring soil moisture content is given in Appendix 3.

2.8 Soil air

When the soil pores are not occupied by soil water they will be occupied by air. This soil air has a characteristic chemical composition and is different to the atmospheric air we breathe.

The living organisms in the soil produce carbon dioxide (CO_2) by respiration but do not use up carbon dioxide by photosynthesis as green plants do. Thus carbon dioxide builds up in the soil air and is present in far greater concentrations than it is in the open atmosphere. Carbon dioxide is also produced in the soil by root respiration and by the decay of organic matter. In the humification process (see section 2.6) organic carbon (C) is slowly oxidized to carbon dioxide (CO_2) by soil microorganisms. Figures for the composition of soil air show the chief differences – there may be as much as 300 times the concentration of carbon dioxide in soil air as there is in the open atmosphere:

	Oxygen content (O_2) (%)	*Carbon dioxide content* (CO_2) (%)
Atmosphere, normal composition	20·9	0·03
Soil air, average range	15–20	0·25–4·5

The carbon dioxide level may rise above the figures quoted, especially where soil porosity is low as in dense clayey soils and in wet weather.

Just as with soil water, soil porosity is a crucial factor in determining the amount and nature of soil air. The soil organisms, roots and decay processes all need oxygen to function. Also the processes all emit carbon dioxide. An 'open', porous soil with large pores allows the carbon dioxide to escape and the oxygen to enter by *gaseous diffusion*. In a soil with dense structures and small pore spaces oxygen cannot readily diffuse to aerate the soil, and carbon dioxide builds up in the soil. The *exchange* of gases is hindered and soil organisms do not flourish, roots do not grow well and decay slows down. Figure 25 contrasts well-aerated and poorly aerated soils. It can be seen that structure will have an important effect on aeration. Since structure affects drainage and water retention the water balance of the soil and the composition of soil air must also be closely linked. A soil with small pores will have increased water retention and decreased aeration while an open porous soil will allow rapid drainage and good aeration. Clearly, a soil with a mixture of small and large pores, as for example occurs in a crumb structure, will be the most advantageous for plant growth. Here some water will be retained in the small pores and aeration will be possible through the large pores.

The production of carbon dioxide shows a marked seasonal cycle. It is most rapid in warm, wet weather when root and microorganism activity are at their greatest (Figure 26).

2.9 Soil organisms

The various soil properties have different effects on soil organisms. Texture and structure are notably important in affecting the porosity of the soil, and therefore water retention and aeration, as described above, and these in turn influence soil floral and faunal activity. There are also soil processes and properties on which soil organisms have a considerable effect and these can be divided into three main groups of processes performed by soil organisms:

1 decomposition processes, e.g. humus decay, pesticide breakdown;
2 transformations and fixation, e.g. nitrogen fixation;
3 structural processes, e.g. aeration by earthworm burrows.

Decomposition

Soil organisms break down complex substances into simpler components. These may then in turn

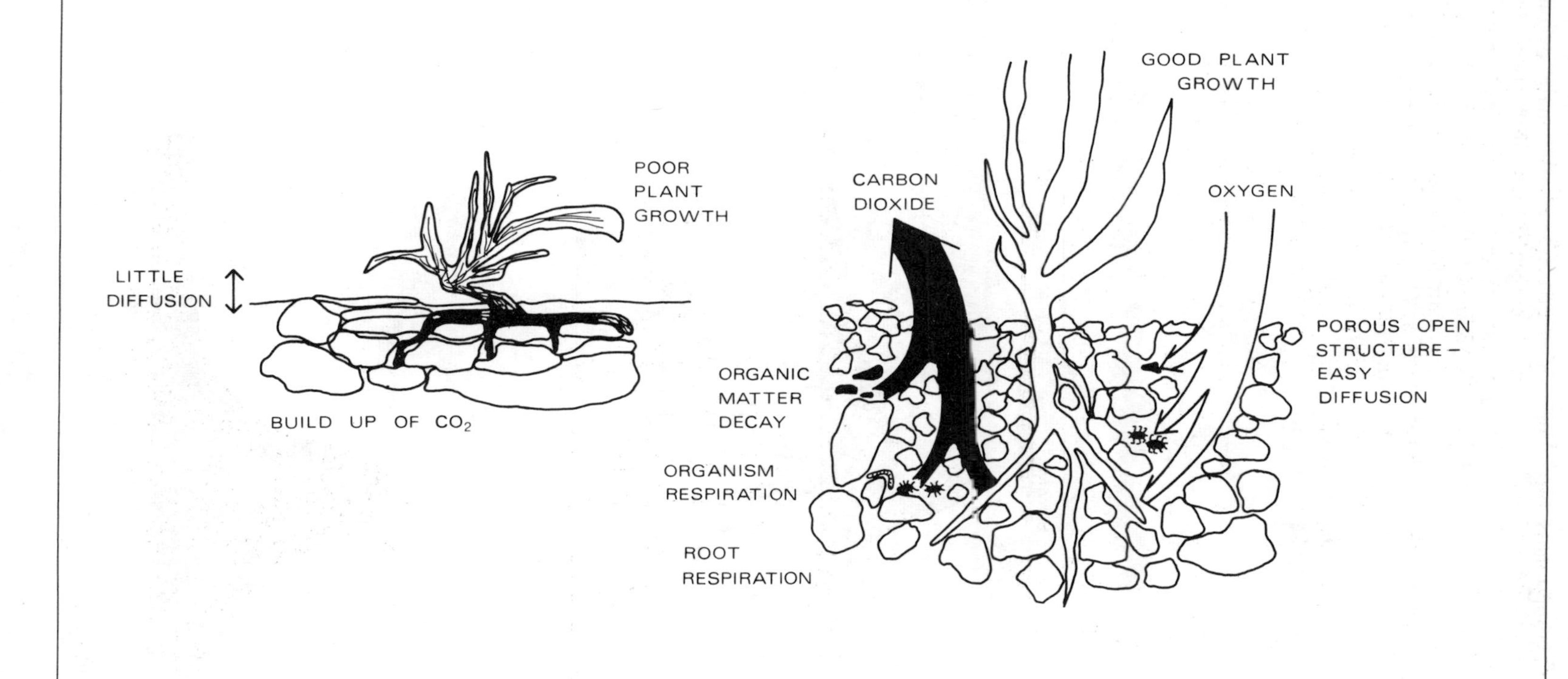

Figure 25 Soil aeration

be synthesized into new soil components, three important examples being:

1 leaf litter to humus,
2 minerals to nutrients, and
3 herbicides and pesticides to simpler (less harmful) compounds.

Earthworms are important in transforming leaves into humus (Figure 22), whilst *bacteria, fungi* and *actinomycetes* are the main decomposers in the latter stages of the humification of organic matter. Actinomycetes are often classified with the fungi by biologists as they can be filamentous like fungi, but they can also resemble bacteria in being single celled. Thus they are a transitional group between the bacteria and the fungi. The activities of the organisms break down the organic matter, releasing organic and inorganic nutrients so that they are available again for plant use.

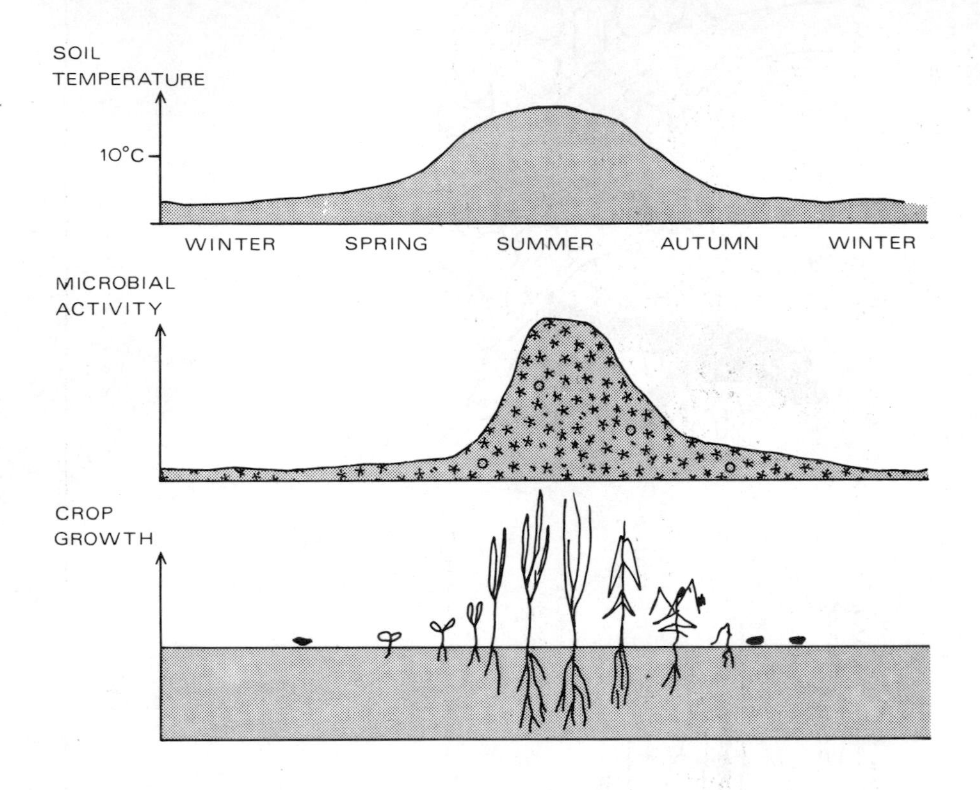

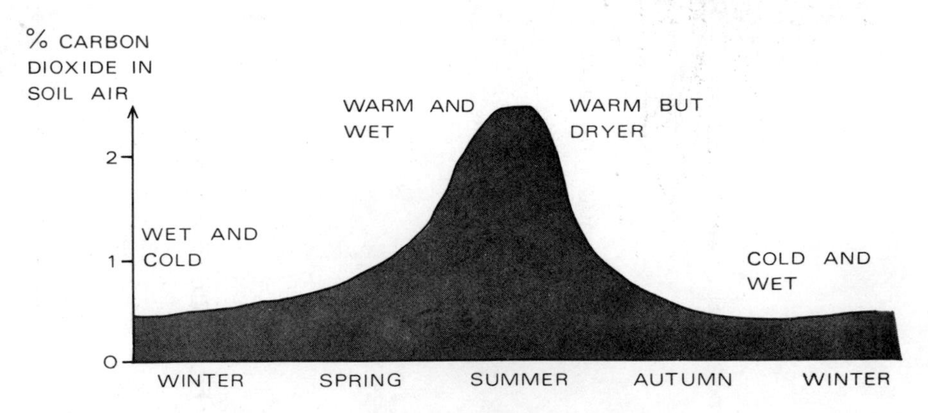

Figure 26 Seasonal carbon dioxide production

Fungi are the chief cause of decay of hard, woody, lignified tissue. The fungal threads, called *hyphae*, penetrate the woody tissue along the cell walls and excrete digestive enzymes (Figure 27).

Soil bacteria and fungi are able to 'attack' many soil minerals, either by producing acids which dissolve them or by directly utilizing chemical elements from the minerals in their metabolism. Their significance lies in the fact that they can attack resistant minerals, like some of the silicates, which are normally almost insoluble in rain water. Thus the nutrients derived from silicate minerals are made more readily available for plant growth by the decomposition of the minerals by bacteria and fungi.

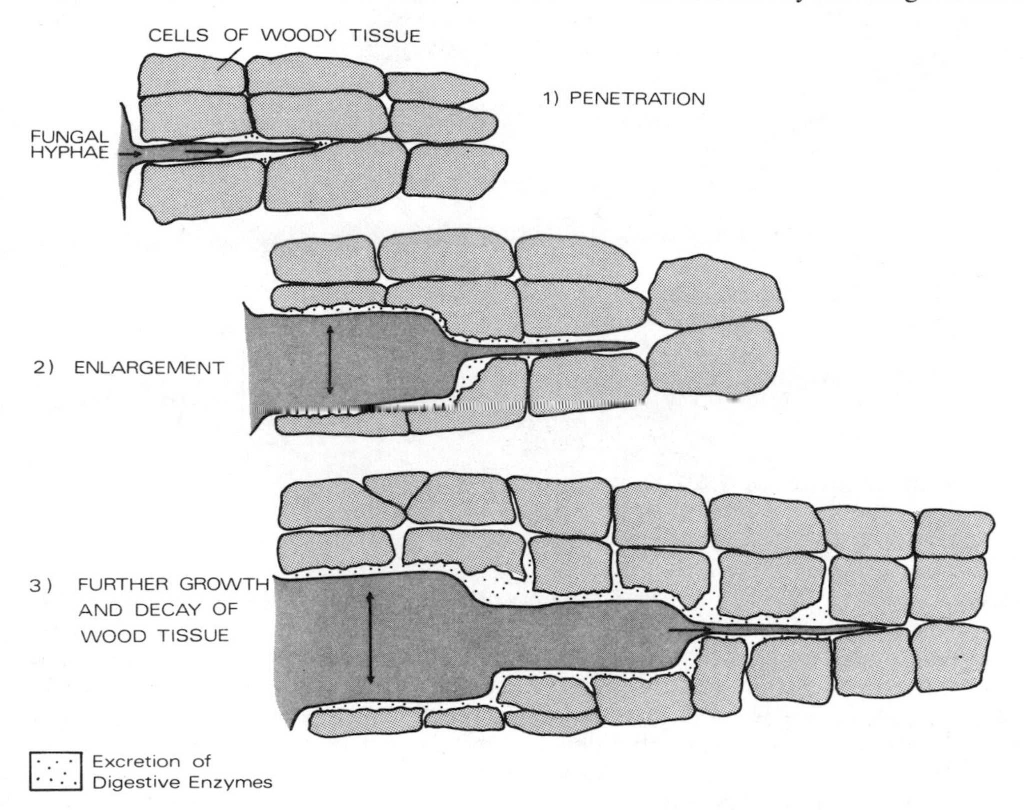

Figure 27 Wood decay by fungi

Modern farming techniques require the use of pesticides and herbicides to control insects and weeds in order to promote agricultural productivity. A significant proportion of these applied chemicals may be left as a residue in the soil and here they may be degraded by soil organisms. For example, the herbicide 2,4 D (2,4 dichlorophenoxyacetic acid) is attacked by the bacteria *Achromobacta, Corynebacterium* and *Flavobacterium.* The basic structure of the herbicide is a benzene ring (a ring of six carbon atoms) with chlorine and organic compounds attached. The bacteria attack the herbicide by cleaving off the compounds and chlorine from the benzene ring (Figure 28). Ninety per cent of the 2,4 D applied to soil may disappear in ten days by this process.

Other pesticides and herbicides are more resistant to attack by microorganisms. These may persist in the soil and if so give cause for concern about their fate in the environmental system.

Transformations and Fixations

One of the most important plant nutrients is nitrogen (N). Soil microorganisms can fix gaseous nitrogen and transform it into nitrate, which can then be used as a plant nutrient. Nitrogen is the basis for all plant proteins and thus is a nutrient vital to growth. While these reactions will be considered in more detail in section 4.3 it will be useful to give an example of a transformation. Nitrobacter transforms *nitrite* (NO_2), which is toxic, to *nitrate* (NO_3), which is used as a plant nutrient:

$$\underset{\text{nitrite}}{NO_2^-} + \underset{\text{oxygen}}{1\tfrac{1}{2}O_2} \xrightarrow{\text{Nitrobacter}} \underset{\text{nitrate}}{NO_3^-}.$$

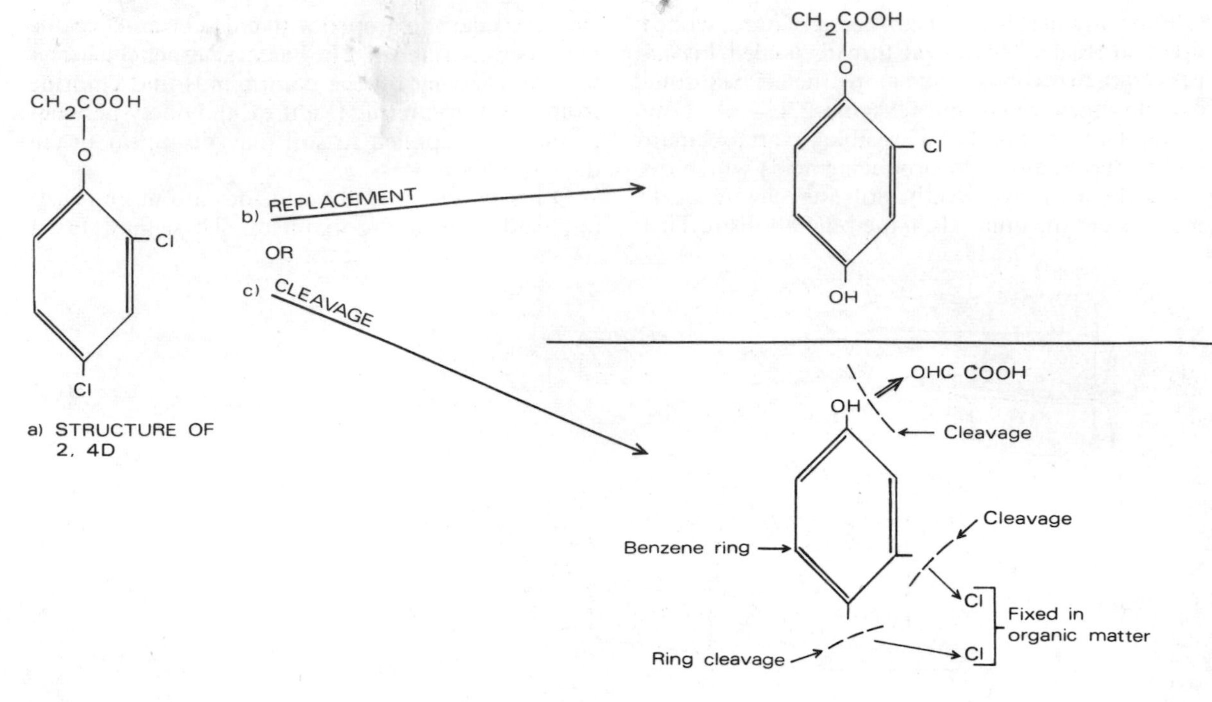

Figure 28 Breakdown of the herbicide 2,4 D by bacteria

Structural processes

Both the micro- and macro-organisms of the soil have a considerable influence on soil structure. Actinomycetes are thought to be important in the binding of individual particles together in crumb structures.

Earthworms, ants and small burrowing mammals are important in altering pore spaces as their burrows allow air and water to penetrate deep into the soil. Plant roots, when they die, also leave channels along which air and water may pass.

Summary	*The effects of soil organisms*
Decomposition	Bacteria Fungi Actinomycetes Earthworms
Fixation	Bacteria
Structural processes:	
Binding	Actinomycetes Fungi
Aeration	Earthworms Insects – ants, grubs, millepedes Burrowing mammals

Many other organisms exist in the soil than have been discussed in this brief section. Both herbivorous and carnivorous organisms are found. The soil is a complex ecosystem where no organism or inorganic process operates in complete independence. This theme will be discussed in greater detail in Chapter four.

2.10 Soil nutrients

Chemical elements found in the soil which are needed for plant growth are termed nutrients. Like soil water, they can be classified in terms of availability to plants (Figure 29):

1 Nutrients in solution in the soil water. These are freely available to plants, but may be washed out of the soil in gravitational water.
2 Nutrients attached to clay–humus complexes. These are the most important reserves as they are available to plants but are not easily washed out of the soil.
3 Nutrients stored in minerals and unavailable to plants unless released by weathering (e.g. by bacterial and fungal attack, see section 2.9).

As the most important reserve of nutrients is found in the clay–humus complexes of the soil plasma it is useful to consider these complexes in some detail. Nutrients can be held in combination with humus substances and attached to the small clay particles. The clays have a *negative* electrical charge at their surface and to *balance* this electrically a layer of *positive* ions (or cations) coats the clay particles. These cations attached to the clay surface are referred to as the *adsorbed ions*. They are not

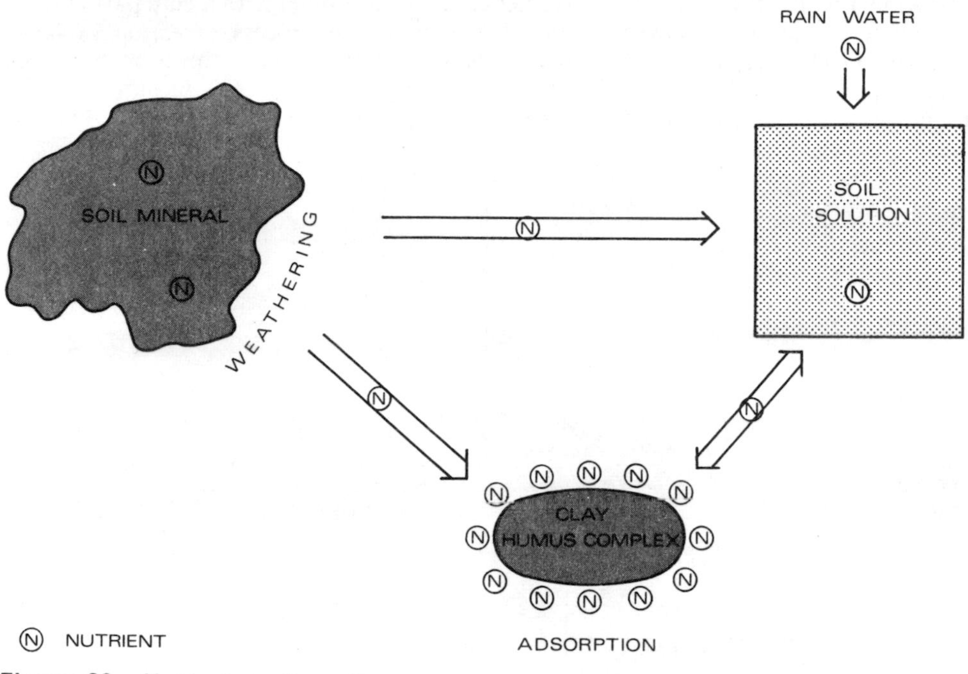

Figure 29 Nutrients in the soil

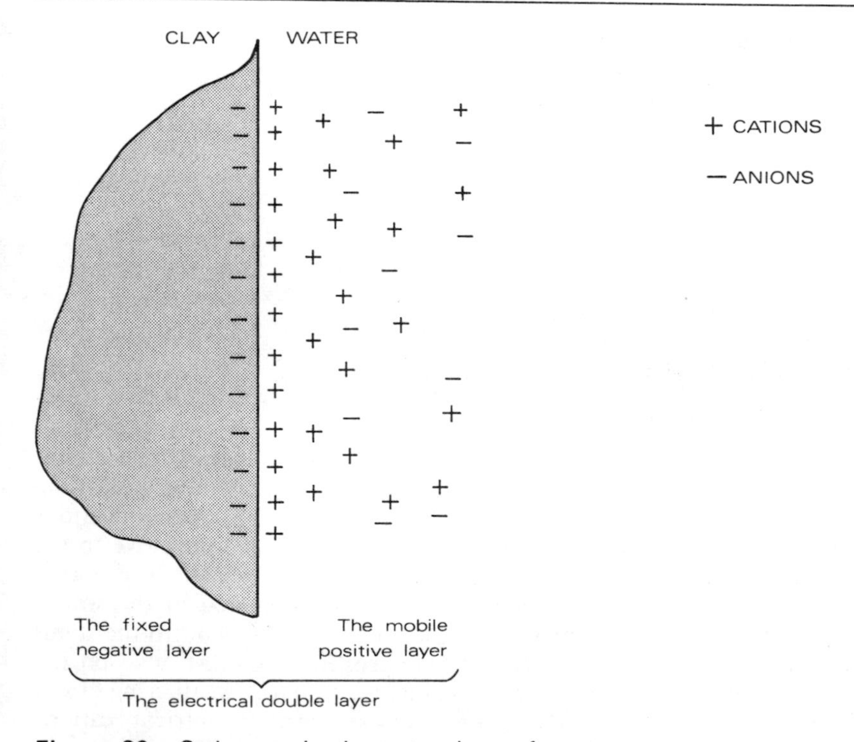

Figure 30 Cations and anions at a clay surface

*ab*sorbed (incorporated in) but are *ad*sorbed (stuck) to the clay. The double layer of negative and positive charges was first described by *Gouy* and is referred to as the *electrical double layer* or the *Gouy Layer* and this is shown in Figure 30.

The amount of cations in the soil water next to the clay increases towards the clay surface and the amount of negative ions (anions) decreases towards the clay surface (Figure 31). The cations are the plant nutrients and include calcium (Ca^{++}), magnesium (Mg^{++}), potassium (K^{+}) and sodium (Na^{+}).

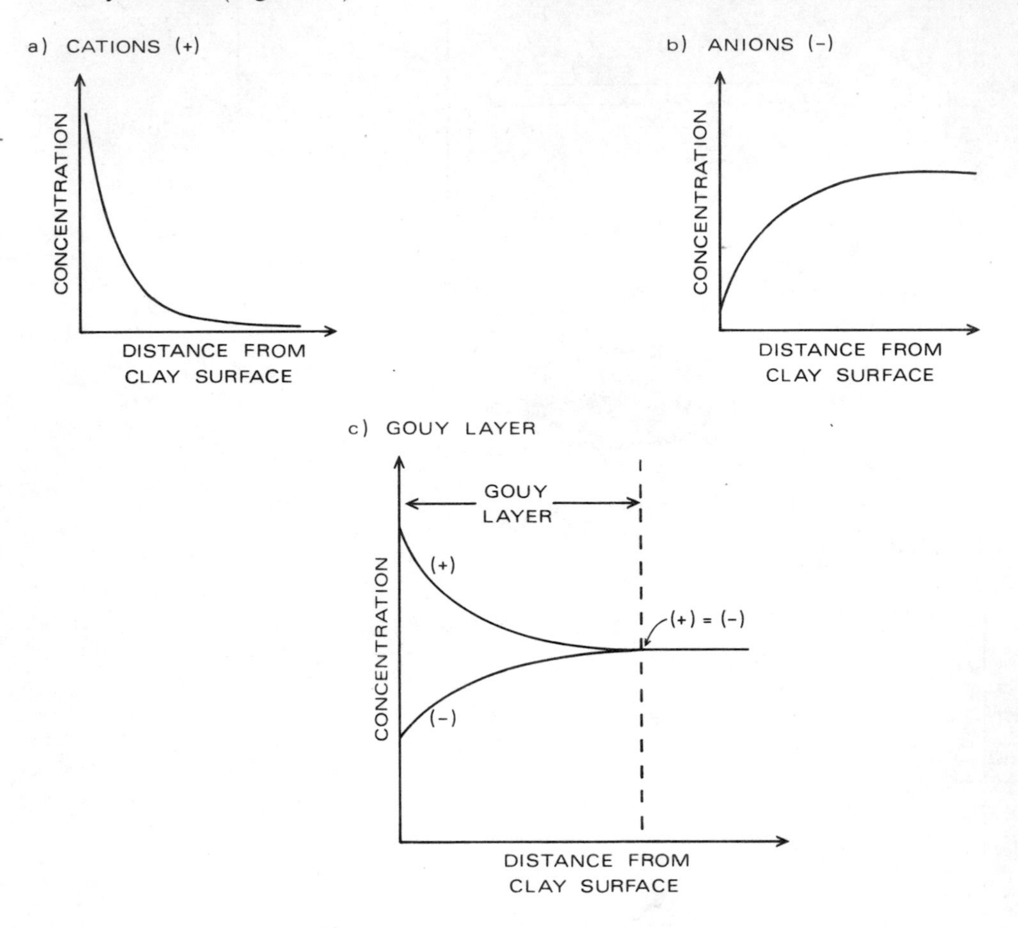

Figure 31 Cation and anion concentration at a clay surface

Calcium is important in the growth of shoot and root tips. *Magnesium* is a basic constituent of chlorophyll, an activator of plant enzymes and is involved in osmosis. *Potassium* is used in the formation and transport of starches, sugars and oils and also in enzyme reactions and osmotic processes. The role of *sodium* is not fully known, but it may fulfil some of the functions of potassium.

Cation exchange

How do the nutrients become detached from the clay and move into the plant root? When a plant rootlet comes into contact with a clay particle with adsorbed cations the process of *cation exchange* occurs. Next to photosynthesis this is probably one of the most important processes in nature as it is the primary mechanism of plant nutrition. The root gives out hydrogen ions in exchange for nutrient ions (Figure 32). Calcium, magnesium, potassium and sodium can migrate from the swarm of cations adsorbed on to the clay surface and move to the plant root; their places are taken by hydrogen ions. The nutrient ions are translocated in the water-transporting xylem tissue of the plant to the stems and leaves. The hydrogen ions substituted on the clay surface are often used in the weathering of soil minerals. This releases further nutrient cations which can be adsorbed on to the clay surface.

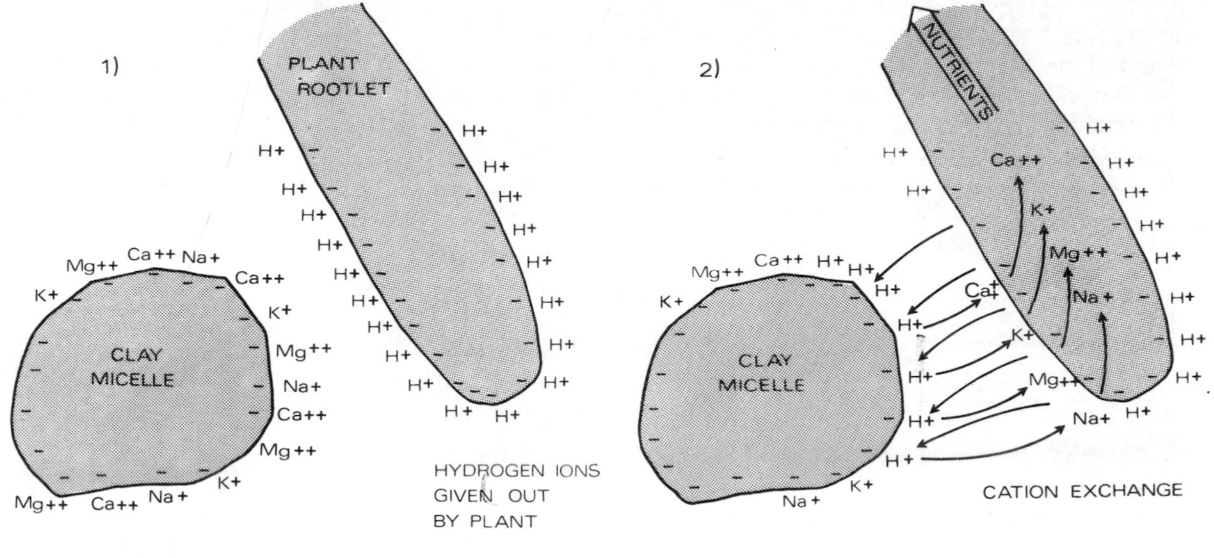

Figure 32 Cation exchange

The ability of clays and humus to yield cations for plant use is *called the Cation Exchange Capacity* (CEC). CEC is measured in mille-equivalents per 100 g of soil. A mille-equivalent is a measure of the ratio in which the element combines with or displaces hydrogen and is calculated by:

$$\frac{\text{molecular weight}}{\text{valency} \times 1000}\,\text{g},$$

e.g. 1 mille-equivalent of calcium

$$=\frac{40{\cdot}08}{2 \times 1000}=0{\cdot}02004\,\text{g}$$

The cation exchange capacity of humus is about twice that of pure clay, which emphasizes the importance of humus in soil fertility. The clay types discussed in section 2.2 differ in their cation exchange capacity, just as they do in their swelling and shrinking capacity (see under Soil structure, section 2.4). Their cation exchange capacity varies according to the surface area available for cation adsorption. Montmorillonite has the highest surface area, possessing both an internal and external surface. With illite the internal surface is already occupied by a fixed layer of potassium ions and therefore exchangeable cations are restricted to the outer surface and thus the CEC is less than that of montmorillonite. In kaolinite the CEC is even lower as it has a much smaller surface area than the other clays (Figure 33).

While cations are readily stored in the clay–humus complexes, anions are less easily stored. Some anions, such as nitrate (NO_3^-), sulphate (SO_4^-) and phosphate (PO_4^-), are extremely important plant nutrients, but they are more readily leached out as they tend to occur more in solution in the soil water. The maintenance of these nutrients often requires fertilizer addition (see section 5.1).

2.11 Soil acidity or pH

Soil pH is a measure of the concentration of the hydrogen ions in the soil water. Since an acid is a compound which dissociates (see section 1.2) in water to yield hydrogen ions (H^+), soil pH is a measure of the acidity of the soil water. Soil acidity is an important factor in the availability of plant nutrients.

Many chemical compounds found in soil are most soluble when they are in a slightly acidic solution, i.e. one charged with hydrogen ions. When the elements are most soluble they are at their most available for plants to take up and use. Thus there is an optimum pH for plant growth and this is when

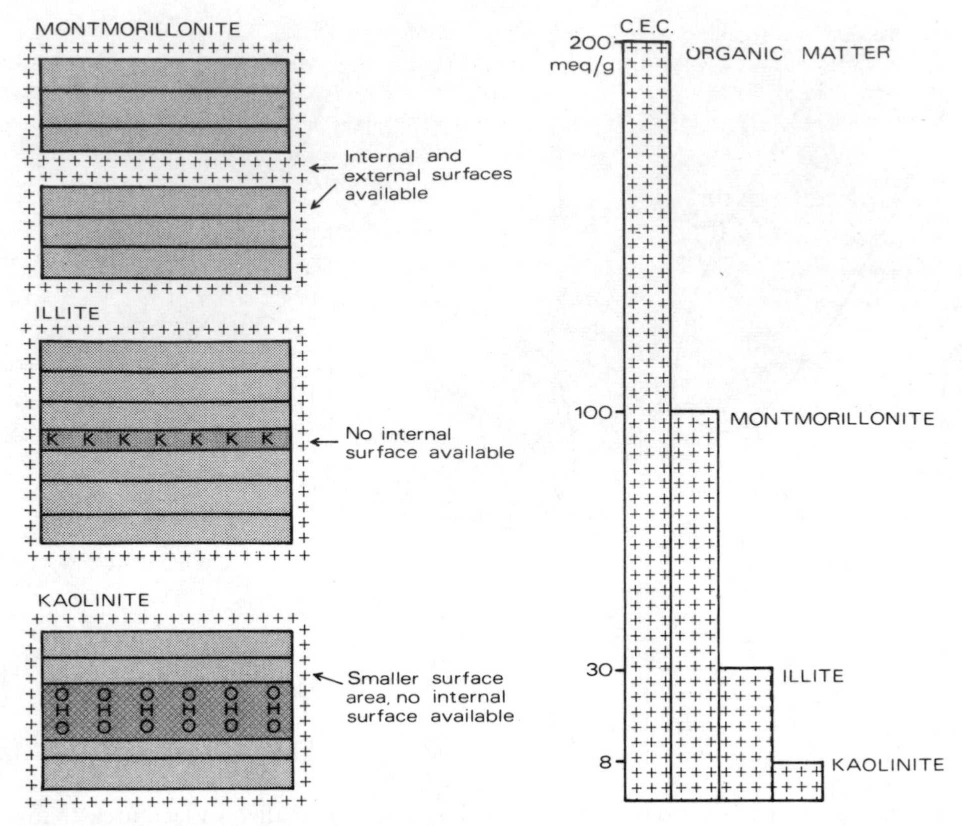

Figure 33 Cation exchange capacity

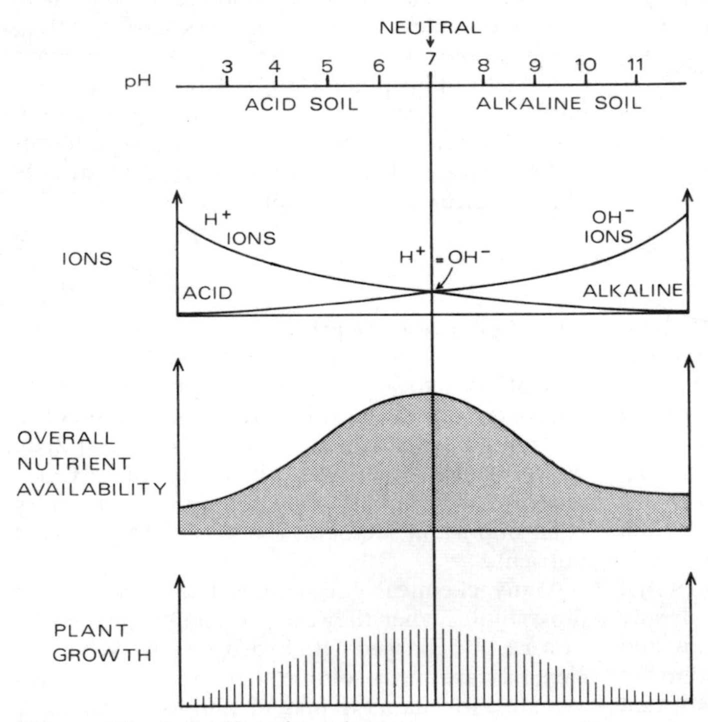

Figure 34 Soil pH and nutrients

the pH is slightly acid. In Britain soils tend to be rather more acid than this optimum and the chemical compounds may be so soluble as to be toxic. The addition of lime brings the pH up to the optimum for plant growth. The liming problem is discussed further in section 5.1.

pH is measured on a scale from 2 to 12 where pH is equal to the negative logarithm of the hydrogen ion concentration in moles per litres (where a mole is the molecular weight in grams, in this case = 1 g/l). Thus at pH 7 the hydrogen ion concentration = 0·0000001 g/l and the negative logarithm of this is 7. pH 7 is neutral and the concentration of hydrogen ions is equal to that of hydroxyl ions (OH^-). Above pH 7 the solution is alkaline. An alkaline substance is defined as one which dissociates to yield hydroxyl ions. Thus above pH 7 OH^- dominates and below pH 7 H^+ dominates.

Figure 34 summarizes the relationships between pH, nutrient availability and plant growth.

Summary

Soil minerals, texture, structure, fabric, organic matter, water, air, organisms, nutrients and pH have been discussed. It has become apparent that many of these soil components and properties are related. These relationships are summarized in Figure 35.

Starting with soil minerals these will influence nutrient status and texture according to grain composition, and the proportions of secondary and primary minerals. Texture will affect structure, controlling water retention and aeration. These are two important controls of soil organisms which in turn influence the nature of the soil humus as well as soil structure. Organic matter influences nutrient availability as does mineral type and water retention. All these factors interact to influence plant growth.

It is emphasized that the soil is a complex system where factors are interdependent and no factor, biological or physical, operates in complete independence.

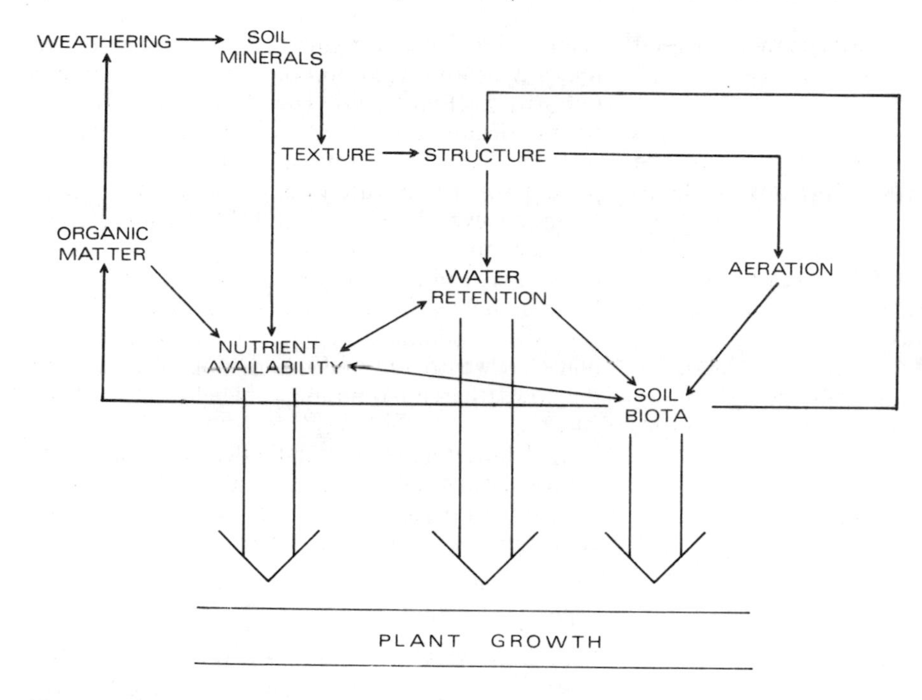

Figure 35 Summary of soil factors discussed

3

Soil types and their development

3.1 Soil processes and the development of soil types

The formation of soil horizons

Soils can be divided into soil types largely on the basis of the horizontal layers or *horizons* which are visible when a pit is dug into the soil. These horizons are a result of the soil development processes discussed in sections 1.1 and 1.2: weathering and oxidation of minerals, leaching, cheluviation and the accumulation of humus.

The horizons are differentiated in detail by several variables, the main ones being colour, texture, organic content, structure, stoniness, acidity and nutrient content (see section 6.4). The major divisions in the soil profile are the A, B and C horizons (Figure 36). The *A horizon* is that of organic accumulation and is composed of leaf litter which is decaying and mixing with the underlying mineral soil. Immediately above the A horizon may be one or several of the L, F and H horizons referred to in section 2.6 (Figure 37). The separate A horizon is the mixed mineral–organic layer beneath the H horizon. Below the A horizon is the mineral *B horizon*, with a minimal organic content. The mineral matter is weathered and this distinguishes it from the unweathered parent material below, the *C horizon*.

A – Mixed mineral-organic horizon

B – Mineral horizon of altered material

C – Little altered parent material

Figure 36 Soil horizons

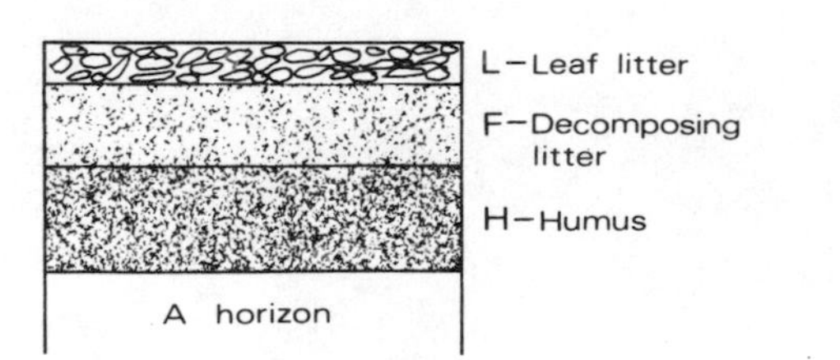

Figure 37 Organic horizons

Each horizon may be subdivided according to its precise nature and it is the formation of individual, distinctive horizons that gives each soil type its characteristic nature (see p. 89).

Inputs and outputs

If we look back at Figure 2 (p. 3) we can see that several soil inputs and outputs are given. We can now use the systems analysis concept of inputs and outputs (introduced in section 1.1) to help in understanding the formation of differing soil types. The three most important inputs and outputs are:

1. *The input and output of parent material* – this controls the depth of the soil B horizon.
2. *The input and output of humus* – this controls A horizon thickness.
3. *The input and output of water* – this controls the solution and leaching processes and also indirectly the accumulation of organic matter.

In many ways the input and output of water is the most important factor, and as the amount of water passing through a soil increases, the amount of solutes moved by leaching also increases. If the water inputs are very large then surface wetness tends to occur. This inhibits the oxidation of organic matter and raw, acid, mor humus will accumulate. As organic matter accumulates, the solution of minerals by cheluviation tends to increase (Figure 38).

Now, if we examine the mineral situation, we see that, as the input of weathered material increases, soil depth also increases. But if the output of weathered material is greater than the input then soil erosion must be occurring and soil depth decreases. This happens on steep slopes and also in some unvegetated areas.

Parent material and soil depth

In some cases soil development may be limited by the lack of weathered parent material. This is the case over very hard rocks, which weather only slowly and do not yield any depth of granular material (such as basalt, well-cemented sandstone and Carboniferous Limestone). Thus the soil depth is limited by the lack of mineral input.

Where a thick deposit of parent material occurs soil formation is not limited by this factor and examples are over glacial tills, clays, sands, gravels, alluvium, colluvium (slope foot deposits), loess and head (periglacial deposits). Thus, in Figure 39, we can describe two main situations, one where the depth of soil formation is limited by mineral material input (A) and one where it is not (B). A further subdivision (A1) is necessary where soil formation is limited by a large output and (A2) where formation is limited by a small input.

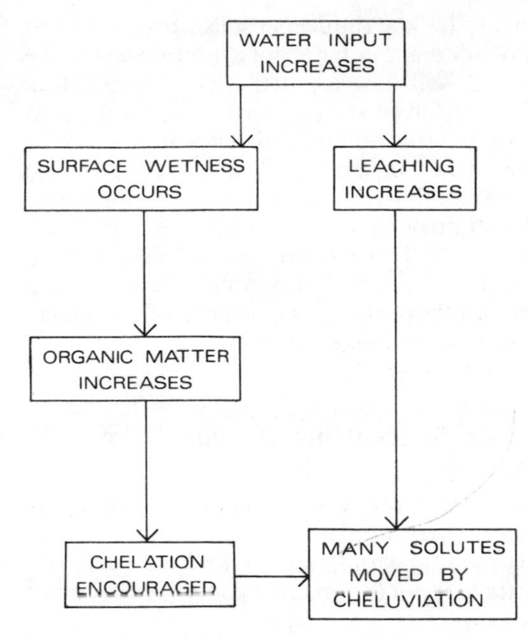

Figure 38 Water input and cheluviation

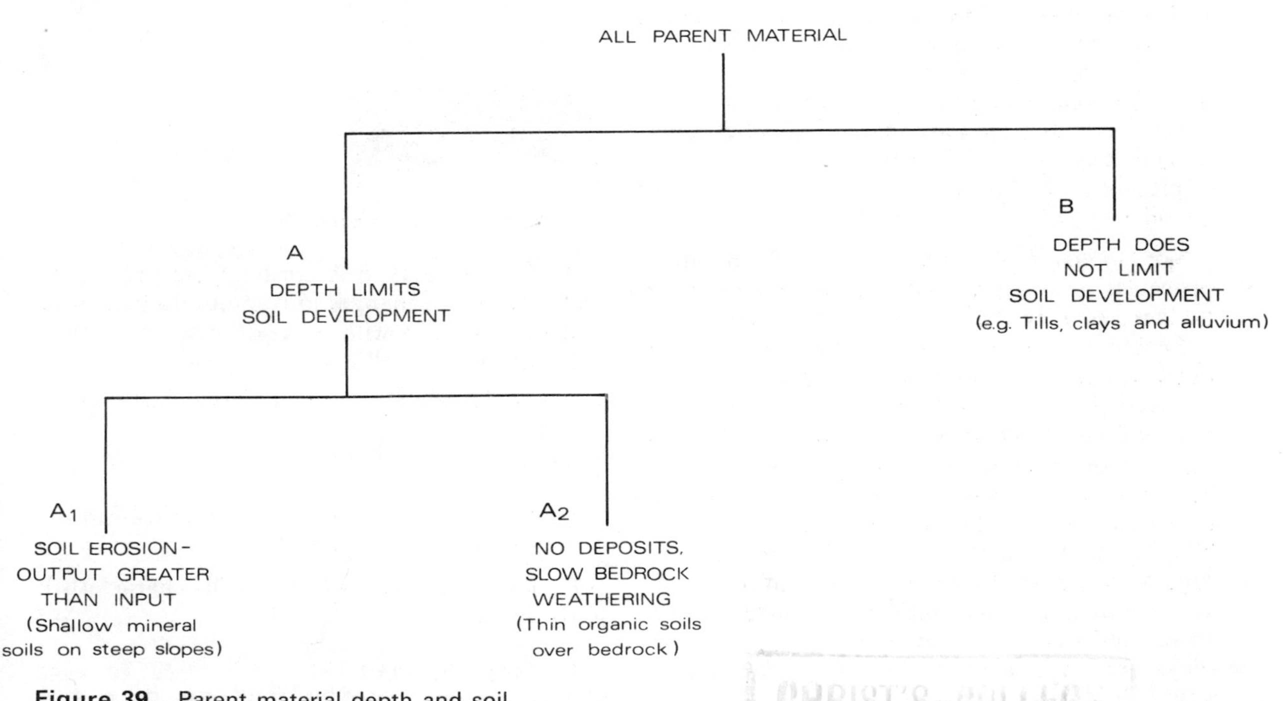

Figure 39 Parent material depth and soil

Soil development on deeper parent materials

We can now look at the deeper soils where the input supply of mineral matter is not a limiting factor. To understand soil development and horizonation attention will have to be paid to the other soil inputs and outputs – soil humus and soil water.

It was shown in section 1.1 that soil developed when a parent material reacted when attempting to achieve a state of equilibrium with its environmental conditions. Three main cases of differing environmental conditions concerning water are important and they rely on the balance of (a) rainfall input, and (b) drainage water output.

The three cases are:

1 where water outputs are equal to water inputs (drainage is good) but the inputs are small (yearly rainfall is not heavy),
2 where outputs are equal to inputs but the inputs are large (high rainfall), and
3 where the rate of output does not keep pace with the rate of input (drainage is poor); rainfall input high or low.

Four soil development processes arise from these three cases – leaching, clay translocation, podzolization and gleying:

leaching is the downwashing of soluble material;
clay translocation is the downwashing of clays; this is also termed *lessivage* (pronounced to rhyme with 'massage');
podzolization is the downwashing of most of the soil solutes by cheluviation, leaving a bleached white or grey upper soil mineral horizon;
gleying is the term given to the soil-forming processes occurring in waterlogged soils where alternate oxidizing and reducing conditions are present according to the level of water in the soil.

We can now look at these four soil development processes in connection with the three environmental cases of water balance described above.

Leaching, clay translocation and podzolization occur where the drainage is good. The differences between the three are that they represent an increase in movement of water down the soil profile due to increasing amounts of rainfall input. However, a crucial step is that as surface wetness increases organic matter accumulation also increases. Thus cheluviation increases as rainfall input increases. Therefore during the podzolization process the accumulation of a thick mat of humus on top of the soil occurs and many more nutrients and clays are washed down the soil profile than in simple leaching (see Figures 8 and 44).

We can see that the simple term leaching is inadequate to describe the overall soil development processes which also involve the movement of clays and chelates. The term leaching is usually reserved for the movement of solutes in waters. The general term of *Eluviation* is used for overall washing out or removal of any material and the term *Illuviation* for washing in to the lower horizons. These terms are similar, but if the analogy with *e*xit is used for *e*luviation and *in* for *i*lluviation they should be easier to remember. Eluvial horizons are termed E horizons (see p. 89).

Soil types

This section describes soil development simply with respect to rainfall and other factors. Soil classification is described in further detail in section 6.6. The different types of soil development processes described above are reflected in horizon variations

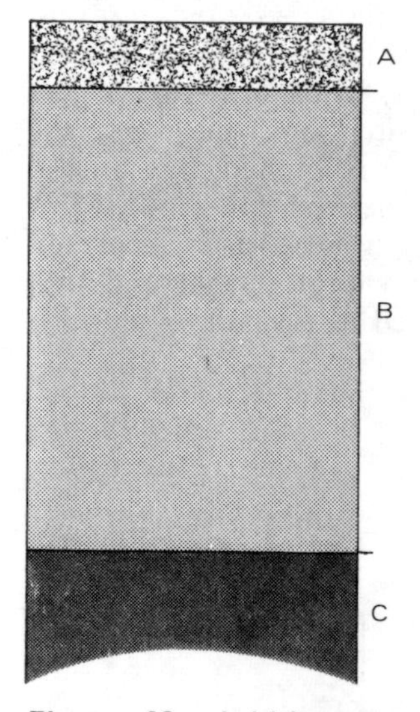

Figure 40 Acid brown earth

and hence in soil type. Accordingly, in Britain, the sequence of soil changes in response to increasing rainfall input (but with good drainage conditions maintained) is as follows:

1 low to moderate rainfall input: *acid brown earth*;
2 moderate rainfall input: *brown earth* (*sol lessivé*);
3 moderate to high input: *podzolic brown earth*;
4 high rainfall input: *podzol*.

Each type can be recognized by its characteristic horizons.

1 ACID BROWN EARTH

The A horizon is capped by a mull or moder humus.

The B horizon shows little signs of differentiation except that it may be slightly lighter coloured in the upper horizons due to the removal of solutes. A typical profile is described in Figure 40. No clay and very little iron has been washed down the profile. Only the most soluble elements, like calcium and magnesium, are being actively removed. The pH will be on the acid side, about pH 5, ranging possibly from pH 4·5 to pH 6·5.

2 BROWN EARTH (SOL LESSIVÉ)
The distinctive characteristic of this soil is that clay has been translocated down through the soil profile. The lower B horizon is rich in clay and therefore can be distinguished from the rest of the profile by its texture. This is termed a textural B horizon or a Bt horizon (Figure 41). Some iron is also moved down the profile. It is thought that the clay is washed down the profile in a jelly-like solution and that the clay is deposited in a skin or *cutan*, which can be seen in thin section (see section 2.5), lower down the profile (Figure 42).

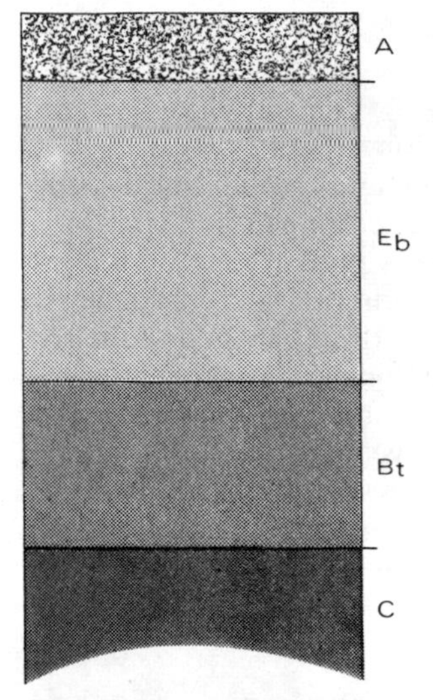

Figure 41 Brown earth (sol lessivé)

3 PODZOLIC BROWN EARTH
This soil is transitional between a brown earth and a podzol. Not only will some of the more soluble elements and clays have been washed down, but also cheluviation is becoming dominant. Thus a fairly thick accumulation of acid organic matter is present at the surface, and if the mineral horizon immediately under the organic matter is carefully examined some bleached, pale coloured sand grains can be seen. This is because much of the iron, which gives the soils a reddish colour when oxidized, is being actively removed in chelates. Furthermore, as well as the light coloured grains in the upper soil, some darker colouring may be visible in the lower profile. This is the redeposition of iron and possibly some of the organic matter that acted as a chelatory carrier. A B horizon rich in iron is termed a Bfe horizon and one rich in downwashed humus a Bh horizon (Figure 43).

4 PODZOL
A podzol can be easily recognized by the sharply contrasting horizons (Figure 44). The A horizon is very dark brown or black, composed of acid, mor humus. Below this is a white or grey bleached horizon of pale coloured mineral grains (often quartz) from which much of the iron has been eluviated. Below this is the redeposited iron in a reddish horizon which may be cemented to form an *iron pan*. The iron pan is probably also rich in organic matter. Iron is washed down under wet conditions (especially in winter) when the iron is ferrous and soluble (see section 1.2). When the soil dries the iron oxidizes and is precipitated. The position of the iron pan thus represents where the iron was exposed to the air, probably at the top of the spring water table in the soil.

The Bfe horizon is the iron pan and the eluvial horizon is termed the Ea horizon (Figure 44). Such a profile is characteristic of porous parent materials and of areas of high rainfall and on sites with good drainage.

Gleying

Where soil water output is not rapid the soil is liable to *gleying*. If water stays in a soil for a long time the soil pores remain filled with water. This does not mean that all oxygen is immèdiately excluded as some can be dissolved in the water (this is how fishes and other aquatic life respire). But the oxygen in the soil water will be quickly used up by the soil microorganisms. Gaseous diffusion is slower in water than it is in air and thus water which is stagnant in a soil soon becomes deoxygenated and the oxygen is not quickly replaced by diffusion from the surface. The iron in the soil becomes reduced because of the lack of oxygen and this gives the soil a greenish or bluish tinge, which is the characteristic background colour of a gley soil (see section 1.2).

However, the most distinctive feature of a gley soil is a scattering of red mottles. These are places where fresh oxygenated water or air have been able to penetrate and oxidize the iron. Thus the larger soil pores, structural cracks and root channels are red with precipitates of oxidized iron. The upper

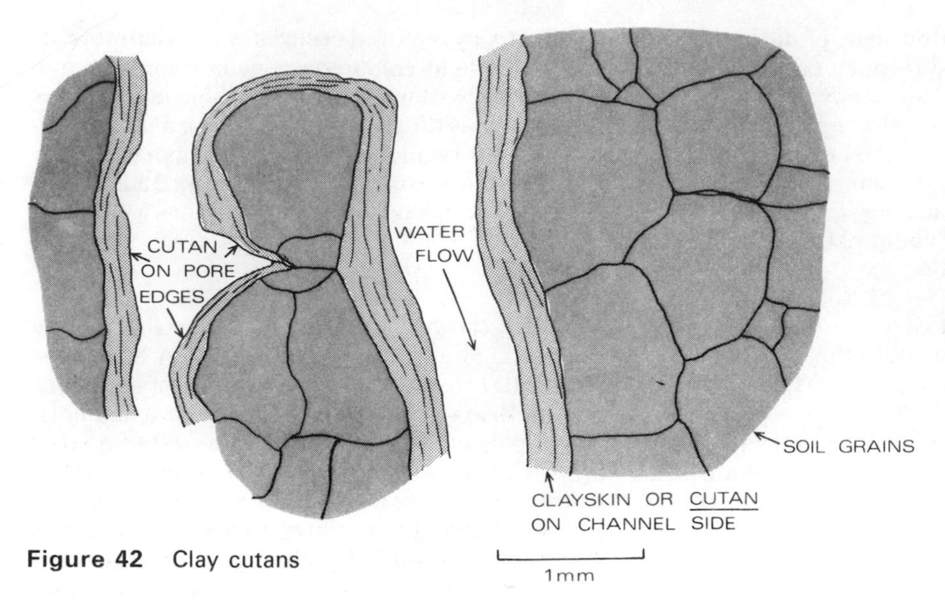

Figure 42 Clay cutans

parts of the profile tend to be redder or browner and the lower parts paler greeny-blue. This is because the position of the water table will fluctuate in the soil according to season and the lower parts will be more permanently waterlogged than the upper.

All soil types may show signs of gleying and thus a gleyed brown earth or a gleyed podzol are possible.

GLEYED BROWN EARTH

The overall appearance is of an acid brown earth, with little movement of clays or solutes, but the B horizon is mottled (Figure 45).

PEATY GLEYED PODZOL

This is one of the commonest soil types in Britain where rainfall is high. High rainfall means two things – that leaching is encouraged and that gleying is encouraged. Thus, a podzol may be formed by leaching, but very high rainfall, especially in winter, may lead to the mottling of some of the soil horizons. A peaty gleyed podzol may show mottling below the iron pan, and in some places the iron pan may become so well developed as to be impermeable to water. Mottling may then occur above the iron pan. Podzols with gleying are usually accompanied by the accumulation of a very thick peaty A horizon and they are often referred to as

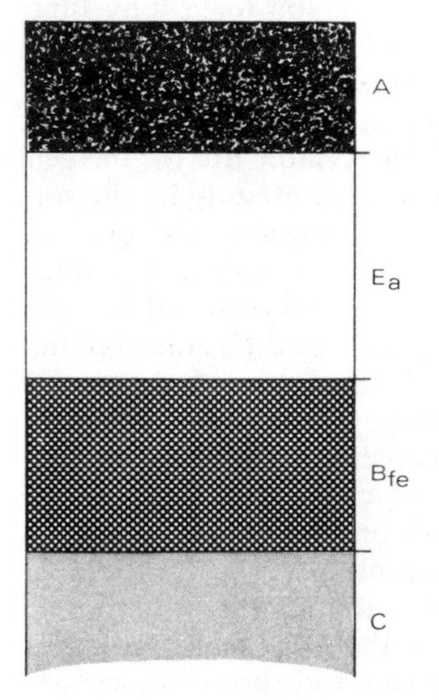

Figure 43 Podzolic brown earth

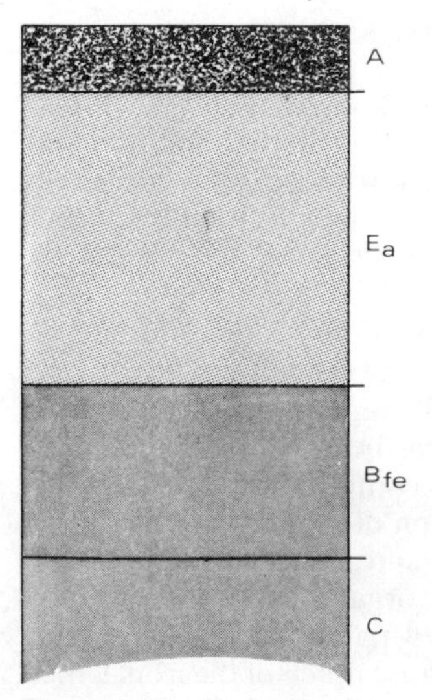

Figure 44 Podzol

peaty gleyed podzols (Figure 46), whereas the simpler podzols without gleying may be referred to as *humus–iron podzols*.

GLEYS

True gleys are usually found in thick clay deposits and the lower profile is simply the unweathered clay. The upper horizons are mottled and merge into the organic horizons (Figure 47).

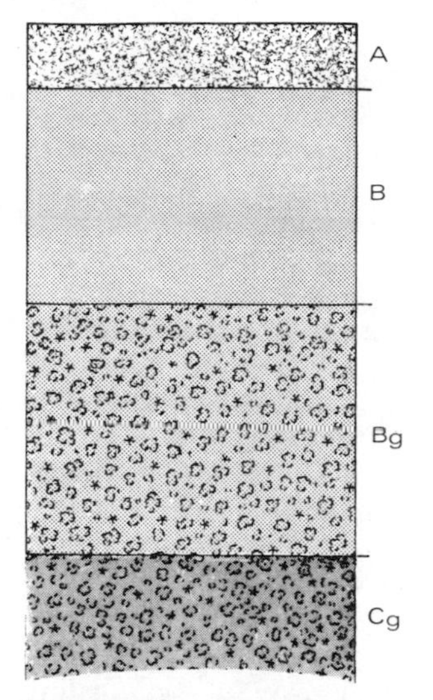

Figure 45 Gleyed brown earth

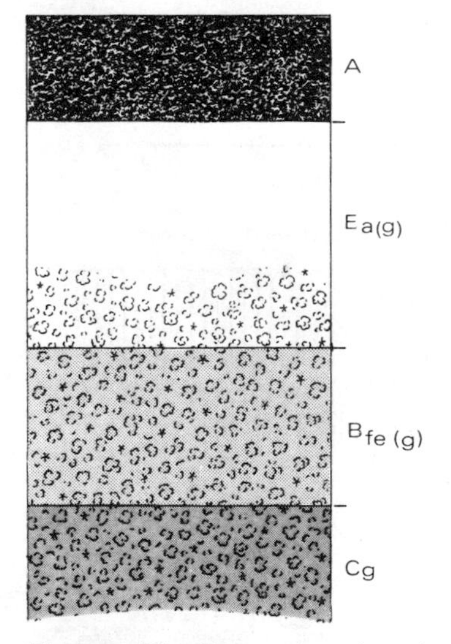

Figure 46 Peaty gleyed podzol

CALCAREOUS SOILS

Calcareous soils are those little affected by rainfall and leaching processes and owe their presence to a limestone parent material in areas of low rainfall. They can be developed *in situ* where the calcareous parent material breaks up and is distributed throughout the soil profile or they can be developed in low-lying areas where calcium carbonate is washed in from surrounding limestone areas. In the latter case a gleyed calcareous soil may develop.

Calcareous soils do not occur on limestone when the rock does not break up easily and where rainfall

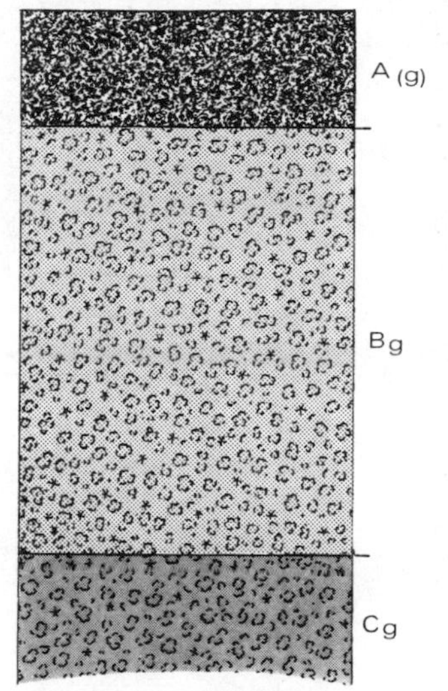

Figure 47 Gley

is high. The calcium is easily leached out of the soil in these cases and is not readily replenished by weathering input from the base of the soil. Figure 48 shows the formation of calcareous and non-calcareous soils on calcareous bedrock according to the nature of the rock and the amount of rainfall. The *calcareous brown earth* and its horizons are shown in Figure 49.

Summary – water balance and soil type

The development of soil types on deep parent material in Britain is summarized in Figure 50. The progression from an acid brown earth to podzol as rainfall input increases is shown and the influence of a low drainage output is shown as causing gleying and peat growth.

The overall distribution of soil types in Britain is shown in Figure 51 in relation to rainfall, the darker shading on the soil map relating to a high mineral content which is concomitant with low rainfall and therefore a low amount of leaching.

Soils developed where mineral input is limited

These soils are generally much thinner than the ones described above and soil development is usually limited to the accumulation of organic matter. Some mineral matter is usually present, but this is incorporated into the organic matter. These soils

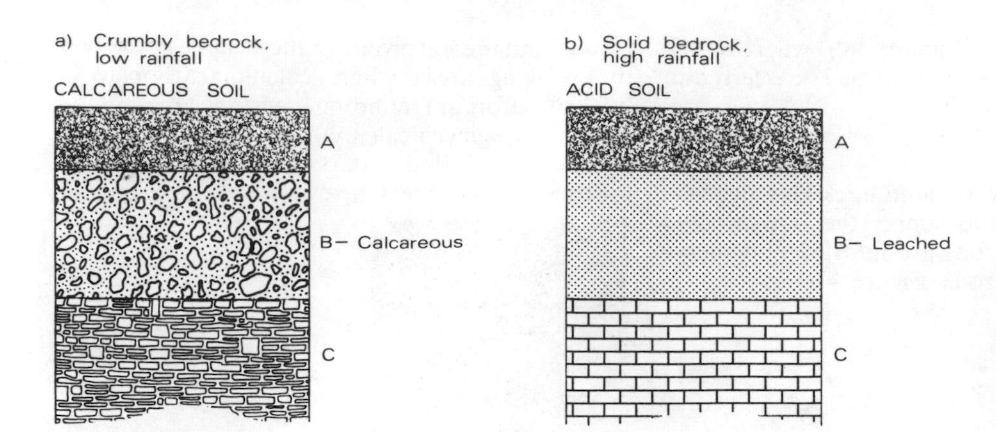

Figure 48 Limestone soils

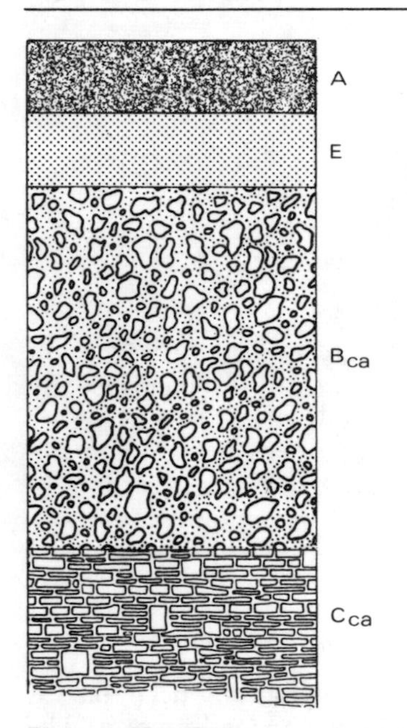

Figure 49 Calcareous brown earth

do not possess a mineral B horizon. A mixed mineral–organic A horizon overlays the hard rock C horizon. On hard limestone a *rendzina* soil occurs and on hard acid rocks a *ranker* occurs.

RENDZINA (Figure 52)
The humus is of the mull type and passes into limestone bedrock below.

RANKER (Figure 53)
The humus is of the mor type and passes into hard rock such as basalt or siliceous sandstone. The humus is usually wetter and more acid than in a rendzina.

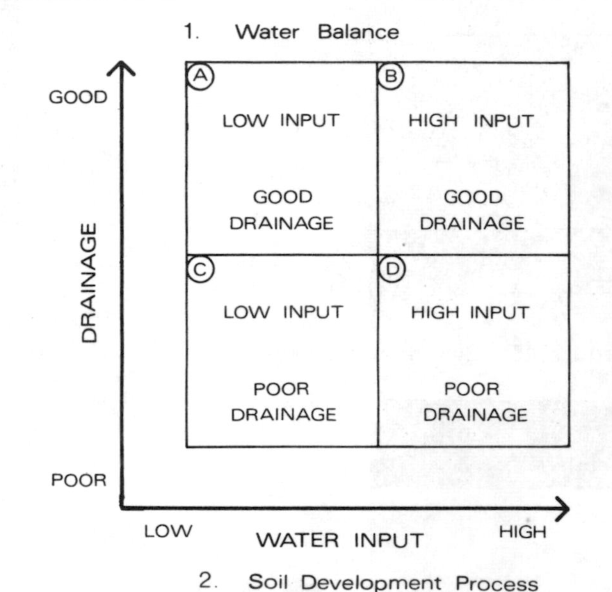

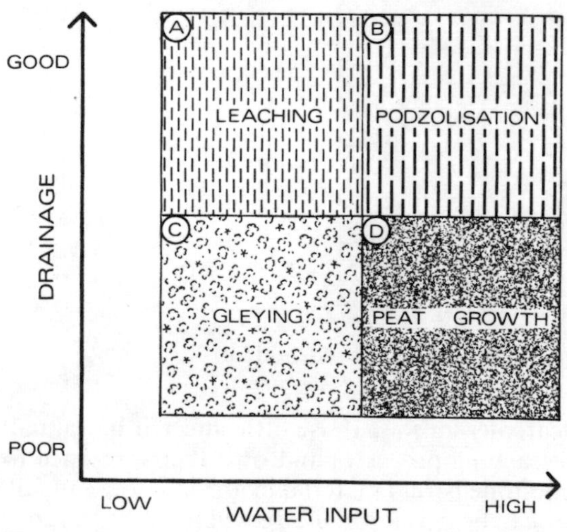

Figure 50 Water balance and soil development processes

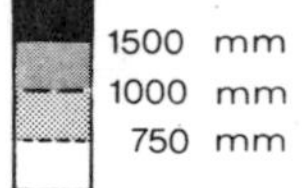

Figure 51 Main soil types and annual rainfall

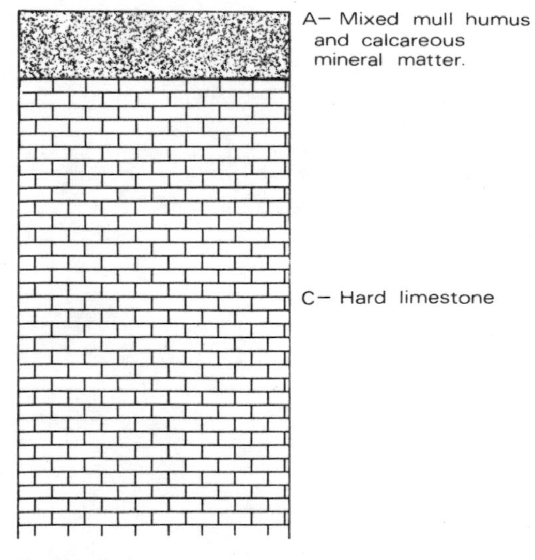

Figure 52 Rendzina

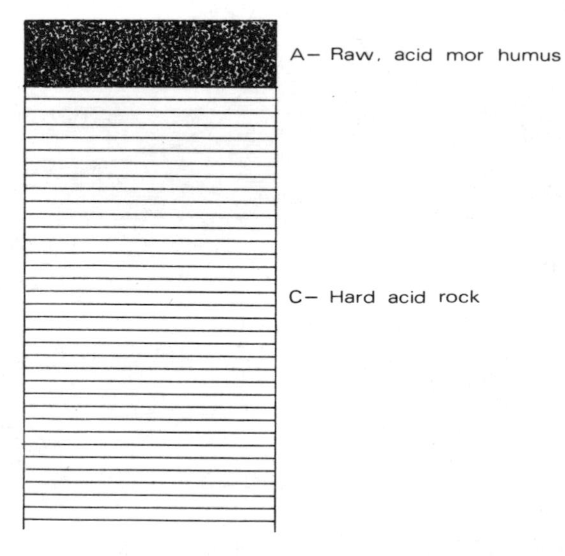

Figure 53 Ranker

3.2 The time factor in soil development

The legacy of the Pleistocene

Between about 1–2 million years ago and 10 000 years ago Britain was covered with a series of ice sheets. These acted to clear away any pre-existing soil and other weathered material from highland areas and to deposit this in lowland areas. Thus in high mountain areas the post-glacial soil parent material was bare rock. In valleys and in lowland plains glacial till, together with sands and gravel, form soil parent materials (Figure 54). The numerous effects of the Pleistocene ice sheets can be seen in the materials which have been left available as soil parent materials:

1 the glacially scoured rock in the highlands;
2 the deposits of boulder clay or till in the lower lands;
3 the deposits of windblown deposits, notably loess and sand dunes;
4 the deposits of fluvio-glacial sands and gravels;
5 the periglacial head deposits (such as coombe rock), frost wedges and colluvium (slope-foot deposits);
6 river alluvium that has accumulated from the Pleistocene to the present day (previous deposits often having been covered or eroded);
7 peat deposits that have accumulated in the wettest times since the Pleistocene.

The significance to soil development of glacial processes is that, wherever deep mineral soils are found in Britain they are almost certainly the result of glacial deposition (boulder clay and loess), coupled with slope-washing and slope-creep processes of periglacial times.

Loess is a little-known component of soil in Britain but is becoming increasingly recognized as an important constituent of soil in southern Britain. Unlike the well-known thick deposits of North Germany or China the loess is only up to about 30 cm deep in most areas. A thin blanket was spread out from the limits of the last glaciation, though thicker pockets can occur (Figure 55). The proportions of the lightest, most easily windblown

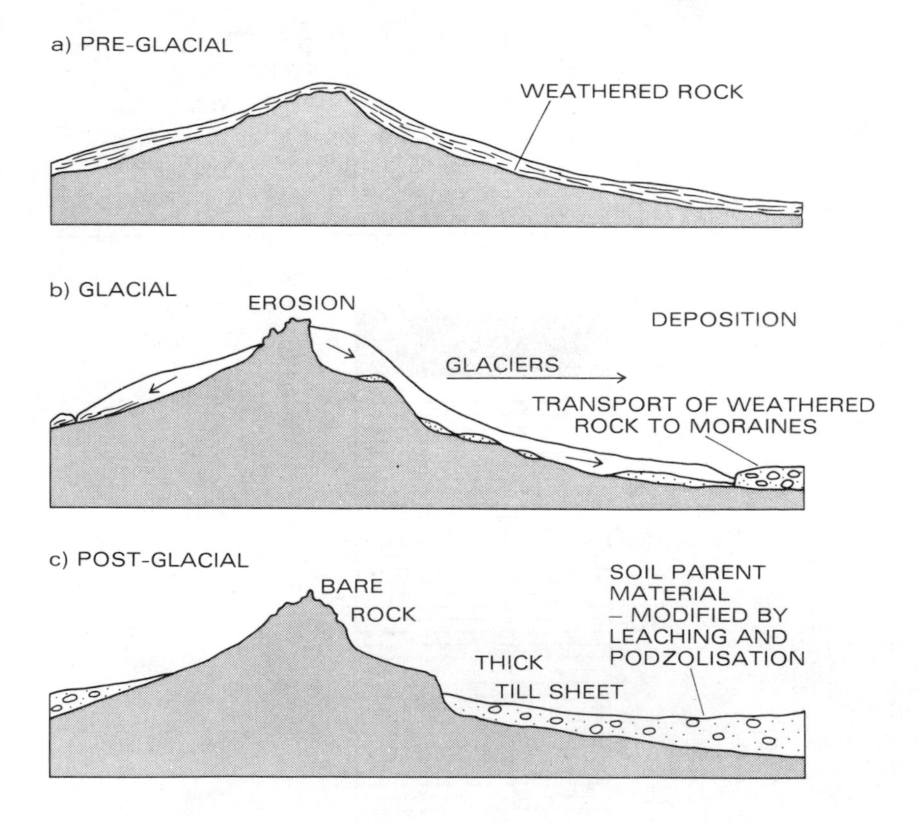

Figure 54 Glaciation and soil parent material

minerals, such as mica plates, increase to the south and west. Not all sizes of mineral grains can be readily blown by the wind. The largest particles are too heavy and the smallest are either light enough to have been blown beyond southern Britain or are clays which, being cohesive, stuck together and therefore were too big to be blown. It follows that most loess deposits in Britain are therefore made

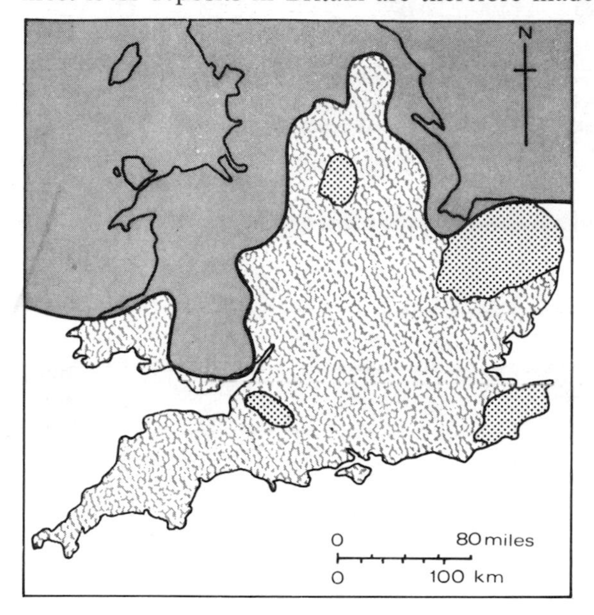

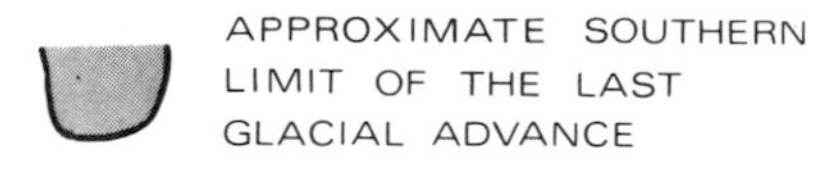

AREA OF POSSIBLE LOESS DEPOSITION

AREAS WHERE DEPOSITS IDENTIFIED AS LOESS HAVE BEEN FOUND

Figure 55 Loess in southern Britain

up of fine silt. A layer of fine silt can be detected in some soils in southern Britain and this can give a Bt horizon similar to that formed by clay translocation. It is, however, formed by a layer of wind-blown loess, not by secondary translocation processes (Figure 56) (see case study 2, section 3.4).

Peat formation in upland Britain dates from about mid-post-glacial times when a wet, cool period of climate (referred to as the Atlantic Period) occurred. Before this time podzolization was taking place and it is common in upland areas to find soil profiles with an old soil or *palaeosol* buried beneath

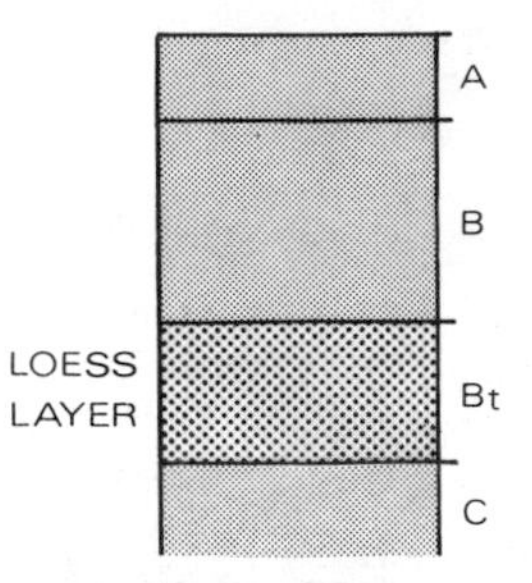

Figure 56 Loess in soil profiles

a thick layer of peat. The tree roots of the former vegetation may often still be seen in the growth position in the buried soil layer (Figure 57).

The rate of soil change

Can one soil type change into another, and if so how long does it take? Does an acid brown earth change into a podzol given enough time and enough rainfall? It is very difficult to answer these questions, as observations on soils have only been made over the last century or so.

However, some evidence concerning the slow rate of soil development comes from studies of river alluvium. In the most recent river deposits horizonation is virtually nonexistent, yet in the older deposits of alluvium (such as those left on river terraces), signs of horizonation can be seen. The most soluble material is being leached from the surface layers of the soil and it is very likely that, given enough time, the constituents of the soil will be redistributed throughout the profile in accordance with the equilibrium theory discussed earlier (Chapter one). The extent of the redistribution will depend upon the rainfall and the drainage conditions.

The evaluation of the rate of soil change is an important problem. Realizing that soil develops in adaptation to environmental conditions it should be noted that the environmental conditions have usually been modified by man. For example, much of the natural vegetation of Britain has been changed by man and soils have long been cultivated by man. Soils originally under oak woods may have changed under cultivation. Structure and nutrient status are perhaps most easily changed by cultivation, while texture is conversely probably the most difficult characteristic to change. Man is changing the soil-developing environment – how far and how rapidly will the soil change in response to cultivation?

Certain soil horizons have been recognized as being virtually manmade. An Ap horizon, for

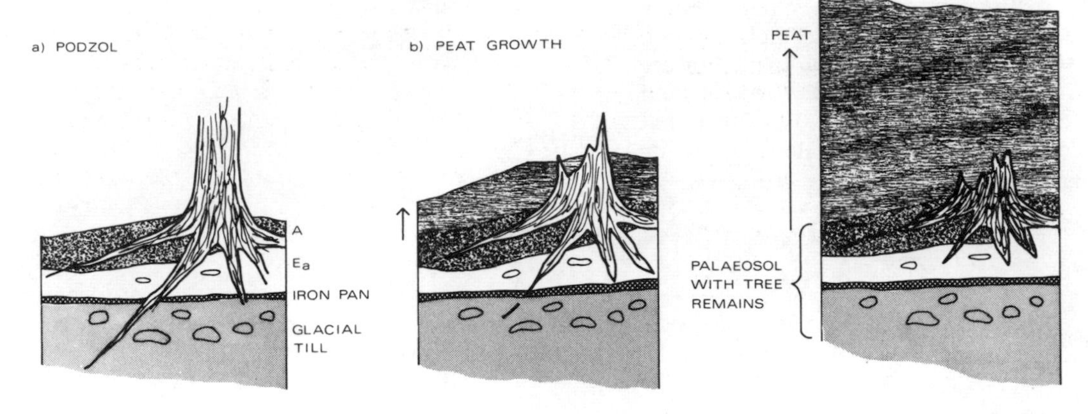

Figure 57 Palaeosol under peat

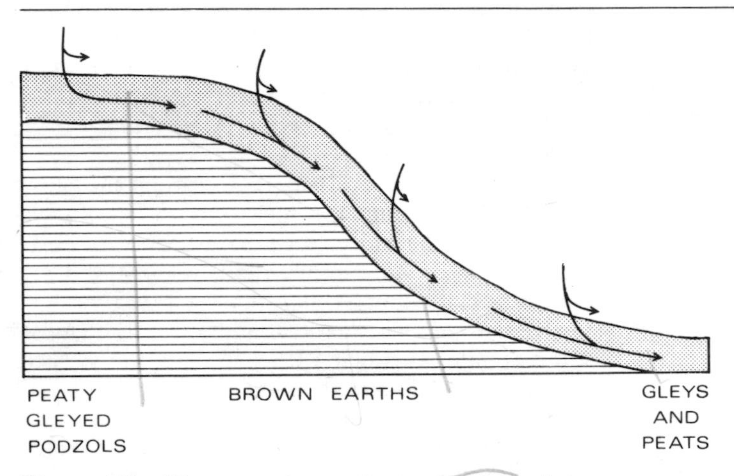

Figure 58 Water on slopes – impermeable rock

example, is one where the structures are changed by ploughing. In America phosphate-rich horizons are recognized. These have developed from a long history of fertilizer application. Clearly structure and nutrient status have been changed in a few hundred years of cultivation.

Soil development is not a process that proceeds to a certain stage and then stops. It is in many ways a continuing process. The weathering of parent material proceeds continually – the components of the soil continually adjusting to each other and to external factors such as climate. While some soil components, like resistant quartz grains, may have survived unchanged since glacial times, other components, especially the soil plasma with its secondary minerals and soil structures and horizons may have changed many times since glacial times in response to changing environmental conditions. Man is now one of the more dominant forces manipulating soil for his own use for the production of food: soil management is the topic for Chapter five.

3.3 Soil on slopes

Just as soil changes over time, it also changes over space. As most of the earth's surface is not flat it is important to understand how slope affects the variations of soil over space.

Slope has two principal effects on soil development. It affects drainage and soil stability.

Drainage

Water will obviously drain more quickly on a steep slope than it will on a less steep slope. Thus flatter areas tend to be waterlogged and here peat grows and gleying takes place. Soils on slope crests tend to be more leached and the nutrients washed from the upper slope pass to the lower slopes, which are consequently richer in nutrients.

It is clear that it is necessary to work out the direction and ease of flow of water movement on a slope before soil development can be fully understood. Two situations can be considered, first an impermeable and second a permeable bedrock (Figures 58 and 59 respectively). On a permeable bedrock vertical water movement is dominant. Water leaches nutrients from the soil and transports them through the bedrock to the slope foot or to the base of the permeable rock. On an impermeable bedrock water has to move sideways through or over the soil. In this way slope tops tend to be waterlogged, slope crests may have podzols, but these quickly pass into brown earths and then gleys and peats downslope.

In lowland Britain slope-foot sites tend to be moister and more nutrient rich while slope tops tend to be better drained but slightly less rich in nutrients. In upland Britain the slopes are longer and as altitude increases rainfall also increases, favouring the occurrence of peaty gleyed podzols and deeper peats.

Sequences of soil changes down slopes, such as in Figures 58 and 59, are termed *catenas* when they are developed regularly in similar topographic positions. The word catena comes from the Latin for a chain; thus the concept implies a series of *linked* soil types. While each soil on the slope is recognizable as separate, each is linked to the next in terms of its relative position.

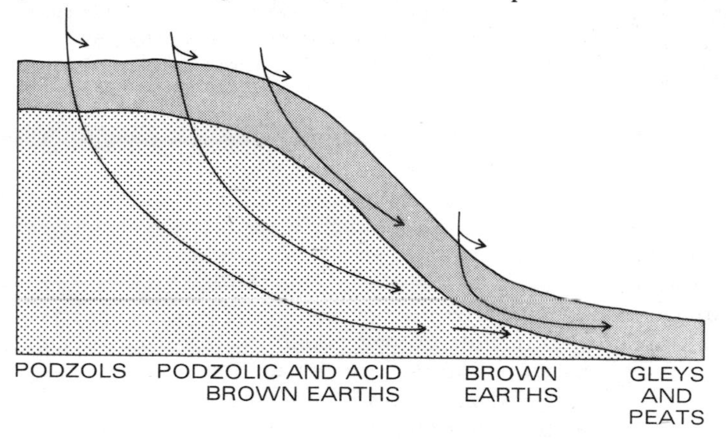

Figure 59 Water on slopes – permeable bedrock

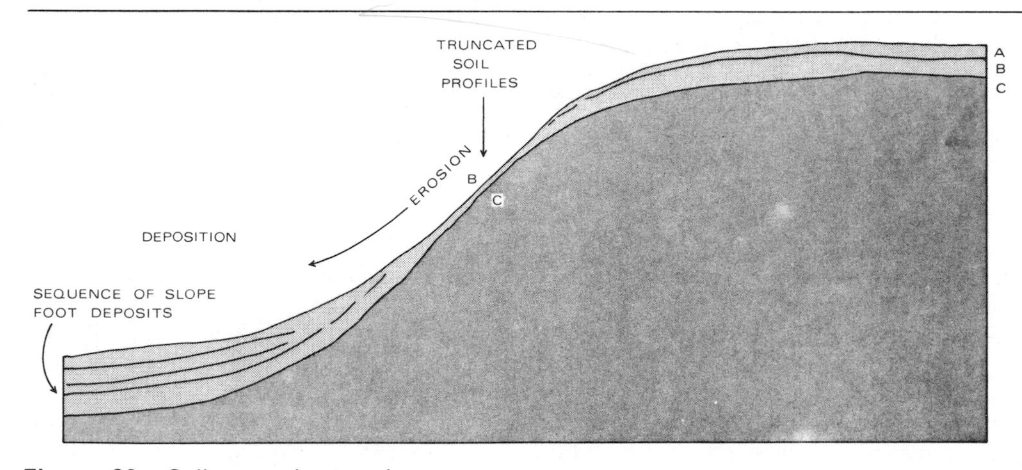

Figure 60 Soil truncation on slopes

Soil stability

Soils on flat areas will be stable but on steep slopes they will begin to slide downhill under the influence of gravity. Soil creep and slope wash are also active on steep slopes and these combined actions lead to a thinning of soils on slopes. Soil profiles can appear 'cut short' or *truncated* when compared with surrounding stable areas. Figure 60 shows fully developed soil profiles at the crest of the hill but erosion of the soil on the slope leads to truncation of the profile. Downslope the eroded material is redeposited as layers and sometimes buried soils can be detected at the bottom of slopes. These have been formed and then buried by deposits from above.

Summary

The factors of slope drainage and slope stability account for soil differences on a slope. In stable areas the direction of water movement influences the course of soil development (soil water is discussed further in section 4.2). In unstable areas on steep slopes the course of soil development is limited by the depth of soil that can accumulate, in other words by the balance of inputs and outputs of soil materials.

The catenary concept is a useful one in the understanding of the evolution of soil on slopes. Once the basic pattern has been identified a detailed picture of soil on slopes can be quickly established, using the idea that each soil type up or down the slope may be linked in a soil catena.

3.4 Case studies of soil development

So far in this book we have been dealing with general principles and theories. It is an important step to be able to put these into practice and to understand how a particular soil may have formed in a particular situation. In this section we shall look at some selected soil profiles, describe their characteristics and attempt to understand the development of the soil in each particular case. It should be emphasized that *typical* soil profiles have been described in section 3.1 while this section deals with *actual examples* of soils which may not necessarily coincide in all details with the typical soil profile.

Soil development can be investigated first by a careful study of the soil profile (as is described in section 6.5). Second, analyses of texture, chemical content, mineral type and amount of organic matter bring lines of evidence which help to piece together a picture of how the soil may have developed. Statements about the development of a soil should always be backed up by responsible evidence and accurate scientific observations on the nature of the soil.

Case study 1 · brown earth, sol lessivé

This soil example is termed an argillic brown earth under the 1973 British Classification (see section

6.6). The profile can be described horizon by horizon.

Depth from surface (cm)	*Horizon*	*Horizon description*
0–25	Ap	Mixed mineral – organic layer; a humose silt loam with crumb structures (influenced by ploughing)
25–60	Eb	Eluvial horizon; depleted of clay; loamy texture
60–80	Bt	Horizon distinguished by its texture; high clay content; clay texture
80+	C	Parent material; clay loam texture; columnar structure

DEVELOPMENT

The Ap horizon has developed from the organic matter supplied by vegetation, subsequently ploughed in, and mixed with the underlying mineral matter. The Eb horizon is an eluvial horizon from which the finer mineral particles have been washed, either in a gel-like solution or as clay particles in suspension in water. The clay has then been deposited again in the Bt horizon as clay cutans (as shown in Figure 42, p. 36).

How and when does the clay become moved and deposited? Analysis of the clay shows it to be dominantly composed of montmorillonite. From our knowledge of clay type and behaviour we know that montmorillonite is very expansive (see section 2.2). Thus in a dry summer the soil cracks as the clay shrinks. In the autumn when it rains clay may be washed down the cracks more easily than at other times. By the time winter has come the soil is wetted enough for the clay to expand and close the cracks up, leaving the skins of clay, or cutans, coating the cracks as shown in Figure 42 (p. 36).

Case study 2 · brown earth, sol lessivé

Depth from surface (cm)	*Horizon*	*Horizon description*
0–20	Ap	Humose silt loam
20–45	Eb	Gritty silt loam
45–60	Bt	Silty clay
60+	C	Bedrock

This is a further example of a similar soil, but although its profile appears comparable to the last example its origin is not necessarily the same.

DEVELOPMENT

In this case clay translocation has occurred. Thin section study (see section 2.4) on samples from the Bt horizon reveals the presence of cutans as evidence of clay deposition. Notice, however, that the Bt horizon has a silty clay texture as compared with the clay texture of the Bt horizon in Case study 1. Detailed textural analysis shows that particles in the size range of 0·002–0·02 mm (silt: International Scale) are common. This texture is comparable with many loessial deposits, which may show a range from the silt to the fine sand sizes (0·002–0·2 mm). It can be suggested from this evidence that the Bt horizon is in part present because of clay translocation but also in part due to deposition of post-glacial windblown loess on which the soil has developed. To be more conclusive mineralogical analysis would be useful in order to detect the presence of foreign minerals, that is those not present in the bedrock and which could have been blown in as loess (see section 3.2).

Case study 3 · Podzol

In this study mineralogical analyses have been used, rather than simply textural studies, in order to understand soil development.

Depth from surface (cm)	*Horizon*	*Horizon description*
0–2	L	Undecomposed litter
2–4	F	Partially decomposed litter
4–7	H	Well-decomposed humus, low in mineral content
7–20	Ea	Eluvial horizon, bleached and ash-like
20–30	Bfe	Illuvial horizon, rich in iron
30–35	B/C	Horizon of weathering bedrock, transitional between B horizon above and C horizon below
35+	C	Little altered bedrock, a mica schist

DEVELOPMENT

In this profile analyses show that organic matter and iron have been washed through the Ea horizon and deposited in the Bfe horizon. In the B/C horizon of the profile the rock is weathered. Mica is decomposed and clay (probably illite) is formed. Quartz, which is relatively insoluble, remains fairly constant through the profile and is almost untouched by soil development processes. Mica, on the other hand, may be attacked by hydrolysis and decreases in quantity up the profile.

In this case mineral, chemical and textural analyses reveal a twofold story of bedrock weathering: the selective decomposition of easily weathered minerals in the bedrock and the movement of iron and humus down the profile by eluviation. The use of several methods of analysis has led to a deeper understanding of the development of this soil profile.

Case study 4 · gleyed brown earth/sol lessivé on chalk

This is a complex soil profile. It is included to demonstrate what steps may be involved in the understanding of some deeper soils with a long history of varied development.

Depth from surface (cm)	*Horizon*	*Horizon description*
0–2	L	Undecomposed litter
2–4	F	Partially decomposed litter
4–15	A	Humose loam
15–30	Eb	Loam, eluvial
30–40	Bt	Clay, illuvial
40–60	Bg	Gleyed clay loam
60–70	CI	Clay loam with flints
70+	CII	Chalk

DEVELOPMENT

Both soil development processes and past geomorphological processes have to be investigated in order to understand this profile. There are two parent material horizons, CI and CII. The CI horizon is a clay deposit with flints (derived from chalk) and is thought to have been derived by Tertiary weathering processes. This separates the chalk (CII) from the rest of the soil profile. For this reason, while the chalk itself is calcareous and porous the soil is in fact acid and poorly drained, deriving its character from the clay deposit rather than from the chalk. Thus the lower layers of the soil are gleyed because water is held up on the relatively impermeable clay loam. Above this gleyed layer is a horizon where clay has been washed in and deposited from the overlying Eb horizon.

In summary, to understand the development of

this complex soil profile it is necessary to include studies of the following:

1 the geomorphological history – the formation of clay with flints over chalk;
2 the gleying process – the perched water table on the clay, and
3 clay translocation – clay being washed down as far as the gleyed layer.

Case study 5 · calcareous brown earth with gleying

This soil profile is developed on glacial till.

Depth from surface (cm)	*Horizon*	*Horizon description*
0–20	Ap	Clay loam
20–60	Bg	Clay loam with gleying, slightly calcareous
60+	Cg	Clay loam with chalk, gleyed

DEVELOPMENT

In this profile the Ap horizon is a mixture of organic and mineral matter. The soil is developed from glacial till which was derived from a chalky area. Investigation of the till below the soil shows it to be composed of clay and chalk fragments. The high clay content means that the drainage is poor in the lower part of the profile and gleying occurs. However, the Bg horizon is gleyed and calcareous below, but towards the top drainage is better and some of the chalk has been leached from the top of the horizon. In the A horizon all the chalk has been leached out and the soil is acid.

Observations on gleying, analysis of calcium carbonate content and pH reveal the story of soil development. In the upper part of the profile, where drainage is adequate, leaching occurs and the chalk is absent. In the lower part of the profile water is stagnant more often. Leaching is not encouraged, the soil remains calcareous and gleying occurs.

While a chalky till shows these features particularly well because chalk is easily soluble in rain water, other glacial tills show the same type of picture of leaching at the surface. The materials deposited by glacial processes are being redistributed by the soil development processes, especially leaching.

Summary

It is often necessary to gather data of a diverse nature in order to attempt to fully understand the development of a soil profile. In the examples selected not only simple data, like texture, are required (for distinguishing Bt horizons, for example), but also more complex mineralogical, chemical or organic analysis may be needed, especially where the more detailed soil profiles, such as the gleyed brown earth/sol lessivé on chalk, are concerned.

4

Soils in the ecosystem

4.1 The ecosystem approach

Systems analysis

Applying systems analysis to soil study is useful because we are able to understand the development of soils by working out inputs and outputs, as explained in Section 1.1. A systems approach is characterized by examination of changes or movements – in this case *movement of materials*, like water and nutrients, from one location to another.

An important aspect of systems analysis is that we can look at small systems or sections of systems (*subsystems*) and combine these in larger systems. In this way we may be able to understand how something works by seeing that the output of one subsystem becomes the input of another system. For example, nutrients from the soil subsystem are taken up by the plant subsystem and a *nutrient output* or loss in terms of the soil becomes a *nutrient*

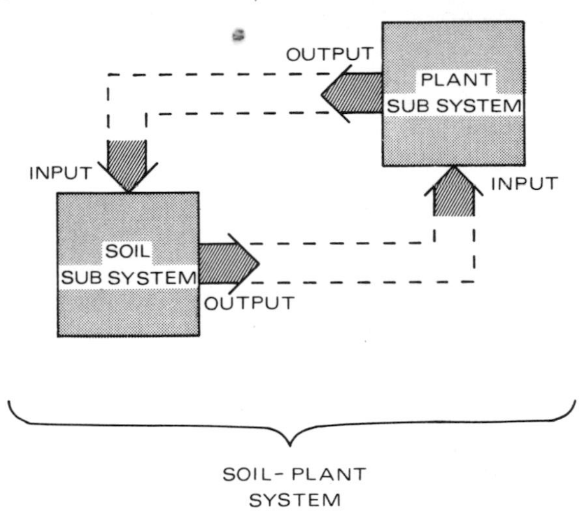

Figure 61 Linked subsystems

input or gain in the *plant* subsystem (Figure 61). We may consider the soil by itself as an isolated system and look at its internal functioning or we may look at the soil and plant systems and study their interactions. Furthermore, we may attempt to place soil in relation to all the factors that interact with it – to study soil in the ecosystem.

Although we may break down subjects like soil study and plant study into various components in order to be able to understand them, it is important to be able to put the components back again to see how the system works as a whole. An everyday analogy is when something goes wrong with a record player. It is necessary to understand how the internal parts work to mend it. It is not sufficient to be able to identify the individual components such as the needle or the turntable; we also need to know how these are all related in producing the sounds we hear from the overall system. Similarly with the soil–plant system. In this case the desired result is a useful crop. If we want to produce the best crop we must understand the individual parts of the system and how they react as a whole to produce the crop. Again, using the record-player analogy, many of us happily use the instrument without knowing anything about the detailed internal workings. We can identify the switch and the needle as the key components. We can make two statements:

1 It is not necessary to know all the details of a system all the time to manage the system in a satisfactory way.
2 It is only necessary to know the details if we want to correct something that has gone wrong with the system.

Obviously there are a multitude of complex factors which influence the overall behaviour of a system. We can divide components into two groups: first *principal operators*, which are the key features to know in order to work the system; and second the *secondary components*, which do not have a major influence on the system output but are important to the detailed working of the system.

Systems analysis uses three levels of detail of analysis:

1 *Black-box analysis* where only the principal inputs and outputs are studied.
2 *Grey-box analysis* where some detail of the internal workings is known.

3 *White-box analysis* where all the detailed components and controls on input and output are known.

In the soil–plant system several levels of analysis are appropriate. In some cases it is sufficient to see a wilting plant and to know that the input of water has been insufficient. The detailed mechanisms of water uptake are, as far as this problem is concerned, secondary and unimportant. Simple black-box analysis is appropriate: the solution is self-evident. In other cases a crop may develop a curious symptom of an unknown disease. In this case a more detailed analysis is necessary. It is necessary to look at soil conditions, pests in the soil, soil nutrients, soil water and a whole host of other inputs and outputs which may be affecting the plant.

While the above examples may have appeared obvious, in other cases it may not be so. It can be said that in essence systems analysis provides a means of solving problems by attempting to resolve inputs and outputs into principal (important) and secondary (less important) factors. The more complex a problem the more a white-box analysis may have to be used, though even complex problems can be 'short cut' by use of a few major factors which may solve the problem.

Ecosystems

One of the largest, most complex systems is the ecosystem. Soils, climate, plants, rocks and animals all interact in a multiplicity of ways. If we are to fully understand how our environment works it is not possible to look at one subject, that of soils, in isolation from the rest of the components in the ecosystem. Nature does not divide herself into neat compartments and therefore to understand her we must

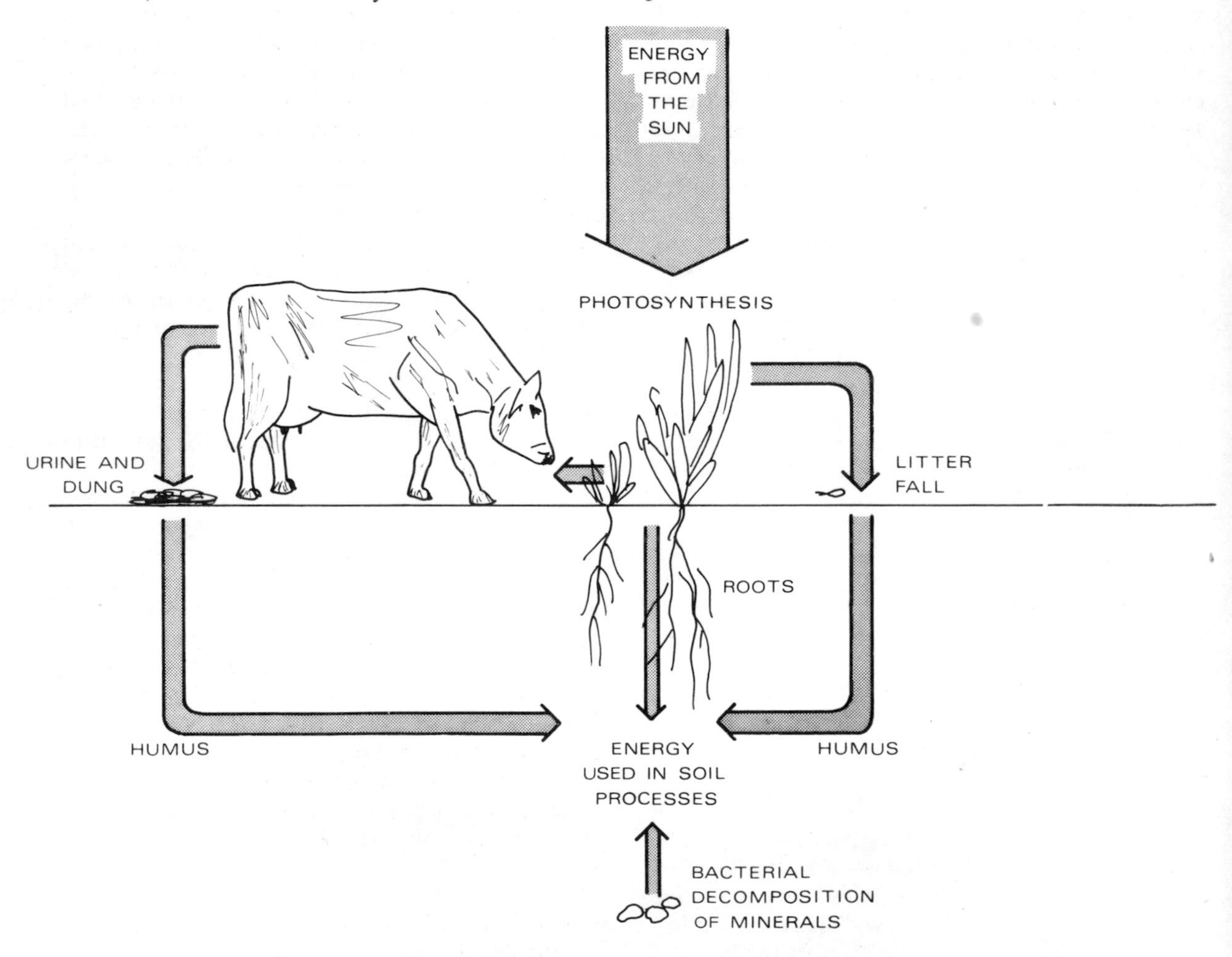

Figure 62 Energy flow into a soil

integrate our separate pieces of knowledge into an interacting whole. We shall study the ecosystem from the point of view of how the soil fits into the overall ecosystem. It will not be necessary to know everything about everything all of the time (even if it were possible!), but different levels of detail will be necessary to solve different problems.

ENERGY

The primary driving force of the ecosystem is the sun's energy. This is utilized in plants during photosynthesis. The energy thus stored may take several routes to become an input into the soil in order to become the driving force of soil processes.

To solve the problem of understanding how soil gets its energy we need to study the links between plants, animals and the soil and also between plants and the soil direct (Figure 62). When an animal eats a plant much of the energy is used in the animal's metabolism, but some remains in the animal's faeces and urine which is deposited on the soil. As far as the cow in Figure 62 is concerned it has finished with the organic matter and it is a system output. But as far as the soil is concerned this is a system input. Soil microorganisms readily obtain energy from the dung and are able to live by decomposing the material. That which remains is incorporated into the soil, where it provides a store for plant roots to tap.

The plant also sheds its leaves. (Even if the plant is evergreen, each leaf may only last for up to two years before it is shed: the plant appears evergreen because all the leaves are not shed at once.) Thus organic matter is continually being added as an output from the plant system but as an input to the soil system. Leaves again supply a source of energy that soil organisms can use for growth and metabolism.

Plant roots provide a kind of motivation force as they push their way through the soil. They help to form structures and they supply a link in the cycling of nutrients by cation exchange. When they die they add organic matter to the soil.

Within the soil a whole series of systems can be identified which are based on discarded organic matter from the plant and animal systems. Herbivorous organisms feed on plant remains and carnivorous organisms feed in turn on the herbivorous ones. Parasites also live on other organisms and on plant roots and so there is a closely linked web of interacting food chains.

The bacteria in the soil are primary operators in this system. They are the ultimate decomposers of all wastes left by other processes and other food chains. They can obtain energy and carbon from organic matter. This type of nutrition is termed *heterotrophic nutrition* (*hetero* – many sources: *trophic* – feeding). Many bacteria, fungi and actinomycetes and all animals feed in this way.

A source of energy for the soil not so far considered is that of minerals. We saw in an earlier section (2.9) how bacteria could decompose minerals. They in fact obtain energy and nutrition from this source. This is termed *chemo-autotrophic nutrition.* The term autotroph (*auto* – self; *troph* – feeding) implies that energy is gained by a primary source like the sun (in which case it is *photo-autotrophic nutrition*), or from the oxidation of inorganic compounds. Carbon is obtained from carbon dioxide. In the case of bacteria deriving energy from minerals the energy has been locked in the mineral for many thousands of years since the initial formation of the mineral in geological time.

Figure 63 shows a simplified food web for the soil. *Phytophages* feed on living plants, *saprophytes* feed on decayed matter and *microphytes* feed on bacteria, algae and fungal hyphae. *Carnivores* feed on soil animals. All organisms decay and provide organic matter. Some organisms can perform more than one function. In terms of systems we have looked at plant, animal and geological subsystems and also at the major ecosystem input – sunlight. We have understood the energy input into the soil subsystem by looking at the principal operators – the bacteria, the animals eating the plants and the plants photosynthesizing. It has not been necessary to look in great detail at the digestive system of a cow, for example, but just at the principal inputs and outputs of the operators (energy and carbon) leaving the details as unknown in a 'black box' analysis.

To summarize, the soil, like any other living system, needs energy. Energy is derived primarily from photosynthesis by plants. It is brought to the soil in three ways: 1) by animals, 2) by litter fall and 3) by plant roots. A further source is the oxidation of minerals by some bacteria. The energy input into the soil is used in driving many reactions involving not only the food chains of many organisms but also the transformations that soil organisms perform on soil components.

4.2 Soil and water

Soil water is fundamental to:

1 plant growth,
2 the growth of soil organisms,
3 the movement of solutes, clays and chelates in the soil.

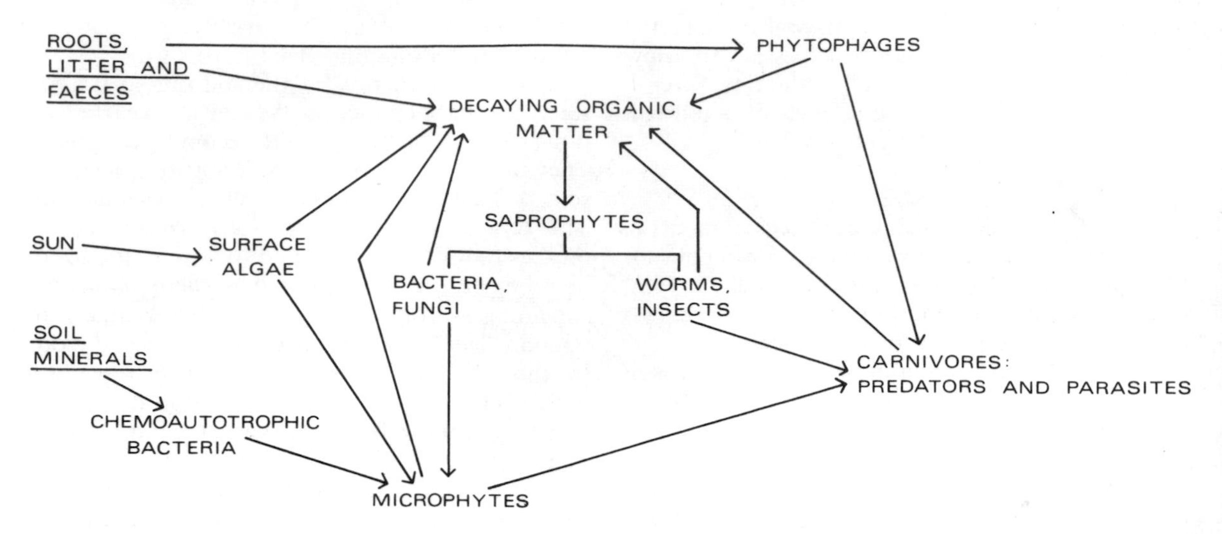

Figure 63 Simplified food web of soil organisms

It is possible to think of soil water in terms of soil alone, but it is more interesting to think of it in terms of a stage in the cycle of water in the whole ecosystem.

The hydrological cycle of water is simply the cycle of water evaporating from the seas, lakes and rivers, forming clouds, falling to the earth as rain and then travelling back to the seas again in rivers. But the soil has a fundamental role to play in this cycle. It, like the bedrock underneath it, may act as a *store* for water. If it does not rain for several weeks you can note that rivers usually keep on flowing, albeit at a possibly lower level. They very rarely run dry. This is because water that falls in a rainstorm is stored in the soil and porous rock beneath it, and is released only slowly (Figure 64).

If you imagine rain water falling towards the earth it may be intercepted by vegetation on the way down or it may reach the soil. Some of the water that has been intercepted on leaves and stems of vegetation may evaporate straight back again into the air, but, especially if rainfall is heavy, the water may drip through the leaves as *throughfall* and some may run down the stem. In Figure 65 *interception capacity* refers to the capacity of the vegetation to intercept rain, which obviously depends upon the leaf size and the arrangement of the leaves. *Interception loss* is intercepted water lost by evaporation. Throughfall and *stemflow* apply to the water that actually reaches the soil as a water input. These terms apply equally to water falling on a forest of trees or a field of grass.

Considering now the water that reaches the soil (by direct fall, throughfall or stemflow) the story is by no means over. The water may enter the soil by *infiltration* into the soil pores. Some soils have a relatively impermeable A horizon. This obviously depends upon the vegetation type, but some plants, such as heather (*Calluna vulgaris*), tend to produce a very greasy humus through which water cannot easily penetrate. Other plants, such as grasses, produce a more permeable humus which water can easily penetrate. If the A horizon is relatively impermeable it will not be able to transmit all the water that is falling on it and the water remains on the surface. If the surface is flat the water is stored in depressions (*depression storage*) from which it will later either evaporate or seep very slowly into the soil. If the surface is sloping the water may still be stored in depressions, but they may spill over into each other until the water runs overland downhill. Obviously the occurrence of *overland flow* requires rainfall to continue until the infiltration capacity of the soil is exceeded (Figure 66).

If the soil A horizon is permeable (i.e. it has a high infiltration capacity), water can percolate into the soil more readily. Its fate will now depend upon the soil characteristics. The general rule applies, however, that as soon as the water meets a less permeable layer it will tend to back up. The water may only pass vertically down through the bedrock if it is permeable. If the bedrock is impermeable it would flow along the top of it. If it meets an iron pan in the soil it may flow over this. This flow of

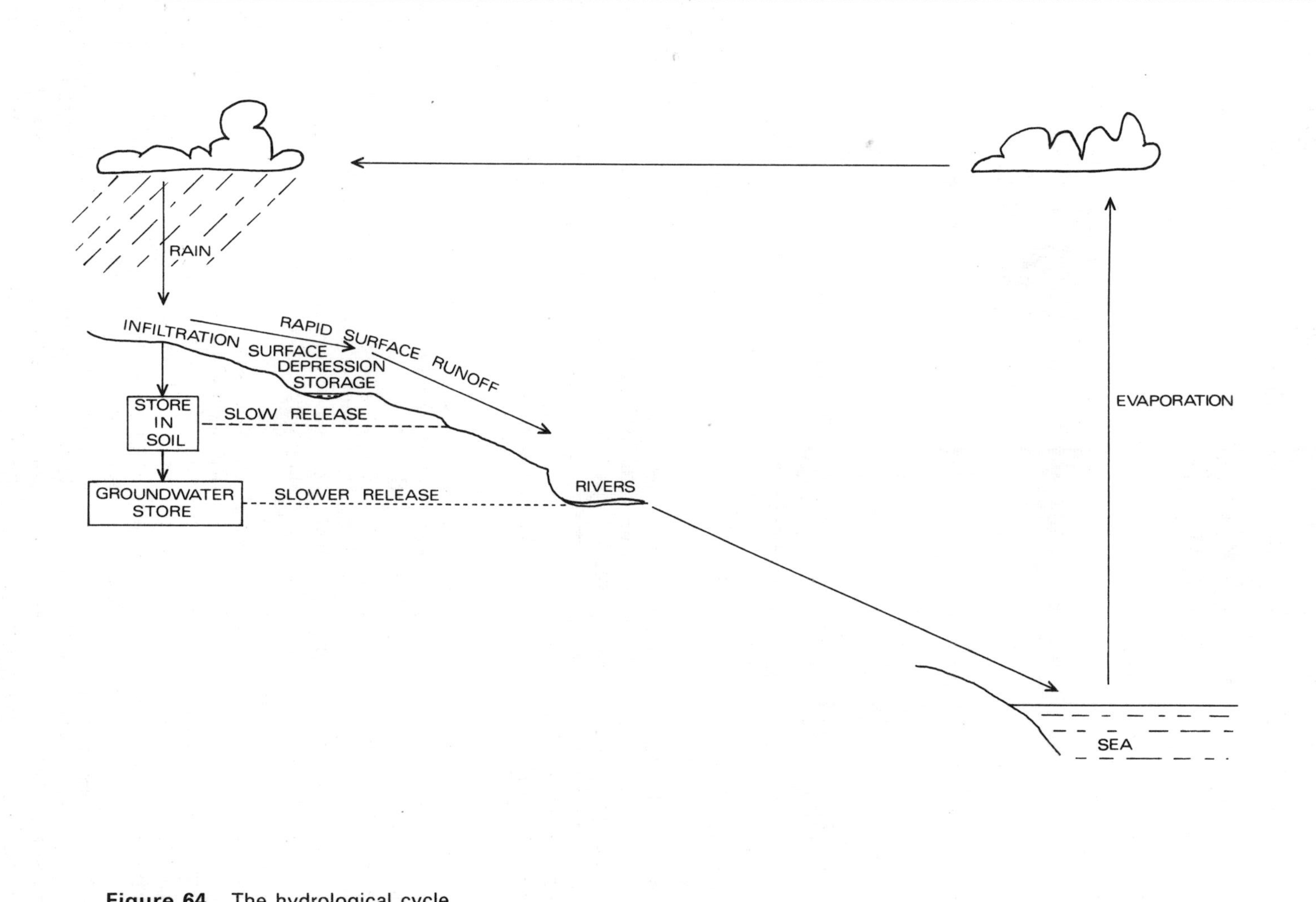

Figure 64 The hydrological cycle

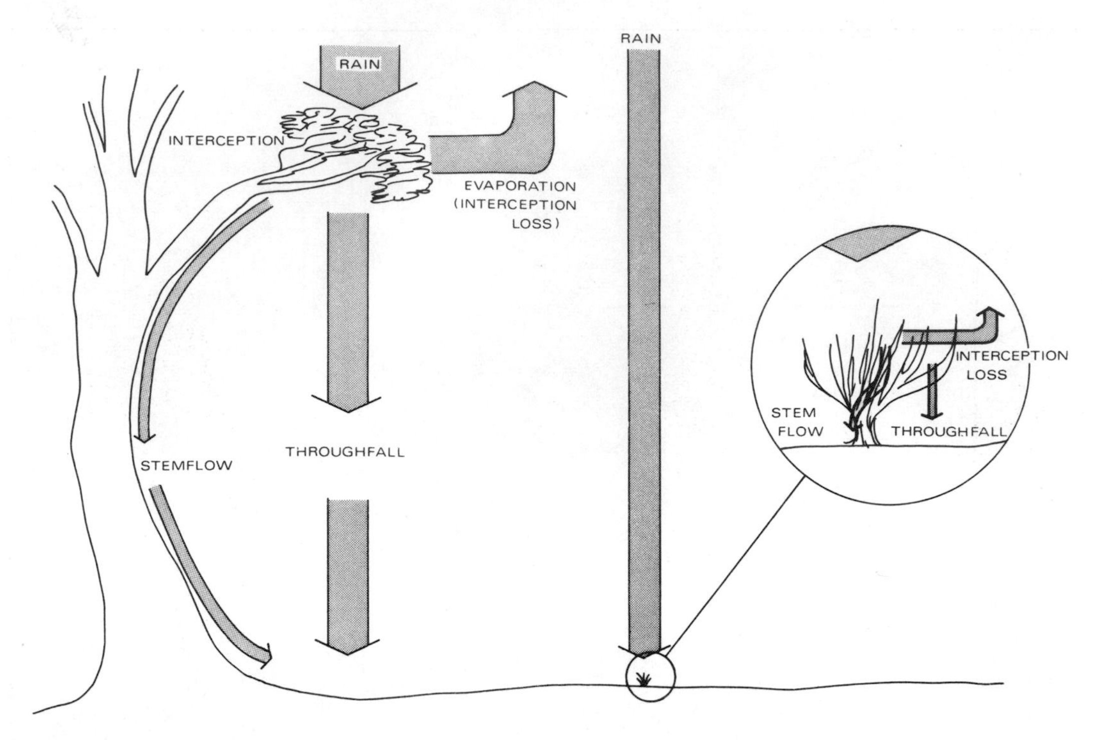

Figure 65 Water reaching the soil

water through the soil is termed *throughflow*. It is produced by gravity pulling down the water in large pores but the water being deflected through the soil because of an impermeable or saturated layer. Throughflow water usually flows downslope, but it may flow upwards through the soil to the surface if the soil is very wet. This is termed *return overland flow*.

Throughflow water is important to the development of soil, as it is this water that can carry away solutes downslope and out of the upper layers of the soil. But what is the fate of the throughflow water?

Throughflow water can be of two types: *saturated flow*, acted upon by gravity and flowing in the large soil pores, and *unsaturated flow*, flowing even though all the soil pores are not filled. The latter water flows gently away from areas of high concentration to areas of low concentration. The more rapid saturated flow is more likely to reach the foot of the slope (where there may be a stream or a marsh) and can carry away soil nutrients and perhaps clay particles as a soil output. The line of water flow, or *percoline*, may be subject to the removal of fine soil grains in the water. In wet upland areas natural *pipes* may form by this process, especially in or under peaty horizons. Pipe flow is extremely rapid and soil mineral matter may be carried away by it. The water generally becomes an output of the soil subsystem to become an input of the river subsystem.

The throughflow that passes at a lower speed through the soil may be intercepted by plant roots. In this case the water is drawn up for plant use and the nutrients that are in solution in the water are kept in the soil–plant subsystem and are not lost to the stream subsystem. The nutrients can be returned to the soil at the time of the next litter fall (Figure 67).

The nutrients and small mineral particles lost to the stream or river systems in throughflow water may go to form alluvial deposits, or they may simply be lost to the sea.

In summary, we have seen that soil water is part of a larger cycle of water. Water falling as rain may enter the soil, depending on the interception capacity of the vegetation above it and the infiltration capacity of the soil A horizon. Once in the soil the water may be lost to groundwater in the bedrock or it may move sideways through the soil. The water may be quickly recycled and reused by plants from which it is transpired back to the atmosphere. It may alternatively flow downslope to a stream or river and be returned to the sea, together with some nutrients derived from the soil.

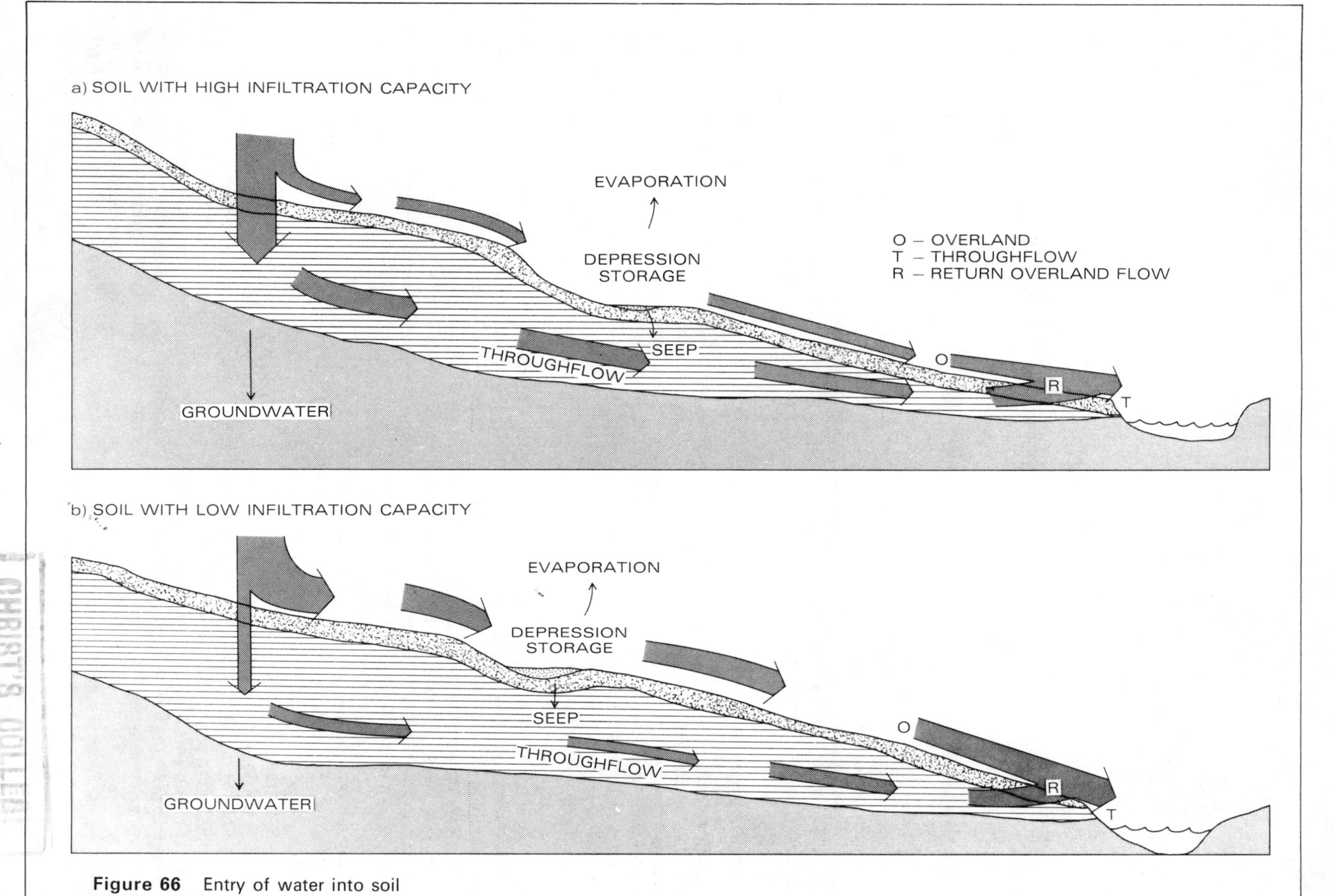

Figure 66 Entry of water into soil

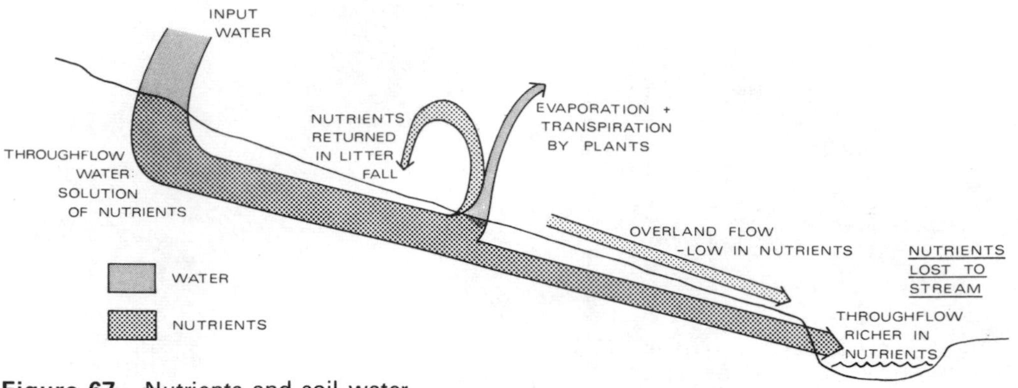

Figure 67 Nutrients and soil water

Soil water is thus part of water in an overall system. The water in the soil subsystem is derived as an output of the atmospheric subsystem. It may then become an output to the plant or river subsystems. No subsystem exists in isolation. All are closely linked by water flow from one subsystem to the next within the overall ecosystem.

4.3 Nutrient cycles

It is a fundamental principle of physics that matter can neither be created or destroyed, but can only change its form. Thus nutrients taken as a soil output cannot be 'lost' but only change their form. They are recycled back to the soil or are moved to another system.

From the descriptions of the inputs and outputs of water in a soil we can begin to build up a picture of how nutrients may be cycled round the ecosystem, moving from plant to organic matter to a clay–humus complex and then back to the plant. Alternatively the nutrients may be lost to the system, for example in a stream. Nutrients lost to rivers in this way are part of a very large geological cycle involving the precipitation of chemicals in the sea, the formation of rocks and the weathering of rocks to release the nutrients again. Nutrients cycled in plants and soils and not immediately lost to streams are undergoing *intra-basin cycling*, that is, cycling within one drainage basin (Figure 68).

Basically nutrient cycling follows the same paths as energy and water flow. Nutrients are released by the weathering of minerals, enter the soil, are used by the plants and animals and return to the soil again. However, some nutrients also enter the soil in rain water and some are fixed from gases in the atmosphere.

It will be useful to consider the nutrients separately, or in small groups, according to the nature of the cycling they undergo. In each case the input and output to the soil is but one link in the chain of linked subsystems in the overall ecosystem. We shall consider two major cycles, the carbon cycle and the nitrogen cycle, and following that the cycling of other nutrients, mainly the commoner cations and anions.

Carbon cycling

We have talked in this book about decomposition of organic matter as one process, soil respiration as another, with photosynthesis and decomposition of minerals as others. As we have already suggested, in the ecosystem these are in fact all linked together. They all form part of a large *carbon cycle* in which the soil plays an important role. We can now piece together the information we already have to produce a picture of the carbon cycle.

Let us begin with the carbon dioxide in the atmosphere. This is thought to have been originally produced from volcanic gases when the world first began and is now used by green plants in photosynthesis to manufacture carbohydrates. The process may be represented by the equation:

$$6CO_2 + 6H_2O \rightarrow C_6H_{12}O_6 + 6O_2.$$

The carbohydrates then reach the soil in plant material using the same paths as were described for energy flow (see section 4.1).

Carbon dioxide is also used in the soil by the autotrophic bacteria that derive nutrition from mineral decomposition. By both this route and the one described in the previous paragraph carbon is 'locked up' in soil humus. One way in which carbon is returned to the atmosphere is by the respiration of the animals that eat the green plants, but a more important step is, however, that of the respiration of microorganisms in the soil. As they decompose

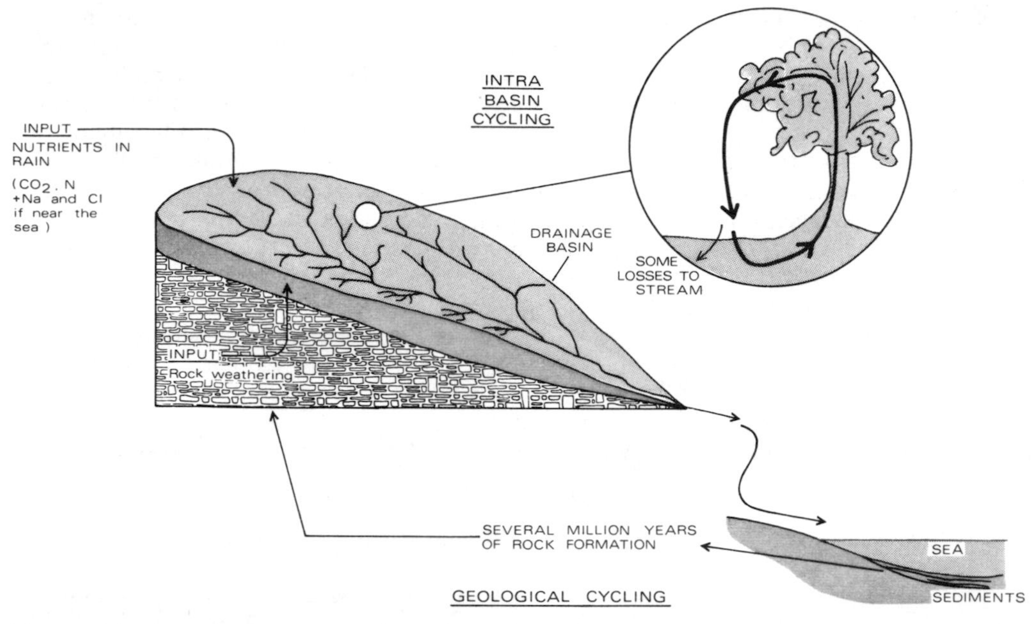

Figure 68 Intra-basin and geological cycling of chemical elements

the organic matter they respire and act to oxidize the organic carbon and return it to the atmosphere as carbon dioxide, where it may be used for further photosynthesis.

As far as the soil bacteria are concerned these organisms benefit from this process in that they obtain energy for growth and also carbon for cell construction. The carbon dioxide is for them a by-product, but this return of carbon back to the atmosphere for photosynthesis is the basis of carbon cycling in the ecosystem.

Nitrogen cycling

The soil has two inputs of nitrogen – fixation and rain water; and three outputs of nitrogen – denitrification, leaching and cropping. Again, these form components of a larger system of circulation and again, to understand the main operations of the whole process, it is necessary to understand something of the main operators in the system, the soil bacteria.

Starting with gaseous nitrogen in the atmosphere, this is fixed in the soil by a free-living bacteria – *Azotobacter* – and bacterial organisms known as *Rhizobium.* The latter occur in nodules on the roots of leguminous plants (such as peas, alfalfa and clovers). The organisms fix the gaseous nitrogen (N_2) to nitrate (NO_3) and it then becomes available as a plant nutrient, either to the plant on which the *Rhizobium* nodule is growing or it may pass into the soil, either as the root dies or when it is ploughed in. Thus the succeeding crop will benefit from the increased nitrogen content. Since nitrogen is the basis of the proteins in the plants this is a vital consideration for plant growth.

The nitrogen may be returned in a number of steps, which may involve a further use by plants but involves three stages of nitrogen transformation from soil humus. The first step is *mineralization* or *ammonification* when ammonia (NH_3) is produced. Under oxidizing conditions this is then changed by *nitrification* by nitrobacteria to nitrate (NO_3) when it may be used by plants. Alternatively *denitrification* may occur where the nitrate (NO_3) is changed to nitrite (NO_2) and is lost to the atmosphere as elemental nitrogen. The denitrifying bacteria, such as those of the genera *Pseudomonas* and *Achromobacter*, obtain oxygen from this denitrification process.

Nitrogen may then be returned to the soil dissolved in rain water or by fixation. An increasing amount is being supplied as fertilizer. The removal of crops clearly takes some of the nitrogen from the soil system, but knowing that nitrate is soluble in water and that water flows as throughflow into streams we can predict that nitrate will also end up in rivers, ponds and drainage ditches. There it may also act as a fertilizer for aquatic plants. This process of nutrient enrichment (*eutrophication*) and the 'blooming' of algae has been noticed in some inland waters in recent years. Farmers are now attempting to minimize this effect by more efficient fertilizer use. Fertilizers are expensive to apply and clearly such losses are uneconomic for the farmer.

Other elements

Cations such as calcium, magnesium, sodium, potassium, aluminium and silica can all be dissolved from soil minerals. They are used as important constituents of plants (grasses, for instance, have a high silica content), and are then returned to the soil in leaf litter. Additions come from nutrients in rain water which can also contain anions, such as chlorine, sulphate and nitrate. Rain is an important source of these nutrients and supplements the supply from rock weathering. A measure of nutrient losses from soil can be gained by measurements of the dissolved load of throughflow waters and also of stream waters.

Summary

It is important to appreciate that soils do not exist in isolation. Nitrate output, for example, 'lost' from the soil, becomes an input to river and bedrock systems. It is important to understand how nutrients are circulated in the ecosystem, especially along the route used by water, both to realize how the nutrients occur in the soil and also to know what effect soil nutrient losses may have elsewhere. It is necessary to place the soil in the context of the ecosystem – the system of linked subsystems – in order to be able to reach this understanding.

4.4 Soil and plants

The soil–plant relationship is a two-way one. Soil conditions considerably influence plant growth but, in turn, plants may modify soil conditions.

The chief soil factors which influence plants are:

1 nutrient supply and soil acidity
2 soil moisture
3 structure and texture (drainage and aeration)
4 depth available for rooting.

(Other factors, like topography, act indirectly to modify drainage conditions and soil type.)

Plants may modify soils by:

1 removing nutrients
2 additions of organic matter
3 increasing aeration by rooting.

Patterns of changing vegetation can often be found which correspond with patterns of soil change. Each vegetation type has its optimum conditions for growth and also limits of tolerance beyond which it will not grow. The actual distribution may also depend upon competition from other plants. Where the conditions are difficult for the growth of one particular species it may be ousted by other species better adapted to the prevailing conditions. Where the conditions are equally suitable for two plant species, the more vigorous one may be the most dominant species.

Two examples will be considered, one on a wet slope on an acid bedrock where the drainage factors are the most important, the other on a slope on a calcareous bedrock where soil acidity is the most important factor.

Drainage

Figure 69 shows a slope with the drainage characteristics associated with an impermeable bedrock (see also Figure 58, p. 42). A corresponding soil and vegetation pattern can be seen. Bracken (*Pteridium aquilinum*) tends to grow in deeper freely draining mineral soils. Heath plants, especially common heather (*Calluna vulgaris*), are associated with acid and slightly less well drained conditions. Other heath plants, such as bell heather (*Erica cinerea*), are found on the drier heaths and moors whereas cross-leaved heath (*Erica tetralix*) is more characteristic of bogs and wetter parts of acid heaths. In the wetter areas rushes (*Juncus* spp.) and mosses (especially *Sphagnum* spp.) flourish.

How are some plants specially adapted to grow in very wet boggy soils? One characteristic of these soils is that under acid, waterlogged, reducing (gleyed) conditions, some elements like iron, aluminium and manganese are very soluble. These elements may be so soluble as to be present in toxic concentrations. It might be thought that plants growing in very wet conditions would transpire very rapidly to get rid of excess water. In fact this does not appear to be the case. If it were so water would rapidly pass through the plant and evaporate from the leaves. In this case the high concentrations of iron, aluminium and manganese that were present in the water would be left behind in the plant, and these concentrations would soon become so high as to be toxic and eventually kill the plant. It would appear that bog plants are adapted to live under wet conditions by having a very low rate of transpiration. This means that they are the most successful colonizers of acid bogs which are rich in iron and other elements. If other plants, not so adapted, were to colonize the wetter areas they would have a rapid throughput of water and

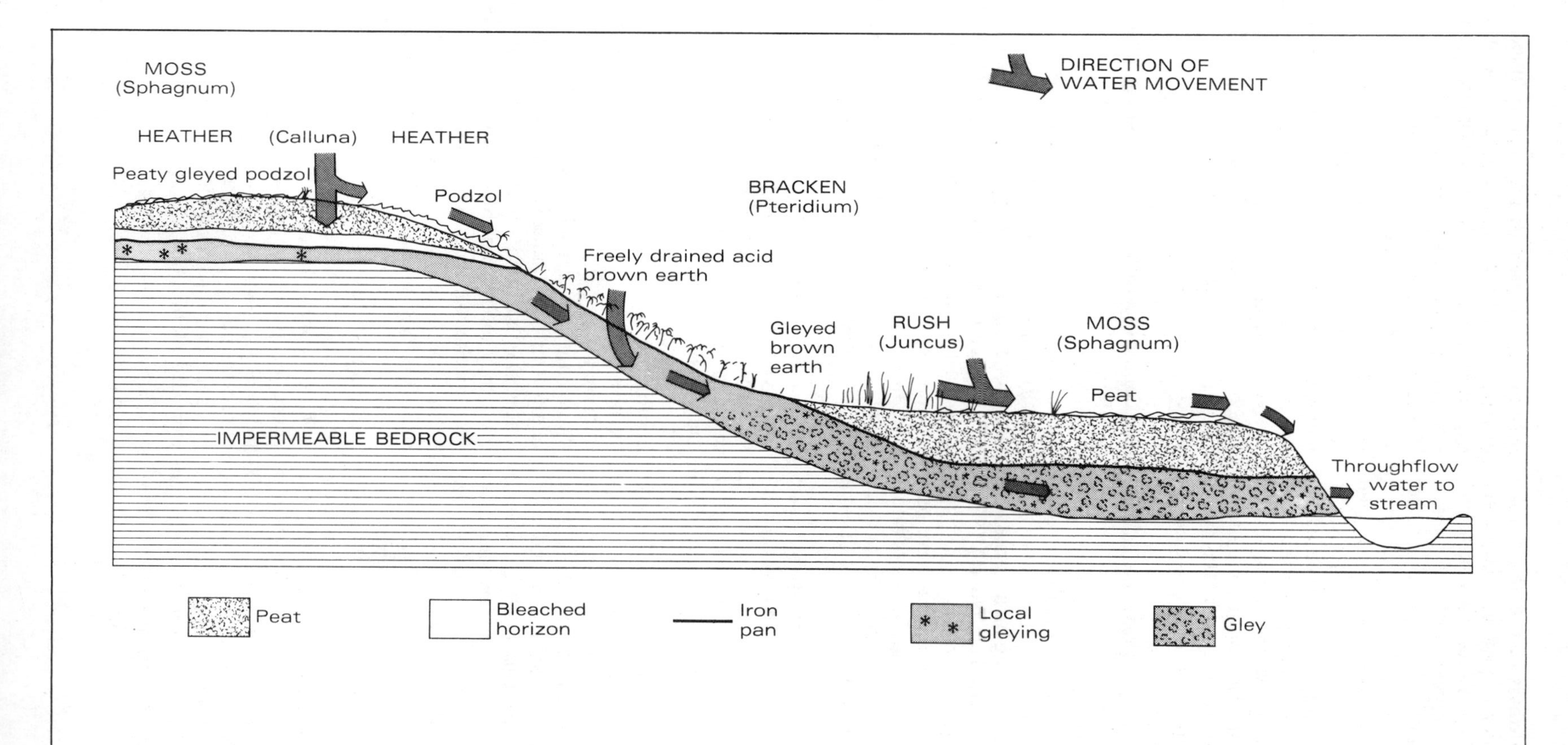

Figure 69 Water movement, soils and vegetation

quickly die from excess doses of the soluble elements present in the water of peat and gley soils.

If you examine some bog plants (e.g. *Juncus* and *Erica tetralix*), the leaves appear hard and glossy – a characteristic of xerophytic plants growing under dry conditions. A thick waxy cuticle and small scattered stomata (transpiration pores) mean that water loss by evaporation from the leaves is small.

The plants act to influence the soil type by the amount and nature of the organic matter that they help to accumulate. Though waterlogging will be the primary control on the rate of organic matter decay, mosses like *Sphagnum* because of their large bulk contribute greater amounts of organic matter than do *Juncus* or *Erica* plants.

Soil acidity

Some plants are adapted to grow in acid soils (pH 3–7) while others are found mainly in alkaline soils (pH 7–9). As alkaline soils are usually rich in calcium carbonate, plants thriving on alkaline soils are usually termed *calcicoles* (or *calciphytes*). Plants restricted to acid soils are termed *calcifuges* (literally 'fleeing from calcium'). These terms have only general application to communities of plants rather than to individuals. Odd plants nominally of calcifuge habit can be found growing on calcareous soils and vice versa.

Figure 70 illustrates a slope on limestone. Again soil drainage is important, but this time we are considering it in the way that it influences leaching. Calcium carbonate will be leached from the upper soil profile except on the slope where the calcium carbonate removed from further above is carried through the soil in throughflow water. In this way the slope tends to be leached and acid on the slope top, but alkaline on the slope itself. A characteristic vegetation pattern is shown with heather (*Calluna vulgaris*) – a calcifuge – on the acid soils and calcicoles on the alkaline soils. These calcicole plants usually include a very varied assemblage of grasses and herbs such as blue sesleria grass (*Sesleria careulea*), wild thyme (*Thymus drucei*), rock rose (*Helianthemum chamaecistus*), carline thistle (*Carlina vulgaris*) and salad burnet (*Poterium sanguisorba*). The presence of just one of these species may not in itself be significant (*Thymus* for instance grows on quite acid soils on mountainsides), but it is the presence of several of these in a *calcicole community* that is characteristic of limestone soils. Calcifuge plants which characteristically avoid lime-rich soils include cotton grass (*Eriophorum vaginatum*), wavy hair grass (*Deschampsia flexuosa*), cross-leaved heath (*Erica tetralix*) and purple moor grass (*Molinia caerulea*).

Some plants may be found distributed in both acid and alkaline soils, though not usually in extremes of either. These widespread species are able to compete in a variety of soils.

The mechanism by which plants are adapted to soils of different acidities is not completely understood. However, if a plant is taken from an acid vegetation community and planted in an alkaline soil, after one or two weeks it frequently turns yellow and appears weak and sickly. The plant is said to be *chlorotic* or suffering from *lime-chlorosis*, which is a photosynthetic disorder. The theory is that iron is needed in photosynthesis and that iron is deficient in lime-rich soils. Just as iron is very soluble under wet and acid conditions, under alkaline conditions iron is far less soluble. Thus, plants moved to alkaline soils are suffering from iron deficiency as the iron is not available for photosynthesis. On the other hand some plants are growing happily in alkaline soils (the calcicoles) and do not suffer lime chlorosis. However, if these are uprooted and planted in acid soils they soon look sickly (this is not just the effect of transplanting – controls can be transplanted to a further sample of alkaline soils and most will survive). The failure of calcicole plants on acid soils is thought to be due to the toxic effects of high aluminium and iron concentrations in the acid soils. Aluminium is very particularly soluble below pH 4 or 5.

It seems that the plants adapted to grow on acid soils, the calcifuges, are able to exclude iron and aluminium from their tissues by a *chelating* mechanism. That is, they produce organic compounds which directly incorporate iron and aluminium and immobilize them in chelate compounds. This enables them to grow in soils where iron and aluminium are present in toxic concentrations. When these plants are transplanted to acid soils they still retain this mechanism and exclude the small amounts of iron and aluminium which are present. Thus they suffer from chlorosis. Calcicoles have no such mechanism because they need to be able to extract the little iron that is available in alkaline soils. It is for this reason that when planted in acid soils they die – they cannot exclude iron and aluminium and they die from the toxic effects of high concentrations of these elements.

This theory is based on reasoning from the evidence of plant success following transplanting plants from one soil type to another. The theory is at present being tested by a search for the chelatory mechanisms which exclude toxic concentrations of iron and aluminium.

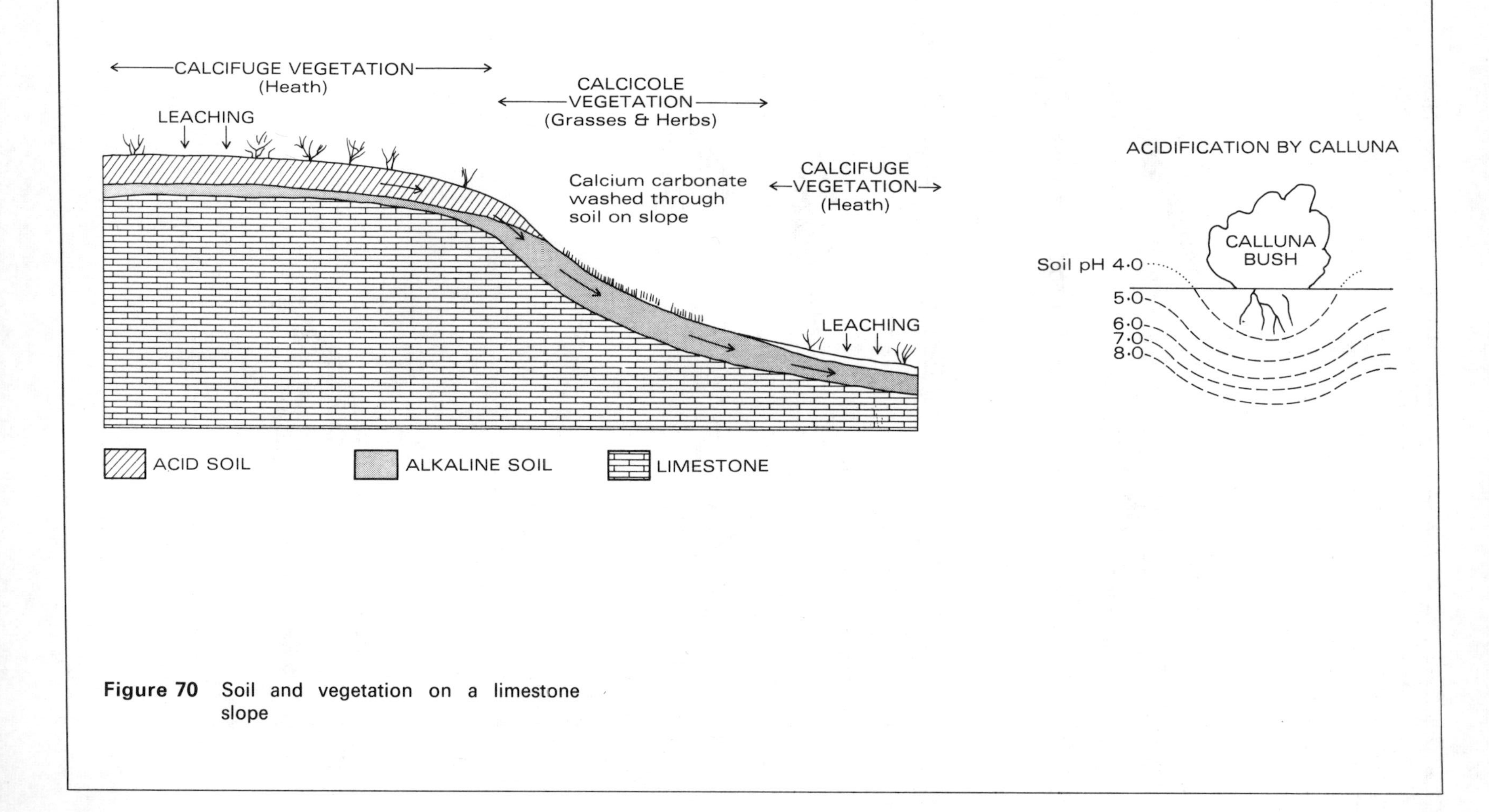

Figure 70 Soil and vegetation on a limestone slope

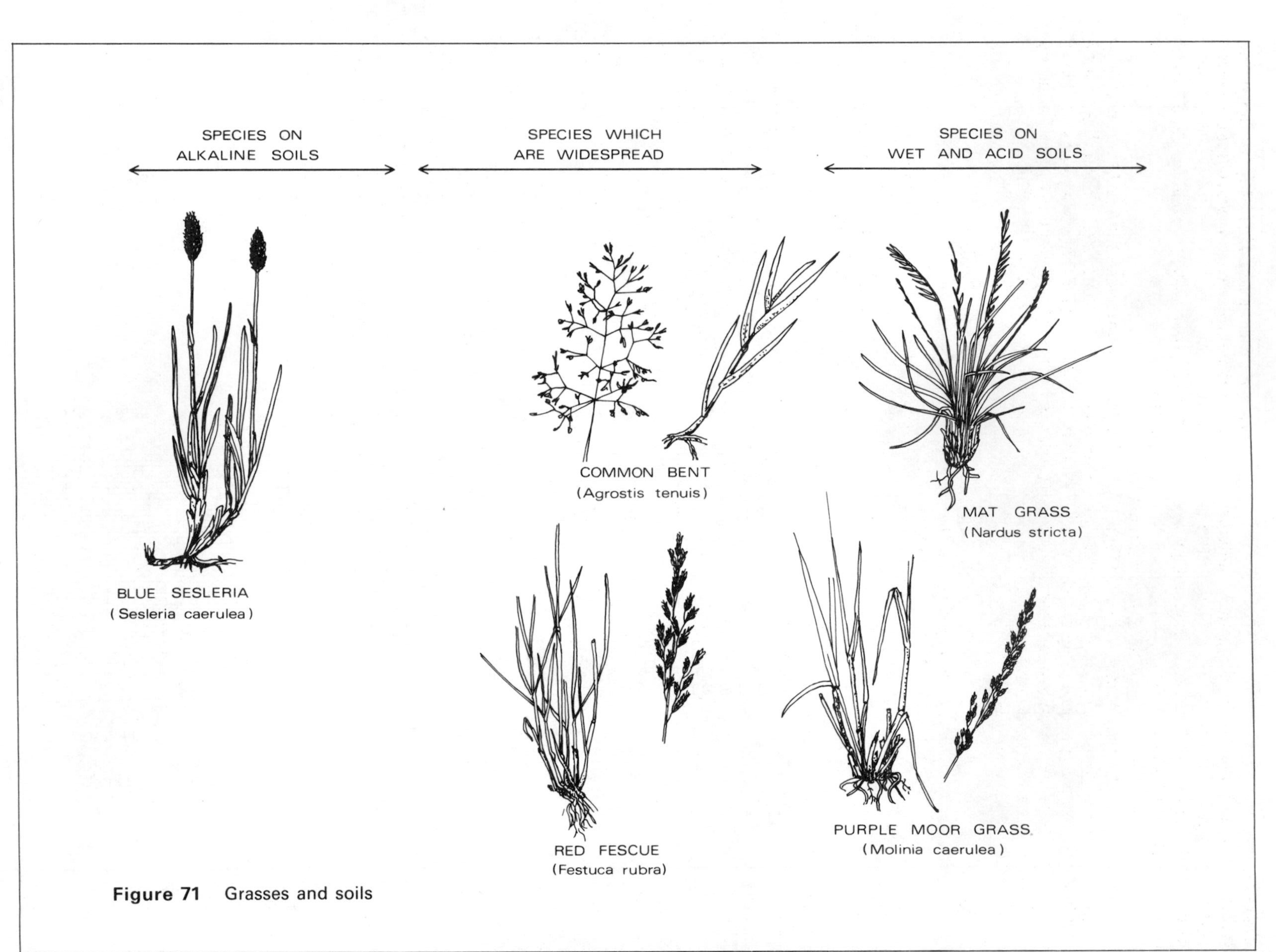

Figure 71 Grasses and soils

HEATHER

Heather (*Calluna vulgaris*) is of particular interest as it can change soil pH by cation exchange. It can locally acidify soil by the uptake of calcium and its replacement with hydrogen ions (see Figure 32, p. 29). In this way pockets of acid soils may be found under heather in otherwise lime-rich soils (Figure 70). Provided the surface soil is acid enough for seedling germination (pH 4–5) the plant can establish itself by soil acidification. Some other calcifuges can also grow on limestone soils by acidification, notably gorse (*Ulex* spp.).

CONIFERS

Coniferous trees may have a similar effect to that of heather, especially through chelation. Chelatory compounds come from the leaf litter of conifers and help to encourage podzolization. If soils are studied under plantations of conifers which have been planted at different times on an originally homogeneous soil it may be possible to detect an increase in podzolization under the oldest trees.

GRASSES

Much of Britain is covered with grasses and they are important economically in grazing of sheep, cattle and horses. Figure 71 demonstrates the tolerances of grasses in terms of soil conditions. Here nutrient status, drainage and acidity are the key controls, and the major point is that some species are very widespread and can compete successfully in a number of soils while others are far more specialized and are restricted to a narrow range of habitats.

CROPS

Crops also show a wide variety of tolerance of soil conditions. Few arable crops thrive really well when the soil is either badly drained or acid or both, and in such places grasses are usually all that will grow well. The relationship between crop suitability and soil pH can be summarized:

Alkaline soils (pH 7–8): alfalfa, sugar beet, clover, some market garden crops (lettuce, peas, carrots).

Circumneutral soils (pH 6·5–7·5): as above and also arable crops (barley, wheat, maize, rye and oats).

Acid soils (pH 4–6·5): potatoes and arable crops.

The optimum pH is about 6·5–7 for most crops, except potatoes which thrive best at a slightly more acid pH (about 5). This places emphasis on the importance of liming to raise the pH of acid soils (e.g. pH 5) to pH 6·5–7 (see section 5.1).

Factors other than soil

One aspect of the science of ecology is the study of the relationships between plants and their distribution to environmental factors. Soil is only one factor to be considered. Frequently it may not be possible to explain a plant distribution with reference to soil factors alone. Other factors, such as the exposure to climatic hazards (e.g. frost and strong winds), rainfall, proximity to salt borne in from the sea, altitude and temperature, together with competition from other plants, the history of the site and the actions of man, may all have to be investigated in order to fully understand plant distribution.

Summary

Soil factors which are important in the study of plant distributions include drainage and acidity. Plants have various physiological mechanisms whereby they are adapted to particular conditions and this gives them advantages over plants which do not possess these mechanisms. Thus it is possible to relate dominant plant communities to soil conditions in many cases. Plants cannot alter major soil attributes, but they can considerably modify some soil properties, especially those related to organic matter and nutrient uptake. Conifers and heather may encourage the podzolization and acidification of soils.

5

Soil management

5.1 Soil fertility

Soil fertility can be defined as the capacity of a soil to consistently produce a desired crop. It is not simply the nutrient supply in a soil. Other soil attributes, such as have been previously discussed (structure, drainage and organic matter content), are also important, both because they influence nutrient availability and also because they are important to plant growth in themselves. Therefore they must all be considered if soil is to be managed satisfactorily to provide conditions of optimum plant growth.

The law of the minimum

The 'law of the minimum' states that the factor which is at a minimum level will control the overall system and this is important for plant growth in soils. For example, it is not useful to have a soil with abundant supplies of nitrogen if the amount of water available is insufficient: crop growth will suffer from a lack of water. Conversely a soil may have a plentiful supply of water but be poor in one or two nutrients. In either case, although the soil may be *generally* fertile, plant growth is limited. *It is the factor in lowest supply (the minimum) which will be the limiting factor in overall plant growth.*

It is therefore important to manage a soil from many points of view and to investigate and understand what the limiting factors are in order to raise the productivity of the whole crop. If water supply is adequate nitrogen supply is usually the most significant limiting factor and therefore nitrogen management is often the most important step in raising crop productivity.

Nutrients

Nine major nutrients are used by plants. These are termed the *macronutrients.* The most important of these are carbon (C), hydrogen (H) and oxygen (O), the others being:

nitrogen (N)
phosphorus (P)
sulphur (S)
potassium (K)
calcium (Ca)
magnesium (Mg)

These macronutrients are the fundamental building blocks of plant tissues and of plant activities. Many other elements are also required in smaller quantities and these are referred to as the *micronutrients* (or *trace elements*). These include sodium (Na), iron (Fe), manganese (Mn), copper (Cu), zinc (Zn), molybdenum (Mo), silicon (Si), boron (B), chlorine (Cl) and cobalt (Co). Other elements, such as vanadium and nickel, have largely unknown, but possibly important, effects. Other trace elements may be picked up by plants, but it is not clear whether they have any important functions and some may even inhibit growth. These include titanium, barium, strontium, chromium, lead, nickel, tin and silver.

In general, since plants take up the water that is present in the soil, they usually absorb the elements that are in solution in the soil water (unless excluded by some mechanism: see section 4.4). Some of the elements picked up are important to the physiological processes of the plant and some are not (Figure 72).

The role of nutrients

Firstly, the plant needs carbon, hydrogen and oxygen for basic cell construction; these being obtained from air and water.

MACRONUTRIENTS

Nitrogen (N) is the basis of plant proteins and is required in large quantities. It promotes rapid growth and it improves both the quality and quantity of leaf crops.

Phosphorus (P) encourages rapid and vigorous growth in seedlings and early root formation. It also hastens maturity, stimulating flowering and seed formation.

Potassium (K) aids the production of proteins and increases the vigour of plants promoting disease resistance and the strength of stems and stalks.

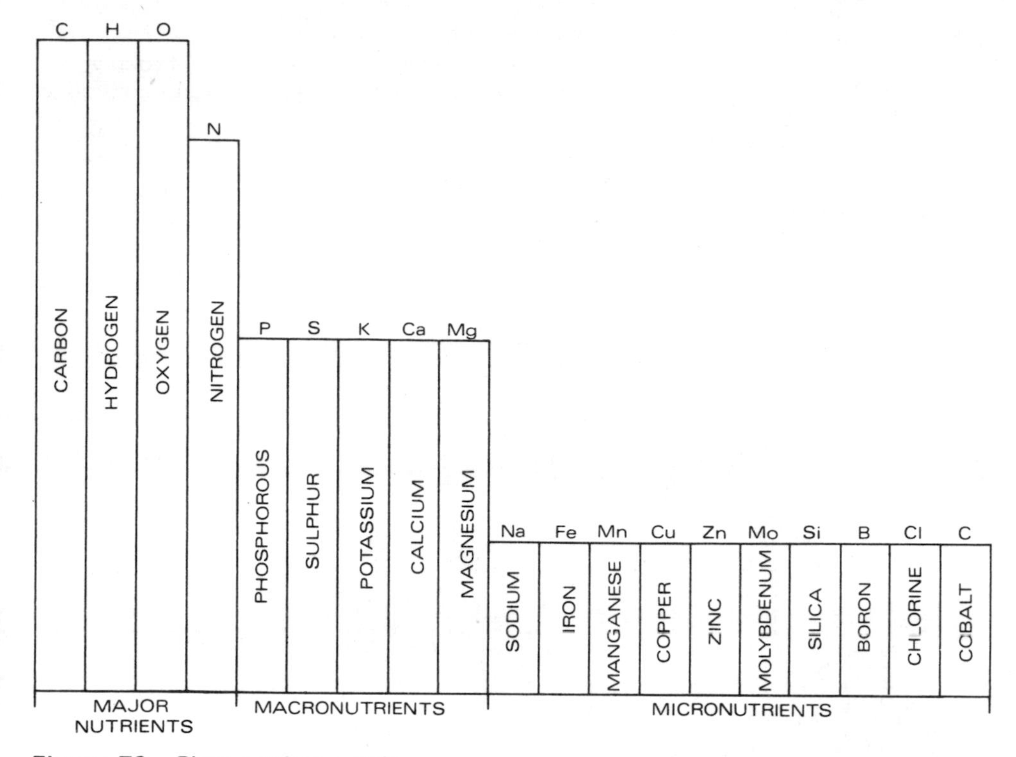

Figure 72 Plant nutrient uptake

It also improves the qualities of seeds, fruits and tubers and helps in the formation of anthocyanin (red colouration) in fruits. It is essential to the formation and translocation of starches, sugars and oils in the plant.

Sulphur (S) is an essential nutrient, more important to legumes, cabbages and some root crops than to cereals and grasses.

Calcium (Ca) is used by root crops. It is most important in the soil for its relationship with soil pH (see below under *Liming*).

Magnesium (Mg) is used in photosynthesis and is a basic constituent of chlorophyll. It is important to arable crops in particular.

MICRONUTRIENTS

Sodium (Na) does not appear to be essential but can increase the yield of grain, roots or fruits produced by a crop. It may be important in osmosis (the transport of water across cell membranes from a dilute to a concentrated solution) and may fulfil some of the roles of potassium.

Manganese (Mn) is used in respiration, protein synthesis and in enzyme reactions.

Copper (Cu) may reduce the toxicity of other elements in large concentrations in the soil and is important in enzyme reactions.

Zinc (Zn) is involved in fruit production.

Molybdenum (Mo) is used in nitrogen fixation.

Boron (B) affects growth.

Chlorine (Cl) may increase the yields of some crops.

Cobalt (Co) appears to be important to some plants, but its function is not clear.

Silicon (Si) is an important constituent of grasses.

Principle of nutrient supply from soils

It is a general principle that for all the essential nutrients there is an *optimum level of supply*. If the nutrient is *deficient* (i.e. there is less than the optimal amount), growth will be stunted, and conversely if the nutrient is in excess, growth will be damaged by toxic concentrations (Figure 73).

The actual optimal value varies for each nutrient. For example, with zinc very small quantities are needed and increasing zinc in the soil from about 50 to 100 parts per million (ppm) helps plant growth, but over 500 ppm zinc is toxic and plant growth suffers. With manganese, up to 500 ppm the plant is healthy but at 1000–2000 ppm the plant's growth is damaged. Below about 20 ppm manganese is said to be deficient because plant growth will be hindered.

Soil pH and the availability of trace elements

Most trace elements (micronutrients) except molybdenum (Mo) are more soluble under acid conditions. Thus in soils of pH 3–5 they are liable to be present in solution in toxic quantities. If the pH is raised to pH 6–7 by liming (see below) they are still available for plant growth but are not toxic. In alkaline soils the elements tend to be less available and deficiency may occur. Iron deficiency (chlorosis) may appear on lime-rich soils (see section 4.4).

Sources and losses of nutrients

From our knowledge of soil water and nutrient

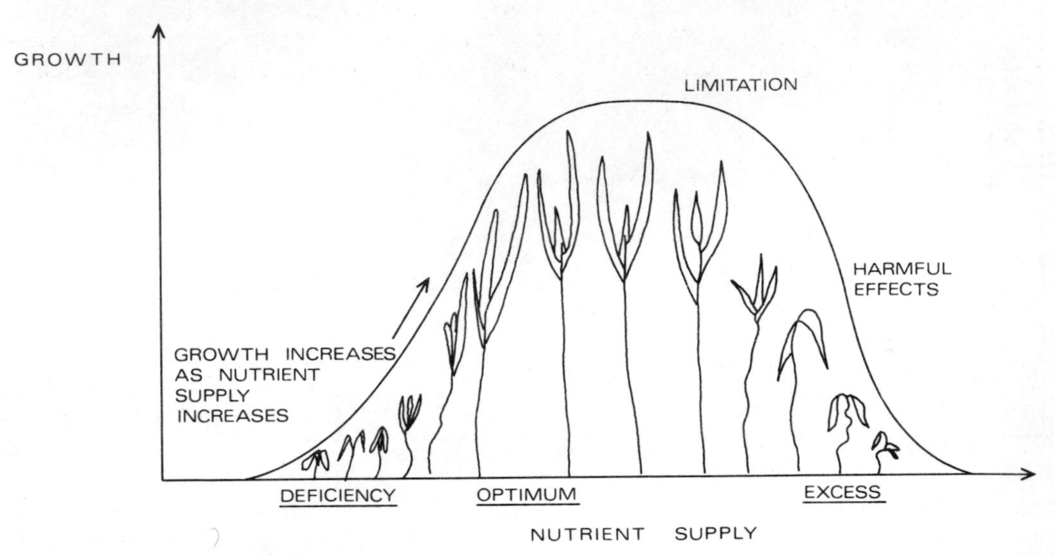

Figure 73 Nutrient supply and growth

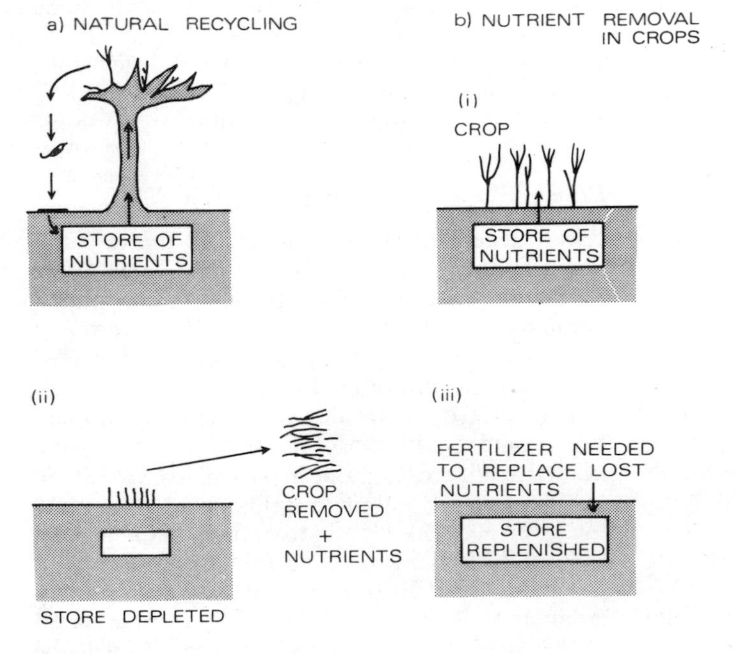

Figure 74 The need for fertilizers

cycling in the ecosystem (section 4.3) we can list the sources of nutrients as

1 rock weathering – release of minerals from rocks, and
2 the atmosphere – (a) gases and also falling dust; (b) rainwater with dissolved elements, e.g. those from sea spray.

In natural systems nutrients extracted from the soil are returned in leaf litter for recycling. When a crop is removed this does not occur and the soil's nutrient store will be gradually depleted (Figure 74). In this case nutrients in the soil must be returned by the addition of fertilizers. These have the dual function of *replenishing* nutrients and also *raising* the nutrient status above natural conditions, thereby increasing productivity.

If a crop is not removed, or only partially removed, some of the nutrients are recycled back to the soil and the organic matter content of the soil is also increased. Whilst the ploughing-in of stubble assists in decomposition of the plant material, decomposition itself often takes a very long time (2–3 years) and this is one reason why farmers often choose to burn off stubble rather than plough it in.

FERTILIZERS

Farmyard manure (FYM) can be applied to the soil and this restores the link in the nutrient cycle (Figure 74), by returning animal excreta to the soil. FYM adds nitrogen, potassium, phosphorus and small amounts of calcium and magnesium to the soil. Other trace elements such as copper, cobalt, manganese, molybdenum and zinc may be supplied, depending on the food source of the animals from which the manure is derived.

If natural organic manures are not added to the soil, fertilizers containing nitrogen, phosphorus and potassium (*NPK fertilizer*) are usually used. These compound fertilizers are common, adding several nutrients at a time. For example 'Nitro-Chalk' adds nitrogen and calcium whilst ammonium sulphate adds nitrogen and sulphur. In each case the soil has to be first analysed for its nutrient content and then the appropriate fertilizer applied to correct for any deficiency.

By replenishing the organic matter in the soil, organic fertilizers add structural processes such as aeration and encourage root growth. Where organic manures are not used it is possible that over several years the organic matter content of a soil may decrease and this may have an adverse effect both on soil nutrient supply and on the stability of some soil structures.

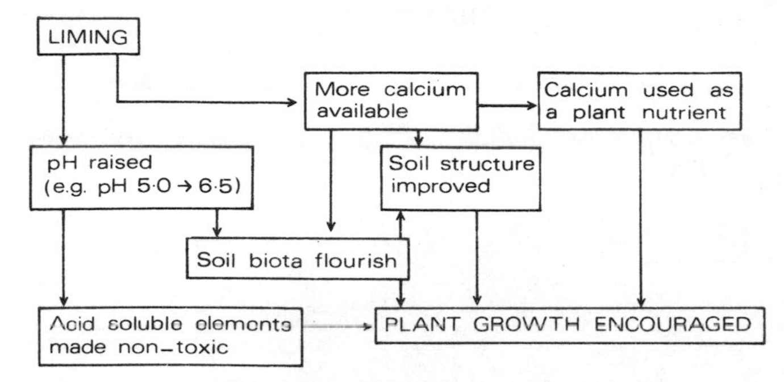

Figure 75 The effects of liming

LIMING

The solubility of many elements, the growth of plants and the activities of microorganisms in the soil are all dependent to some extent upon the pH of the soil, with the optimum situation being a pH of about 6–7. The addition of lime (calcium carbonate, $CaCO_3$) raises the pH of the soil (see Figure 34, p. 30). The *lime requirement* of a soil is *the amount of calcium carbonate needed (per hectare) to raise the pH of an acid soil to a value of between 6 and 7.*

Liming has the following effects (Figure 75):

1 it reduces the toxicity of acid soluble trace elements,
2 it improves soil structure, and
3 it increases the availability of calcium as a nutrient for soil organisms and plants.

These effects all help to improve plant growth.

Lime may be needed to counteract the acidity produced by the repeated addition of nitrate fertilizers. Nitrate (NO_3^-) is an anion and as such can combine with nutrient cations such as calcium (Ca^{++}). This may be carried off in solution in mobile soil water in combination with nitrates, and liming helps to counteract this process.

From a knowledge of the relationships between soil pH and crops (section 4.4) it is evident that liming is not a necessary treatment for *all* crops. While arable crops such as barley and sugar beet benefit from a pH corrected to 6–7, other crops such as rye grass and potatoes thrive in a much wider pH range from below pH 5 to above pH 7, with an optimum in the pH 5–5·5 range (see p. 61).

They are removed by leaching output in mobile soil water and, in agricultural systems, by output in crop removal. The output is offset by management procedures which recycle some nutrients, especially ploughing in unwanted parts of the crop (e.g. stubble), and the application of farmyard manure. The major source of replenishment is the addition of fertilizers. Nitrogen-fixing plants can be grown which increase the soil content of this important nutrient. Fertilizers not only restore the nutrient content of the soil after crop growth but can also be applied to areas of infertile soils in order to reclaim them for agricultural use.

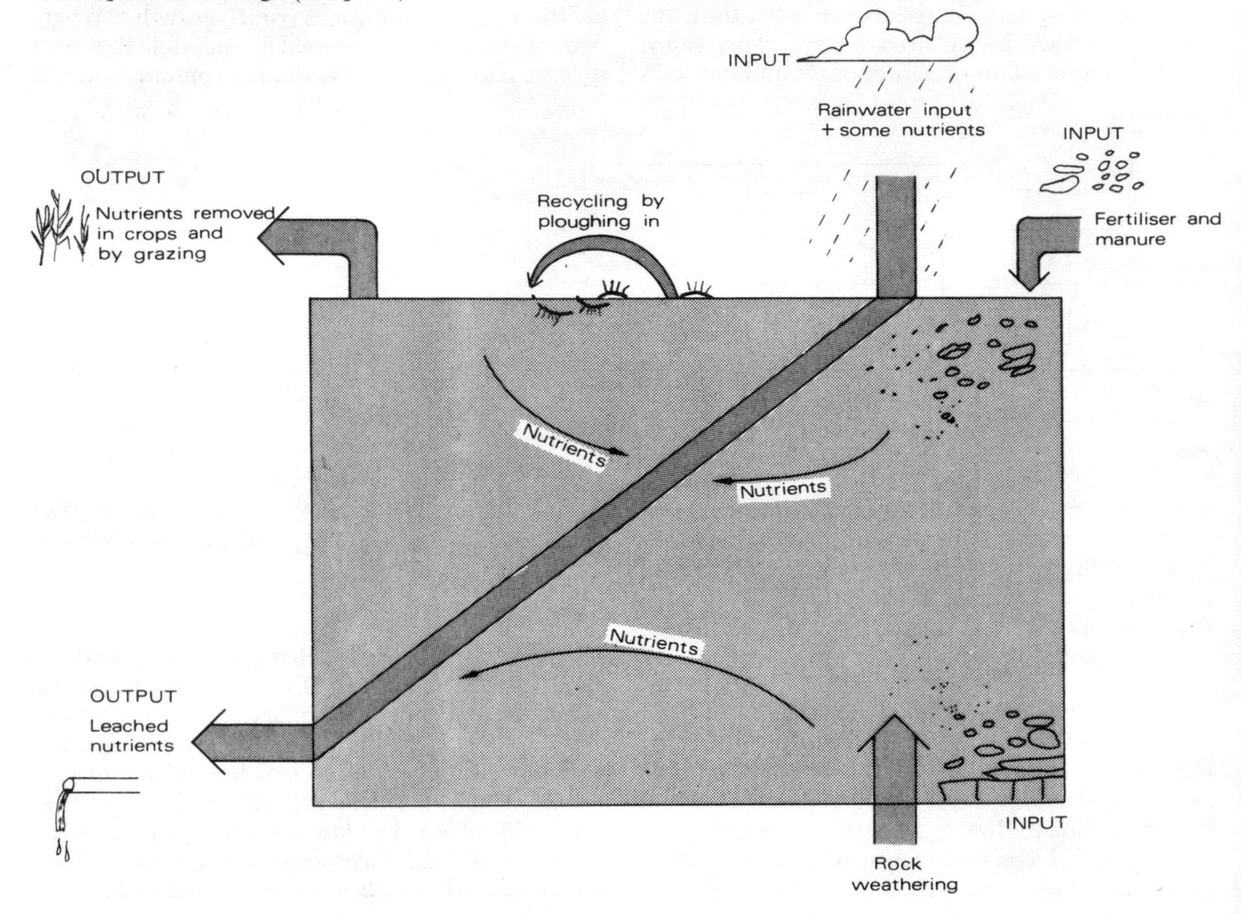

Figure 76 Soil nutrient balance under agriculture

Summary (Figure 76)

Nutrients are supplied to a soil by rock weathering, atmospheric and rain-water inputs. The nutrients perform three main groups of functions:

1 Physiological functions – as a constituent of, or an effect on, enzyme activities and growth.
2 Affects other nutrients, e.g. copper appears to affect availability of other nutrients.
3 Affects soil properties; especially the effect of calcium on solid acidity and structure and also the effect of sodium on the dispersion of clays.

5.2 Soil structure

Basic concepts

The aim of soil-structure management is to produce structures which are suitable for plant growth. As with plant nutrients there is an *optimum level* of structuring. Large, cloddy, persistent structures are as unsuitable for farming as are unstable structures which easily break down into a finely dispersed structureless state. In the former, soil structures are too well established to permit easy root growth, while in the latter the structures are too weak to aid aeration or to permit drainage.

Cultivation and artificial drainage can modify soil structure. The ideal result is a layer of stable crumb structures at the surface, which allows seedling establishment, with a prismatic or blocky subsoil which aids good drainage and helps to prevent waterlogging (provided that the local ground water table also permits soil drainage) (see p. 16).

Two concepts of structure are important:

1 structure *form* (shape and size), and
2 structure *stability*.

Form determines packing and pore space and therefore aeration and water retention, whilst stability determines how the structures will behave under ploughing or under pressure from livestock hooves and also the degree to which the structures will break up. The structures should be able to be reduced to a tilth suitable for seed growth and plant establishment, but they should be stable enough not to be able to be reduced to a structureless soil.

Structure form and stability are both related to texture, organic matter content, water content and also, to some extent, nutrient content (see pp. 16–17).

Structure form

In order to understand soil porosity (and therefore aeration and water retention) it is necessary to understand how the texture components (sand, silt and clay) are packed together. Figure 77 shows the unstructured porosity of sand, silt and clay. It can

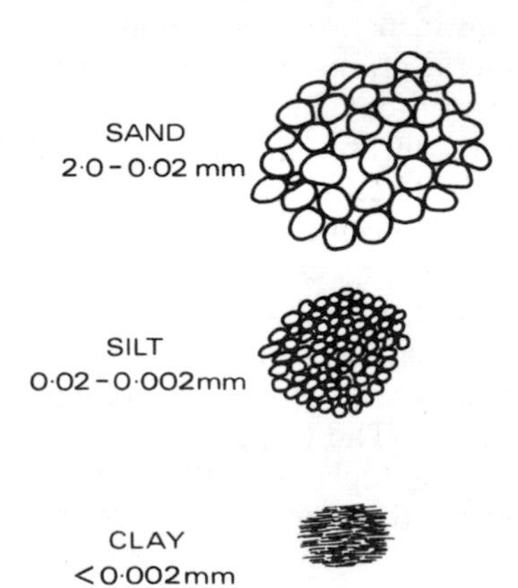

Figure 77 Packing of sand, silt and clay particles

be seen that it is chiefly the size of the pore space between the grains that varies, rather than the total pore space. Indeed the total pore space may be greater in a clay than in a sandy soil, but from a knowledge of how soil water behaves (section 2.7) we know that it is the size of the pore that is most important in determining water retention.

In a soil the particles will be aggregated together by various organic and inorganic cements. They will also shrink and expand according to water content. This causes the organization of the particles into structures, of which four main types can be recognized (Figure 78):

1 granular or crumb structure;
2 platy;
3 blocky; and
4 prismatic (or columnar if the tops are rounded).

The crumb structures tend to be associated with mixtures of particles in all texture classes and the blocky and platy structures with soils having a slightly higher content of silt and clay. The shape of the overall structure tends to reflect the shape of the particles that make up the structure. Thus clays tend to be organized in the longer, narrower structures formed by cracking and compression with alternating shrinkage and swelling.

Variations in the combinations of these structures account for differences in porosity. Two types of pores can be recognized: structural pores and textural pores. The former includes those pores occurring between soil structures and the latter those within soil structures, between the individual grains (Figure 79). Often it is the structural pores which determine the rooting of the plant and the drainage of superfluous water, the inside structures being smaller and holding the reserves of water (see p. 17).

Structure packing will determine the amount and size of the pore spaces between the structures. Platy structures will obviously pack very closely and leave little pore space, but crumb and blocky structures can be packed together more loosely given a larger pore space.

Structure stability, drainage and cultivation

Structure stability and soil drainage are closely linked. Structure stability is a limiting factor in plant productivity when either the structures fail to allow good drainage or alternatively where the water table is too high. In either case the soil is liable to be wet when ploughed and a smeared layer of broken-down structures can appear which may lead to the formation of a hard 'plough pan'. The

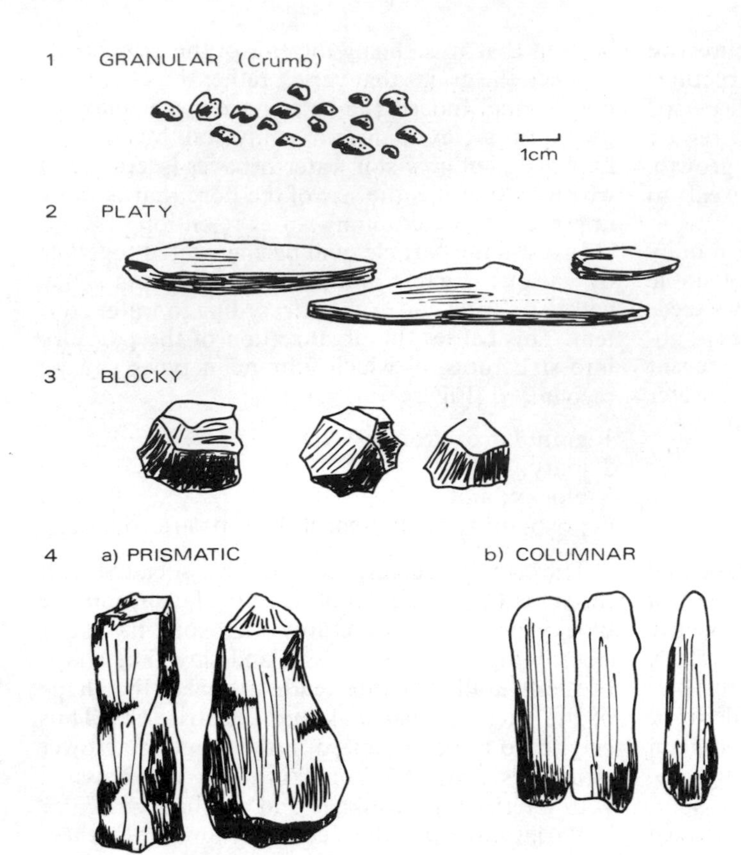

Figure 78 Soil grains grouped in soil structures

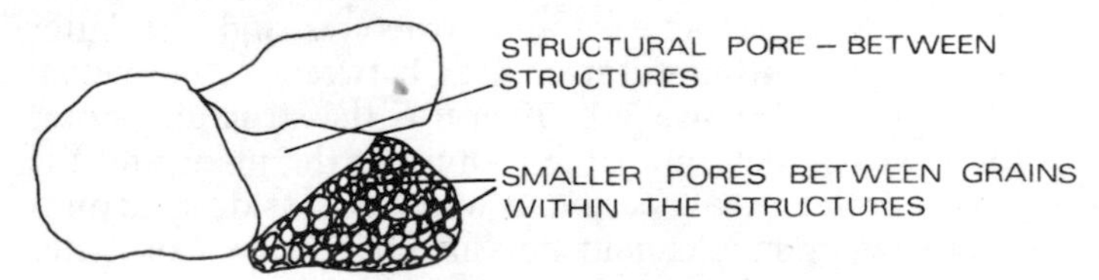

Figure 79 Pore spaces and structures

pre-existing structures are unstable when wet and ploughing at the wrong time can break down the structures to give a compact structureless soil layer (Figure 80).

The weight of tractor wheels and livestock hooves can have a similar effect in compacting surface layers of the soil and in the latter case this is known as *poaching*. Platy structures are produced if the soil is subject to pressure when it is wet and the structures are unstable (Figure 81). Basically the weight is causing *compaction* and in both cases of plough smearing and of weight compaction the soil pore space is reduced. (Figure 82).

Why are some structures more stable than others and why is water content important in determining structure stability? Organic matter, calcium carbonate, aluminium and iron hydroxides and silica all act to cement mineral grains together. Clearly, if a soil has a high content of these then the structures will be more durable and the soil will tend to be more stable. The importance of water content is that many of the cements are soluble in water. Thus, when the soil is wet the cements dissolve and the structures are unstable. When the soil is drier the cements reprecipitate and recement the grains into stabler structures. This is why plough pans form as a hard layer. The cements are weak when the water content is high, but after the soil has dried in dry weather the cements reform and the structures can harden again. Silt soils are particularly prone to structure deterioration as the particles are not cohesive enough to stick together, like clay, nor are they coarse enough to prevent close packing, as in sand. Silt-soil structures may easily collapse as the particles pack closely under pressure. Such soils are usually found over silty parent materials.

The addition of farmyard manure and lime can obviously help to form more stable structures as they will help to aggregate individual particles. But on some soils, for example silty clays prone to structure deterioration, drainage may provide an answer. Drainage can be the most important factor because if the soil is ploughed when wet the soil is more likely to suffer structure deterioration than if the soil is ploughed when drier. Drainage may also influence the amount and nature of iron in the soil, and iron precipitation may be an important factor. Iron is soluble when it is in chemically reduced (waterlogged) conditions, but when the soil is aerated the iron precipitates as ferric oxide, a strong cementing material. Hence drainage of difficult soils may induce the cementation of structures by precipitated iron oxide.

Subsoiling – the breaking up of the subsoil by deep ploughing – greatly improves the overall soil drainage. A well-drained subsoil means that a soil will have few problems of structure stability.

Treatment of the soil may assist in some situations, but the most important step in solving the problem of cultivation on soils with unstable structures is the *timeliness of cultivation.* The right time to cultivate the soil is when the surplus water has drained off sufficiently for the consistency and strength of the structures to be correct. This is when they can withstand the pressures put upon them and will not smear. This time will vary with different soil textures but will be earlier in the year if the soil is well drained (Figure 83).

Soil-structure management can be a delicate problem. The aim is not to try to produce rock-hard cemented structures (similar to those produced

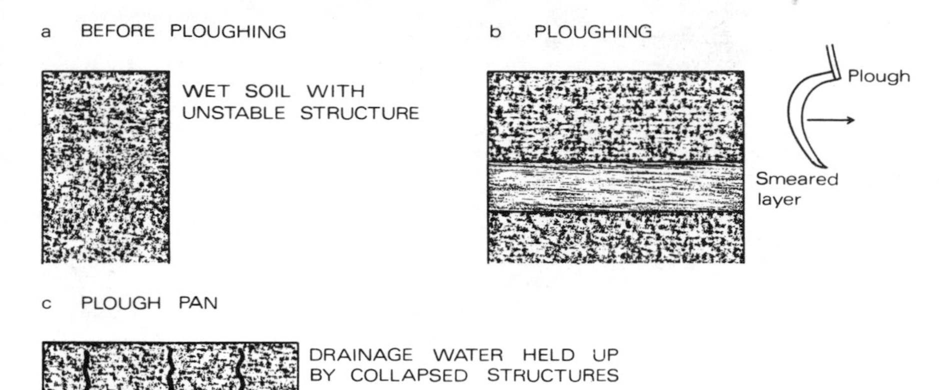

Figure 80 Formation of plough pans

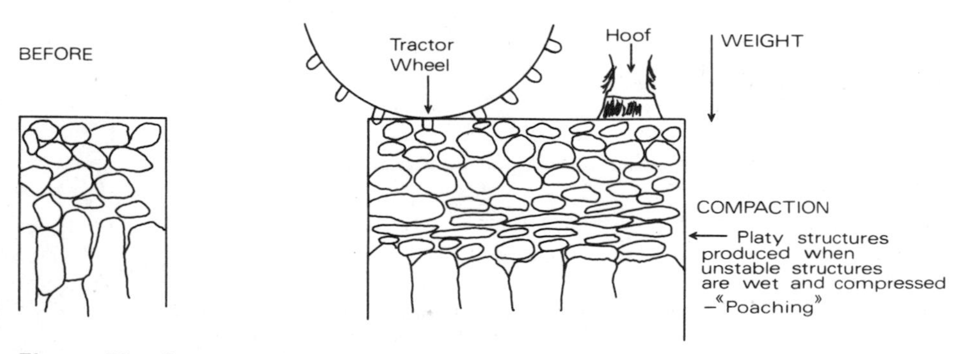

Figure 81 Structure alteration by weight

when unstable structures are broken down when wet and then set when dry). The aim is in fact to attempt to work out when the structures have their optimum stability for forming a useful tilth for seedling growth.

No one measurement of structure stability is completely reliable in all situations, but a useful idea of structure stability can be gained by applying a force to the structure until it breaks. We can work out, for instance, how many kilograms per square centimetre a cow or a tractor may exert. We can then apply this force to a structure to see if it breaks down. Figure 84 shows one such experiment. By such testing it is possible to estimate both the ideal water content, when ploughing should take place and when to wait for appropriate field conditions.

Structure stability is a much debated topic. Some farmers maintain that ploughing when wet does their soil no harm (using powerful 'crawler' tractors with caterpillar tracks). Such soils usually need to be high in organic content and other cementing agents. The discussion of structure management will continue as new facts relevant to the problem emerge through scientific investigation.

Summary

Some soils have structures which are unstable and prone to deterioration under cultivation or intensive stocking with animals. These soils often have closely packed mineral grains and are badly drained. Drainage and the addition of organic matter usually assist in solving the problem, but the timeliness of cultivation is the main way of overcoming the problem.

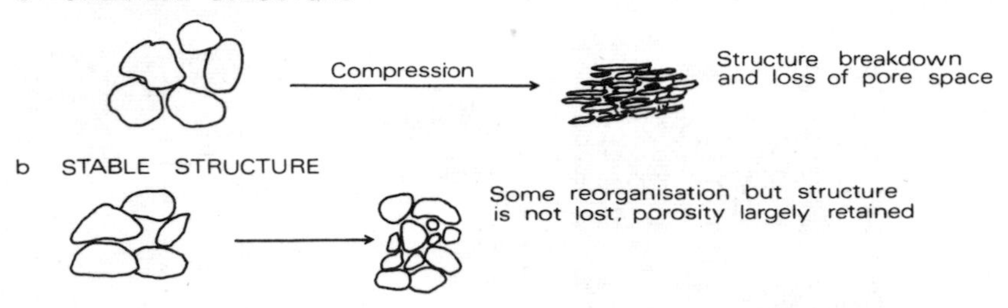

Figure 82 Pore space and compaction

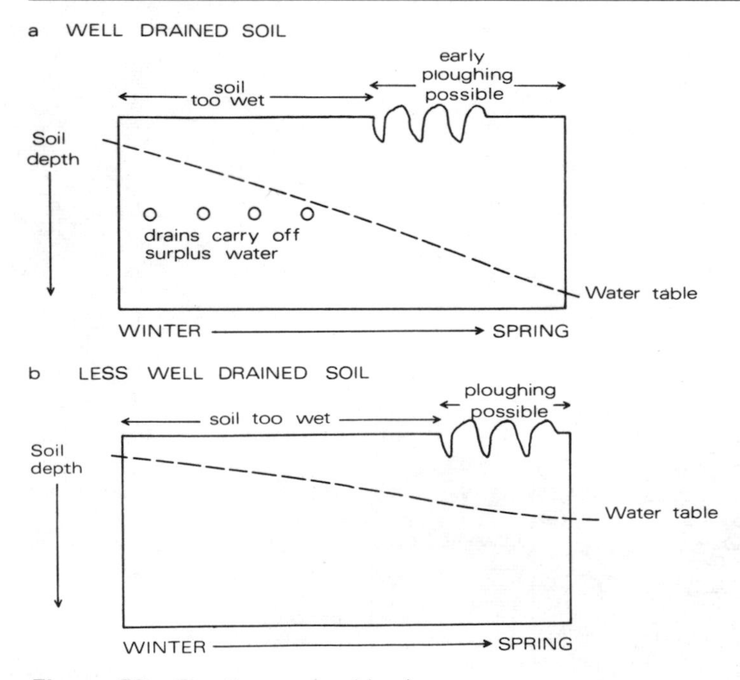

Figure 83 Timeliness of cultivation

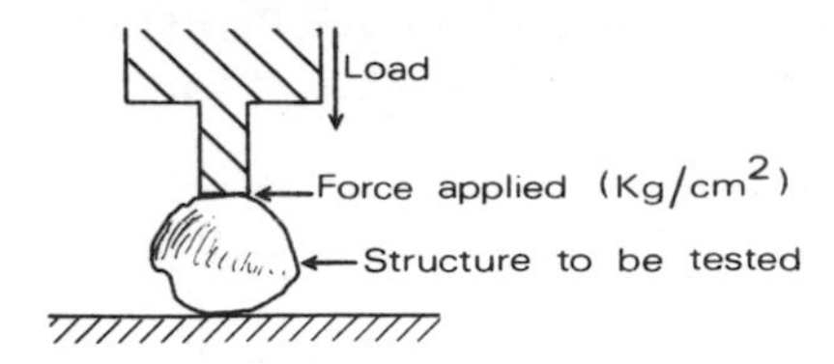

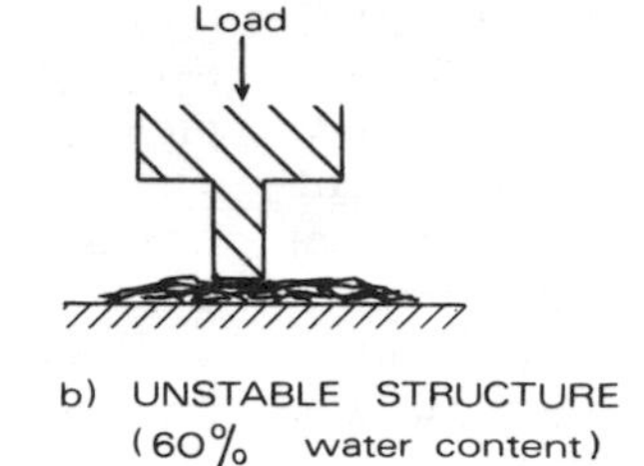

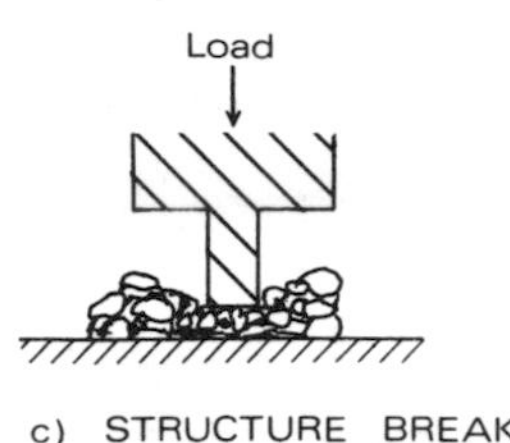

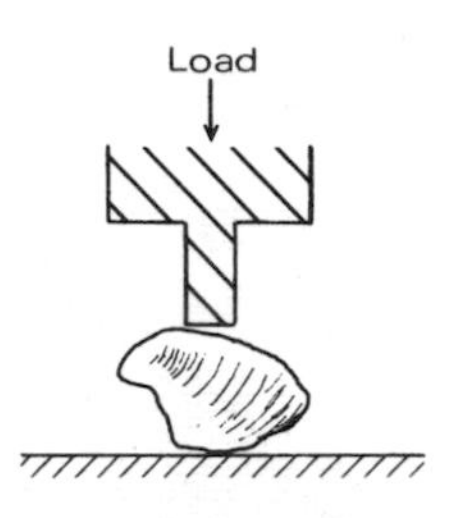

Figure 84 Structure stability

In many areas of Britain, particularly those with a loamy texture with a high organic content, structure stability is not a problem. Ploughing is used to return the unwanted part of the old crop to the soil and to expose the soil so that frost and other weathering processes can help to break down the larger structures. This and other treatments, such as harrowing, will render the soil suitable for seedling establishment and crop growth.

5.3 Soil mechanics

The subject of soil mechanics deals with the properties of soils which are important for engineering purposes. Of particular importance are the *load-bearing* properties of soil. When a road is to be built or the foundations of a building are to be laid, it is important for the engineer to know the mechanical strength of the soil and the load which the soil can bear, lest the road slides down a hillside or the foundations of the building collapse.

Also when a dam, or other construction involving water, is to be built the amount of seepage through the soil has to be assessed. Therefore the *permeability* of the soil is often an important characteristic to measure.

Moreover, the load-bearing properties of soils and their water contents are frequently closely related. A wet clay soil is both more liable to flow downhill under pressure, and also to have lower load-bearing properties, than is a drier clay soil.

Soil mechanics, then, concentrates on the mechanical strength and stability of soils which are subject to physical forces. This involves the detailed study of such factors as soil water content, soil permeability, soil plasticity, soil compaction and consolidation, soil particle cohesion and soil movement on slopes.

Engineering definition of soil

Before we can elaborate on these detailed considerations it is first necessary to define what a soil engineer means by the word 'soil' because it is not the same as what, say, an agriculturalist might mean by the term. For the purposes of engineering, soil is considered to be any loose sedimentary deposit; the emphasis being on the word loose, to distinguish it from solid rock. Deposits such as gravels, sands and clays, or mixtures of these, are included. Moreover, it is generally the case that the topsoil is removed before any engineering projects are started. Thus the agricultural and the engineering definitions of soil can be thought of as being almost mutually exclusive. To the agriculturalist the engineer's 'soil' is generally known as the subsoil: to the engineer the agriculturalist's soil, usually distinguishable because it is richer in organic matter than the material beneath, is often removed as a thin layer of unwanted 'topsoil'.

FRICTION AND COHESION

Of prime importance in the study of soil mechanics is the degree of cohesion between the solid soil particles. Particles with highly angular edges and those cemented together by calcium carbonate, iron oxides or clays will tend to be locked together strongly and to be resistant to movements between grains. On the other hand, smooth, rounded individual grains not cemented together, or those well lubricated by a film of water, will tend to slide over each other, especially when subject to a physical force such as gravity or the weight of a building.

In studying the movement of soil particles it is useful to discuss the two important controls on the physical strength of soils: *friction* and *cohesion*. Friction between particles is greatest, of course, when the surface area of grain contact is largest and when the surfaces themselves are dry and rough. In practice, however, cohesion may be the dominant control. A strongly cohesive soil is one where the individual particles are most firmly stuck together by the cementing agents mentioned above; most important are clays, but iron oxides and calcium carbonate are also significant. But the role of water is perhaps the most crucial. A film of water between particles, if it is thick enough, will act to cushion the particles from one another and thus the lubricated particles can more easily slide over one another with a minimum of cohesion. But a very thin film of water acts to bind the particles together by cohesion. Therefore, in practice, it is necessary to specify the water content of the soil in order to predict its stability.

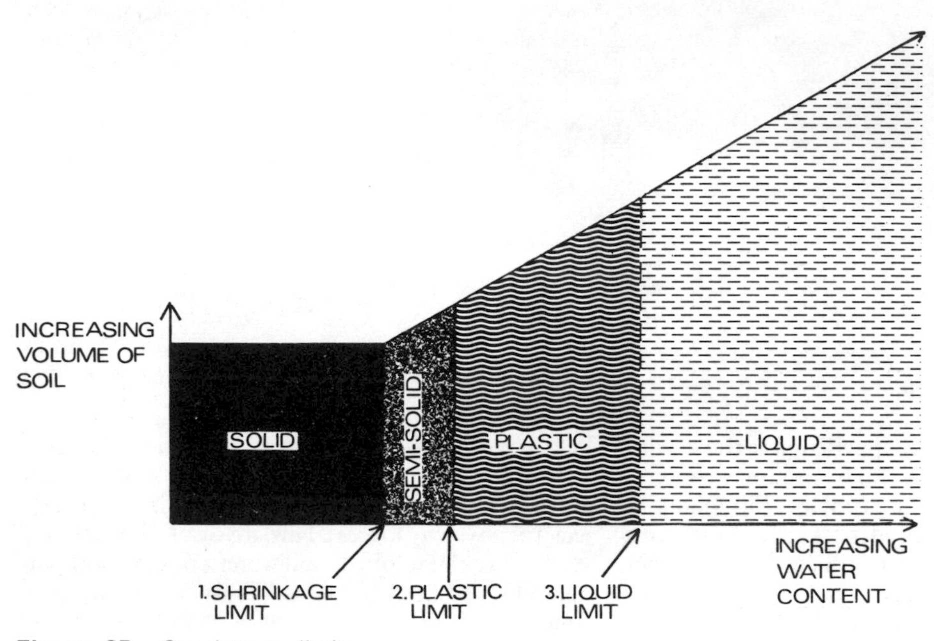

Figure 85 Consistency limits

CONSISTENCY LIMITS

Every small child knows that when earth is puddled with water it forms a delightful mud for making mud pies. But if too much water is added then the consistency of the mud will change till the mixture will flow and the mud pies collapse. Soil engineers attempt to be rather more sophisticated about these matters by the precise specification of *consistency limits*. As water is added to dry soil it passes through the stages of *solid* to *semi-solid* and then to *plastic* and finally to the *liquid* state. Between each of these states a boundary, or consistency limit, is specified and these are defined as follows:

1 *The shrinkage limit* – the limit between the solid and the semi-solid states. Above this limit the soil expands in volume when water is added and is then in the semi-solid state. Below this limit the removal of water causes no further decrease in volume and the soil is in the solid state.
2 *The plastic limit* – the limit between the semi-solid state and the plastic state. As water is added to the soil it becomes plastic and is able

to be moulded into shapes. The plastic limit is usually defined more precisely as the minimum moisture content at which the soil can be rolled into a thread of 3 mm diameter without breaking up.

3 *The liquid limit* – the limit between the plastic and the liquid states. At this moisture content the soil will flow under its own weight.

These consistency limits are illustrated in Figure 85. They are also frequently called the *Atterberg Limits* after A Atterberg who first specified them. Details of the procedures for the determinations of the limits can be found in most standard soil mechanics books (see Further Reading, p. 115).

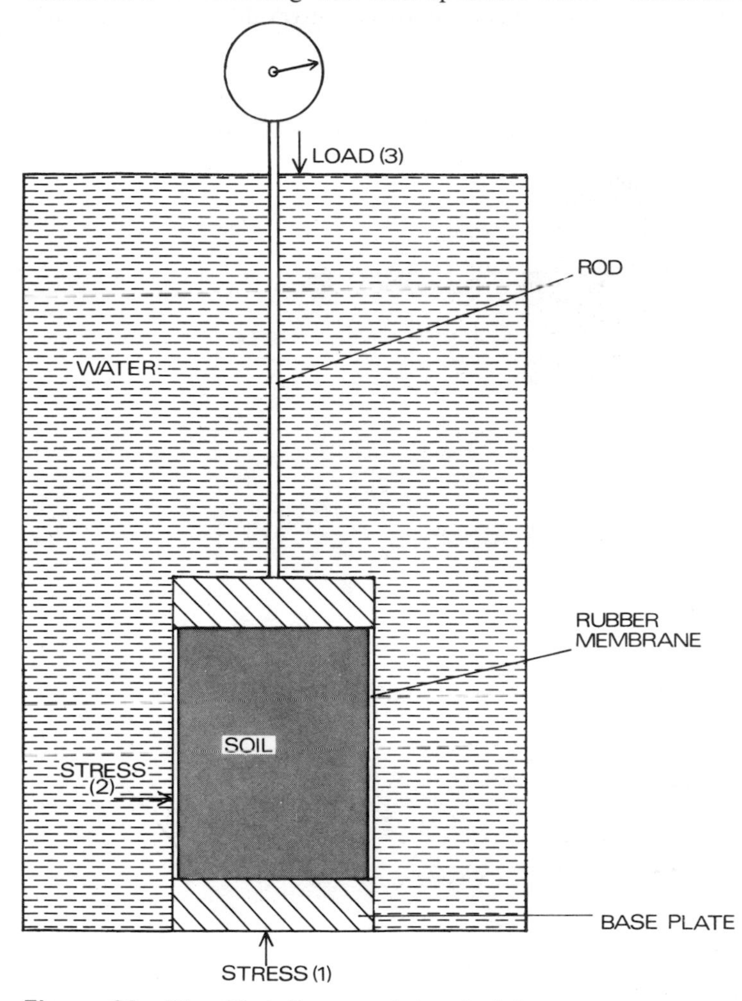

Figure 86 Simplified diagram of the triaxial test

SHEAR STRENGTH

The load-bearing properties of a soil are usually estimated by measurements of the *shear strength* of the soil. A soil can be subjected to a load and at a certain stage will give way and *shear*, one part of the soil sliding over another. The maximum resistance a soil has to such a shearing stress is called the shear strength.

In coarse-grained soils it is the friction between the grains that is important in determining shear strength. A high proportion of grains which are rough, jagged and interlocking will give a soil a high shear strength, able to bear a heavy load before shearing. In fine-grained soils, however, the cohesion between the grains is the most important factor. Because of this the water content is a major factor in determining the load-bearing properties of fine-grained soils. Thus fine-grained soils in the semi-solid state will exhibit stronger bonds between the particles and have a higher shear strength than similar soils in the plastic or liquid state.

A common test used in soil mechanics labora-

tories to find the shear strength of a soil is the *triaxial test*. The soil sample is stressed from three directions (or in three axes, hence, 'tri-axial') as shown in the simplified diagram, Figure 86. The soil is subject to pressure from 1) the basal plate which is stationary, 2) the water surrounding the soil which is enclosed in a rubber membrane as shown, and 3) the load applied vertically downwards, which is increased until the sample fails, or shears. The load at which the sample fails is recorded. The sample, upon shearing, appears as in Figure 87.

A second test is the *shear box test* which is shown in simplified form in Figure 88. The soil is placed in a two-part box, the upper half being subject to a force in the opposite direction to the lower. The shear force is increased by cranking a handle until the soil shears. The box is supported on rollers and held in place by a load above. The force being exerted when the soil fails is recorded on the proving ring on the left of the apparatus.

A simpler, cruder, apparatus can be made to demonstrate the principle of shear strength by loading soil into two boxes (e.g. seed boxes that do not have large gaps between the slats) and fixing the lower one on to a bench (Figure 89). The upper box should be held down by a large weight (experimentation should show the correct weight, but at least 2–5 kg may be necessary to prevent the upper box lifting at the back and hinging forward at the front). String should be attached to the top box, to which is attached a receptacle for weights. The box should be placed some way back from the bench so that the stress in the top box is as horizontal as possible. Weights can be added to load the upper box until it shears and the load at which shear took place can be recorded. The results gained from such an apparatus will be very crude, and cannot be compared with shear-strength results recorded in the literature, but it can be used to demonstrate the differences between very sandy soils and clayey soils or between wet and dry soils.

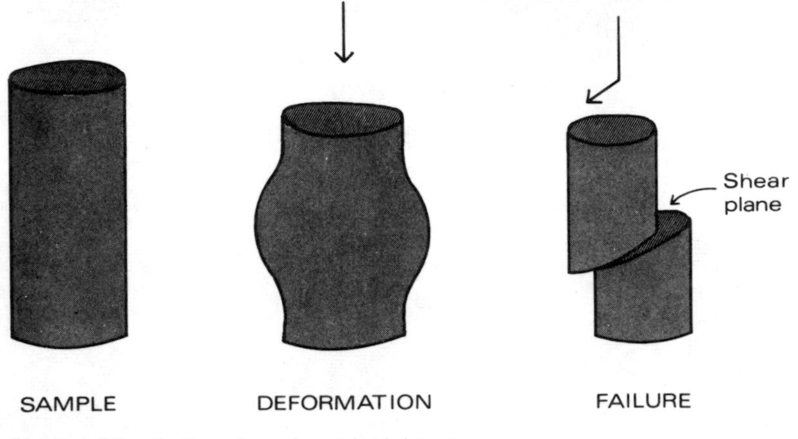

Figure 87 Soil undergoing triaxial test

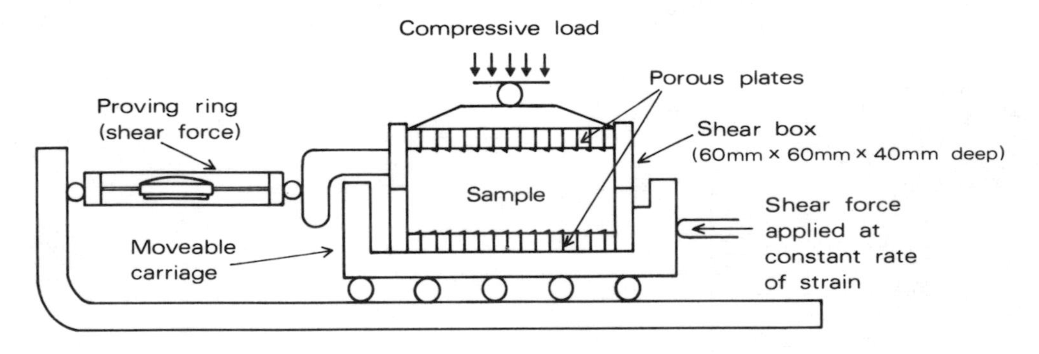

Figure 88 Diagram of a shear box

PRACTICAL APPLICATION

Many problems of the practical application of soil mechanics can be found in problems ranging from dam building to laying the foundations of buildings

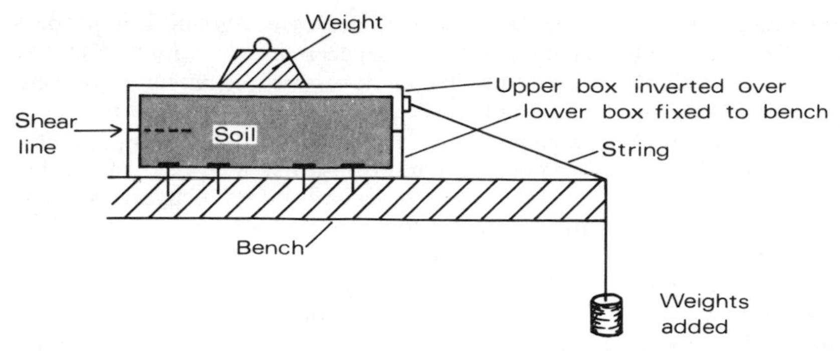

Figure 89 Diagram of a simplified shear-box apparatus

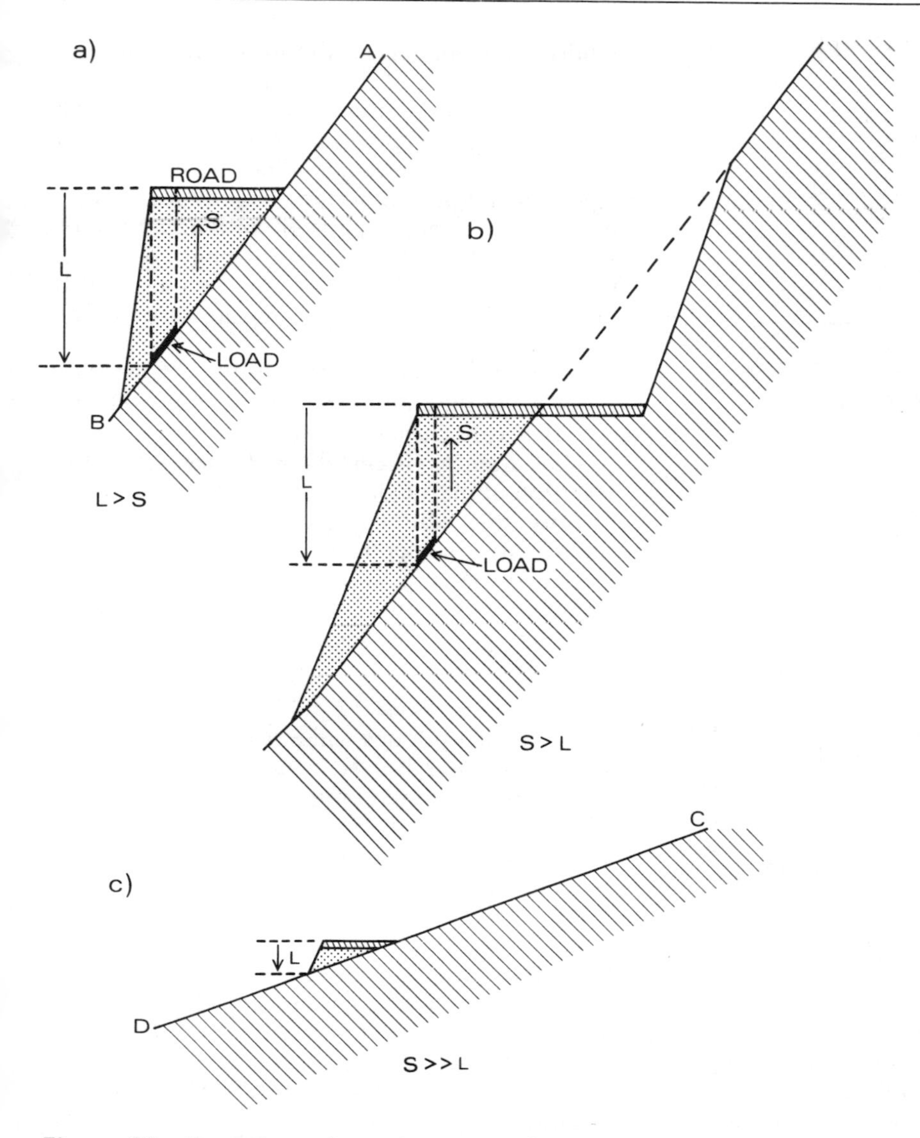

Figure 90 Examples of road construction on slopes

and road construction. As an example, imagine a slope A–B (Figure 90), on which a road is to be built. If a road embankment was built as shown in a) then the load per square centimetre of the original slope is represented by 'L'. If 'L' is greater than the shear strength 'S' opposing it then the road will slide downhill, especially if the soil becomes wet. However, if the road is constructed as in b) so that there is less depth of soil loaded on to each part of the original slope, a decreased load is seen as represented by 'L'. The road is now stable as 'S' is greater than 'L'. This is also true of c) where the slope C–D is much gentler than the slope A–D. This is one

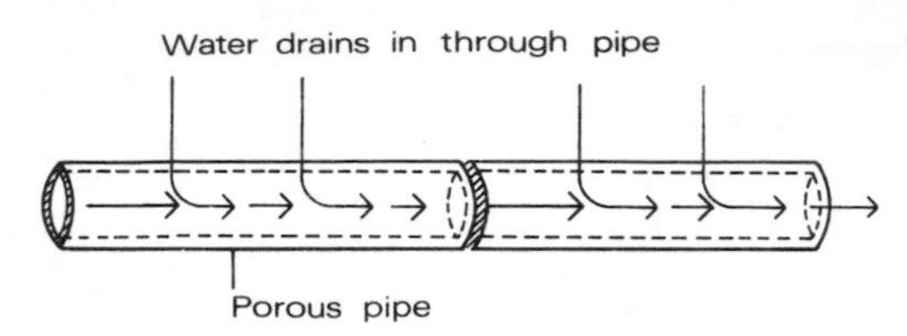

Figure 91 Tile drains

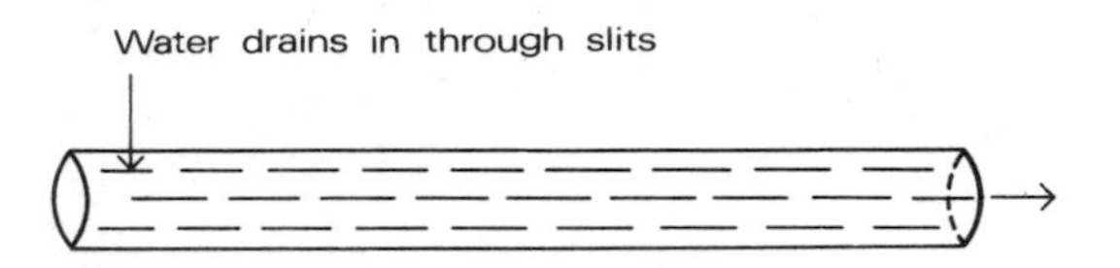

Figure 92 Plastic pipes

hypothetical example of the way in which the measurements of shear strength may be used in the calculation of the slope angle of the embankment necessary to make the embankment stable.

Summary

Soil mechanics deals essentially with the load-bearing properties of soils, defining soils as any loose, unconsolidated material occurring naturally on the earth's surface. These load-bearing properties depend largely upon the properties of each class of soil texture and the water content of the soil because both have a strong influence upon the cohesion between the particles bearing the load.

5.4 Soil drainage and soil reclamation

Drainage

Areas of soil can suffer from excess moisture if:

1 they have a clay texture and poor structures so the water is held in the soil, or
2 they are low lying and the water cannot drain away, or
3 they are in high rainfall localities.

Excess moisture decreases agricultural productivity primarily by decreasing aeration. This, in turn, will have an adverse effect on soil organisms and on plant root development. Usually when a soil is gleyed, drainage could be used to improve the land for agriculture. Earlier methods used either ditches alone or raised ridges between shallow ditches which thus encouraged water to run off. Tile drains, plastic pipes and mole drains are now used widely.

DRAINS

Tile drains are cylindrical pipes of porous clay which have been hardened by firing in a kiln. Each section is laid down adjacent to the next in a trench in the soil and then the trench above the pipe refilled. The depth of drains depends upon a number of factors, including the nature of the soil profile and the desired effect on the water table. The pipes will drain off gravitational water from the soil pores. Various diameters of pipes are used according to the amount of water it is required to draw off, but a common type is 30 cm long and about 10 cm diameter, the porous sides being about 1·5 cm thick (Figure 91). The runoff can be increased by 'backfilling' the drainage trench with porous gravel.

Plastic pipes are a modern substitute, with slots or holes cut in the sides (Figure 92) and it is claimed that these are cheaper and more effective.

Mole drainage (Figure 93) is effected without the insertion of plastic or tile drains. A moling implement is attached to the tractor and pulled through the soil at some depth. This leaves a small tunnel in the soil through which water can drain. The tunnel usually lasts for two or three years, but this depends upon many factors, particularly the soil structure.

DRAIN SPACING

The spacing of drains will obviously influence the amount of water drained from a field and the water table can be lowered more by a closer spacing of drains, as shown in Figure 94. The height of the water table (H) is directly proportional to drain spacing (L). Thus

$$H \propto L.$$

To work out the relationship between (H) and (L) accurately it is necessary to study the rainfall (R) of an area, the permeability of the soil (P) and the depth of the water table to the impermeable layer which is holding up the water (D). Then using the formula

$$H^2 - D^2 = \frac{R}{P} \times L^2$$

the amount by which spacing (L) affects the water-table height (H) can be predicted.

The drains are usually laid out in a 'herringbone' pattern with the open end of the pipe being lower down the slope (Figure 95).

Reclamation

Land reclamation in Britain became important during the ploughing-up campaigns induced by food shortage in the Second World War. Marginal land (clay wetlands, scrubby lands and heaths) came under cultivation. Today, although some of the wartime ploughed-up land has reverted to uncultivated land, reclamation of marginal land still continues in the form of drainage of wetlands and the improvement of hill pastures.

Upland soils, which include peaty gleyed podzols, podzols, rankers and peats, obviously suffer from two main factors inhibiting crop growth – wetness and acidity. A management policy of drainage and liming has to be undertaken to overcome these.

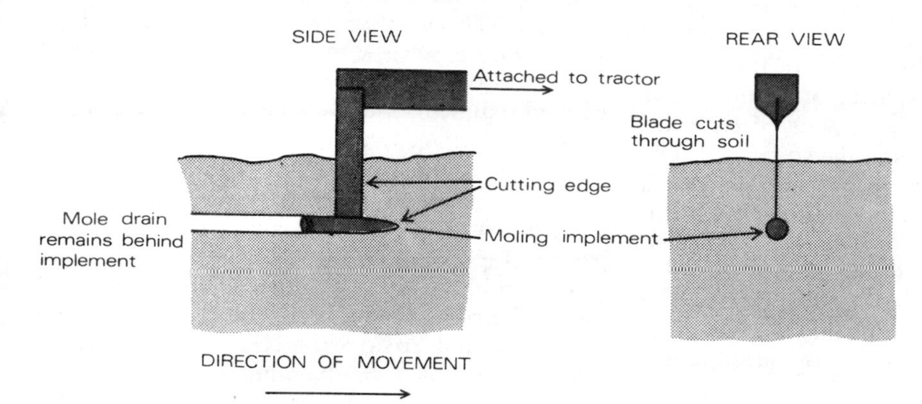

Figure 93 Mole drains

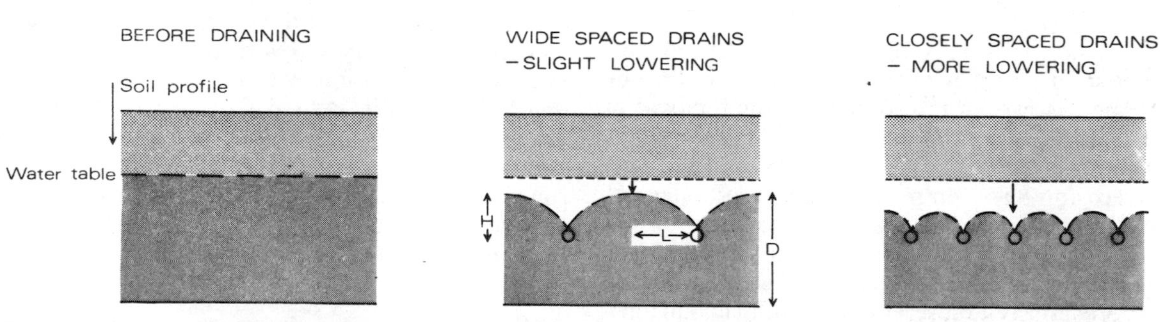

Figure 94 Drain spacing (in section)

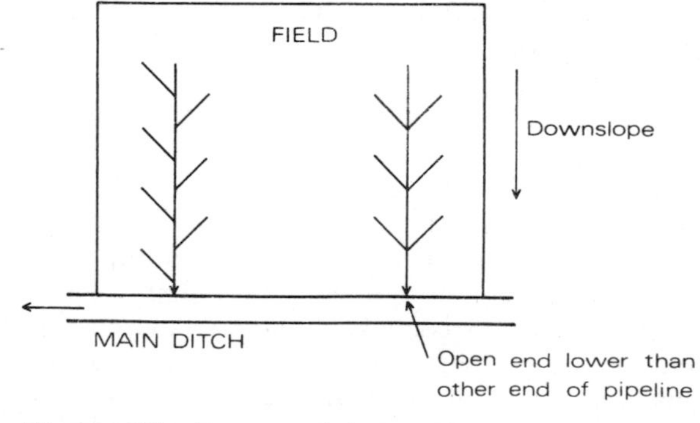

Figure 95 Pattern of drains (in plan)

However, the problem is not generally a simple one. Stoniness and the persistence of bracken rhizomes can be almost insuperable. In spite of the fact that bracken may be suppressed by constant crushing and also that chemical sprays have been developed which kill bracken, no method is really wholly effective. The use of sprays is still experimental and not entirely satisfactory. Ploughing may actually encourage bracken growth by cutting its fleshy rhizome into two or more pieces, each of which may grow into a separate plant.

Summary

Drainage can be achieved by tile drains, plastic drains or mole drains. The type of drain, the manner in which it is laid and the spacing of the drains are all important considerations in their effectiveness at reducing the water levels in the soil.

Reclamation of soils in upland areas needs a knowledge of soil drainage and fertilizer application, but problems may arise from stones and bracken. The crops that can be grown are often more limited by climate than by soil factors, but grassland can usually be profitably improved by fertilizers and drainage where topography will allow.

5.5 Soil erosion and conservation

Basic concepts

Soil erosion is usually thought of as the removal of the solid particles of the soil, either by wind or by water. However, it also includes the removal of nutrients and clay colloids in solution in the water which flows within the soil matrix. While solid particle erosion has been recognized as a problem for some time the loss of nutrients as a form of soil erosion has only recently received much attention. The latter is referred to as 'chemical erosion' or 'fertility erosion' as it represents a potential loss in agricultural productivity by a depletion of the store of soil nutrients. Furthermore, awareness of this problem is necessary to tackle the problems of eutrophication of inland waters (see section 4.3), because the loss of nutrients from the soil may represent a gain in nutrients in ponds, ditches and rivers. This may lead to the 'blooming' or sudden multiplication of algae detrimental to the aquatic life.

Soil erosion in Britain can occur in one of three ways:

1 the physical washing of topsoil downslope,
2 the physical blowing of topsoil away in strong winds, and
3 the chemical solution of soil nutrients and loss by leaching.

Slope processes

If old fields on slopes are studied carefully it may be noticed that the soil is at different levels on either sides of hedges and banks running across the slope. The soil is higher on the upslope side and lower on the downslope side (Figure 96). On the upslope side soil has accumulated by surface wash and soil creep. These processes are known as *colluvial wash* or *colluvial movement* and the resultant material is *colluvium*. These are natural processes and will occur to some extent even under a close grass cover, especially where the slopes are steep and the rainfall is high. The processes may be accelerated by ploughing on steep slopes. Runoff may be rapid after rainstorms, and the accumulations of water-washed sediments may be seen at the foot of fields after storms. In general the overall effect may be to damage the soil as the topsoil is lost from the upper part of the field and buried in the lower part of the field, but over all, the slope will tend to become gentler (Figure 97).

The problem, which is only a slight one in many parts of Britain, can usually be avoided by not ploughing on steep slopes and leaving the soil under grass. The problem can be minimized, and the soil conserved, by ploughing across the slope, along the contours, rather than straight up and down the slope. With contour ploughing the water seeps through the ridges and does not build up enough momentum to carry soil with it (Figure 98).

This type of erosion, with rapid runoff, is a particular problem in upland Britain where forestry drains are installed. Here the drains are put in up and down the slope in order to achieve maximum runoff. However, this also gives the water maximum erosive power and some soil erosion may result, bringing sediments down into streams and rivers.

Wind erosion

Wind erosion of topsoil occurs in eastern England and some other areas where very large open arable fields exist. Fine sands and silts may be winnowed out of a dry soil if a bare exposed surface is left after ploughing.

Not all soil grains are as prone to wind erosion as others. The larger sand grains are too heavy to be moved, except by the strongest winds. Clays, on the other hand, although in themselves small enough to be blown away, possess the important property of cohesion (see section 2.4), sticking together and thus being resistant to wind erosion. Silts are the most prone to wind erosion as they are small enough to be blown and yet possess no cohesive properties. The strength of wind needed to

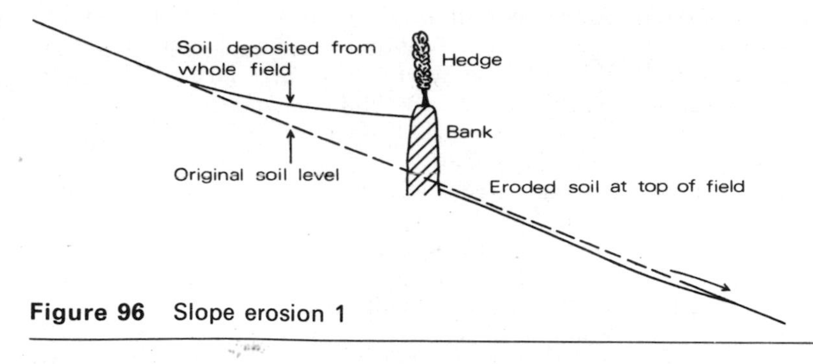

Figure 96 Slope erosion 1

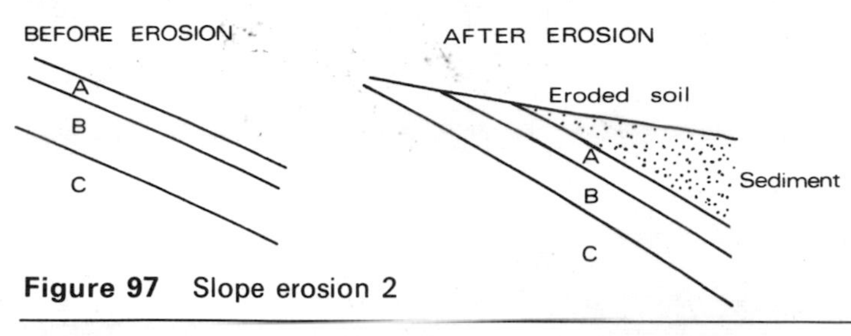

Figure 97 Slope erosion 2

a Ploughing up and down slope encourages erosion

b Contour ploughing minimises erosion caused by running water

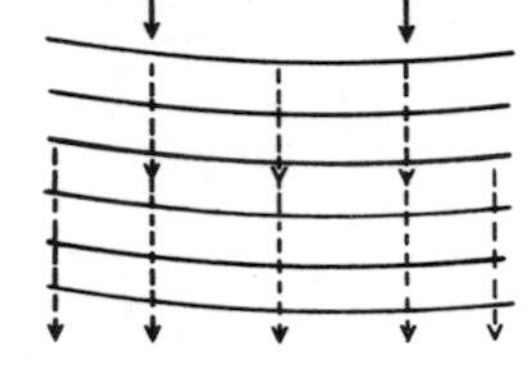

Figure 98 Ploughing on a slope

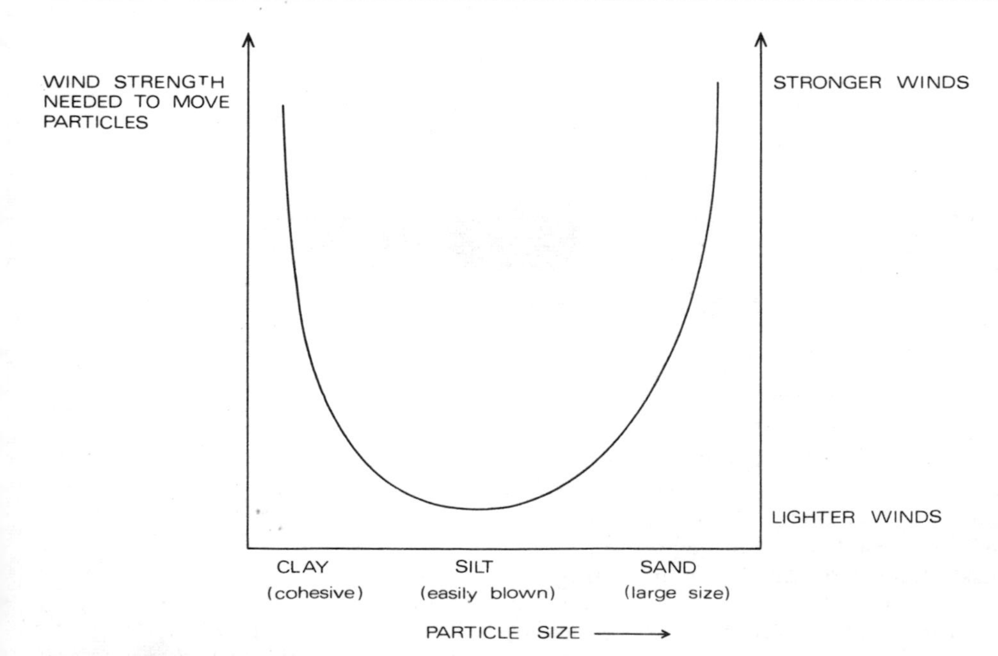

Figure 99 Wind erosion

move a soil particle of a particular size is shown in Figure 99.

In open fields wind can select the erodable soil fraction (mainly the finer silts) and leave behind the coarser soil structures and stones. Silt may accumulate on field edges and banks (Figure 100). This is particularly common in parts of eastern England where hedges have been removed.

In parts of the Fens of eastern England organic peaty silts overlay clays. The fertile peats and silts have been quite extensively eroded away by wind action (Figure 101). At present, a management policy is to plough the remaining peat into the clay and thus promote a more stable topsoil of mixed mineral and organic matter.

Well-planned shelter belts are an important way of managing soils liable to wind erosion. Shelter belts can reduce wind strength, depending upon the type of belt used. Trees with a low barrier of bushes are the most effective (Figure 102). Closely spaced shelter belts are the best protection against physical soil erosion in arable areas. On the other hand it is often more economic to have larger fields and thus be able to use efficiently large agricultural machinery, such as combine harvesters. Whether or not a hedge is grubbed up or a shelter belt is planted must be a matter of balancing the economics of having large mechanized field systems against losing money by crop deterioration resulting from soil erosion. It is also important that the shelter belts act as reserves for wild life, which itself has some significance to crop growth. (For example, hedges act as habitats for birds which are important in reducing insect pest numbers.) Also one needs to balance long-term costs of the possible alternative management schemes (open fields *v.* shelter belts) and decide the best policy.

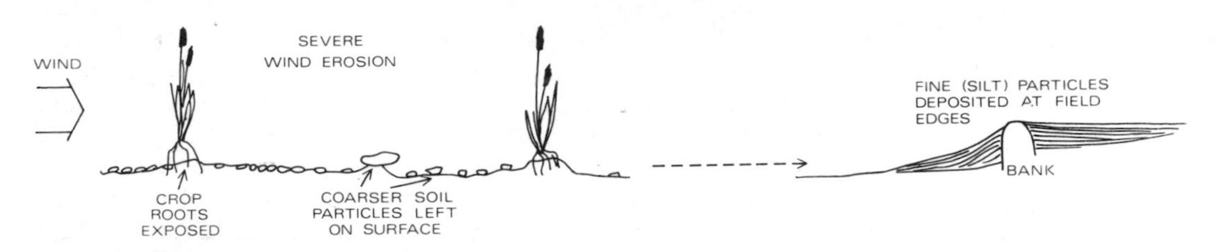

Figure 100 Wind erosion in open arable fields

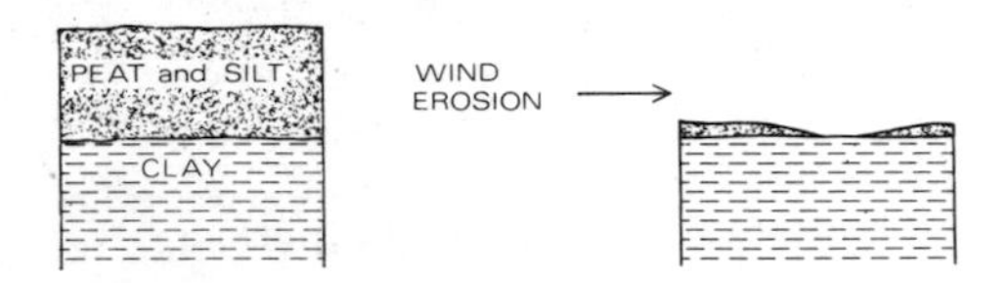

Figure 101 The Fens

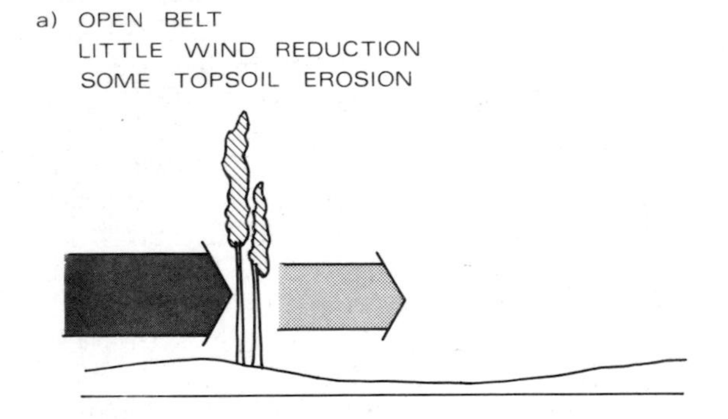

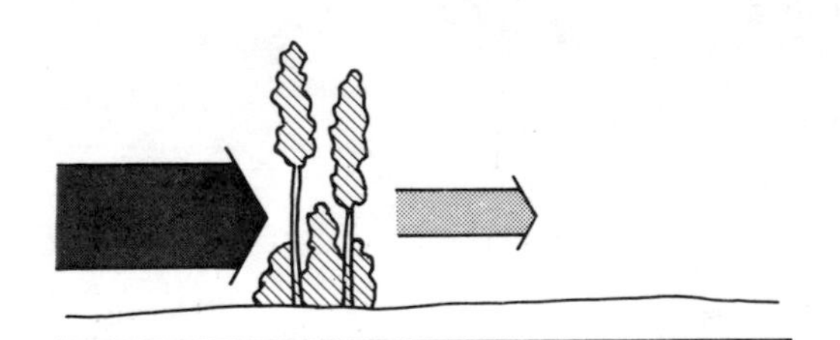

Figure 102 Shelter belts

Chemical erosion

Leaching has already been discussed under the headings of soil development, nutrient cycling and soil fertility (sections 3.1, 4.3 and 5.1). Leaching not only represents chemical loss in itself but it may also lead to the physical weakening of soil structures. This is because chemicals which cement soil structures may be lost. If large structures break down they are more susceptible to wind or water movement.

Where the dominant water movement is downwards, under adequate rainfall (i.e. where rainfall exceeds evaporation), nutrients will be dissolved and washed down from the topsoil. Natural pipes, such as are found in and under peat in upland areas (see section 4.2), also encourage this process.

Chemical analysis of waters flowing out of tile drains shows that calcium, sulphate, chlorine, nitrogen (in the form of nitrate) and sodium are the most readily lost nutrients. Magnesium, potassium, phosphate and nitrogen (in the form of ammonium compounds) are also lost, but in lesser amounts. Much of these nutrient losses may, in fact, come from applied fertilizer, although in some cases the origin is from rock weathering or, especially in the case of chlorine, from rain water. It is possible that in many cases too much fertilizer is being applied to the soil and the excess is being lost in drainage waters. Clearly, in marshy areas the nutrients would be retained in the soil if the tile drains were not installed, but the soil would be too wet to be of much value.

In high rainfall areas with natural free drainage there is little that soil management can do to prevent leaching losses. It must ensure, however, that:

1 artificial drainage is not too excessive, i.e. runoff is not more than it need be for adequate crop growth, and
2 excess fertilizer (which costs money) is not applied.

Summary

The agents of physical erosion are water running over the surface and wind. The problems of soil erosion can be curbed, or avoided, by not ploughing on steep slopes and by the use of shelter belts to protect those soils with a texture prone to blowing (chiefly those having a silt texture). Chemical erosion includes natural leaching and the leaching of applied fertilizers. Careful calculation of correct fertilizer application minimizes the chemical erosion of fertilizers into inland waters.

5.6 Soil resources

With a high population there is an incessant demand on the land to produce food. This is made more difficult in an urbanized country like Britain by the demand for building land which encroaches upon agricultural land, but some land is obviously more suitable for agriculture than others. Soil can sometimes withstand intensive cultivation; in other situations it may not. The proper and full use of our soil resources needs the understanding of how a soil works and a knowledge of the capability of soil in agricultural production. In this way the land available for agriculture is made the most productive to meet the demands of society for food.

Land capability

Soil erosion can be avoided, soil fertility and structure maintained and yields improved if a soil is not cultivated beyond its capability. For example, a valuable, deep, fertile loam, with a high organic matter content, a neutral pH, stable structures and adequate drainage in a region with moderate rainfall, can grow a very wide variety of crops with profit and without any damage to the soil. It is a versatile soil with a high capability. On the other hand a less fertile soil on a steep slope has a low capability and the crops that can be cultivated with profit are limited. Attempts to grow wheat on such a soil might show little profit and may cause damage to the soil by slope-wash erosion. The capability of the soil would probably limit its agricultural use to grassland.

While the physical capability of the soil is a major constraint on soil productivity it should always be remembered that economics usually govern what is grown. Therefore, if, in the case of the steep slope mentioned above, there was a tremendous demand for wheat and it fetched an extremely high price it might possibly be profitable to terrace the slope and contour plough to stop soil erosion. In this case, wheat could be grown with profit because of the economic demand for it. This is obviously an extreme example, but physical limits are only important when related to economic factors. The crucial factor is whether the cash benefits gained from managing difficult lands are greater than the cost of improving the land for agriculture.

Agricultural land classification

Land in Britain has been grouped by the Ministry of Agriculture, Fisheries and Food into capability grades in an *Agricultural Land Classification.* The highest grade, Grade 1, is extremely versatile and is capable of growing a wide variety of crops. Lower grades are more restricted in use, but their capability can be improved by, say, irrigation or

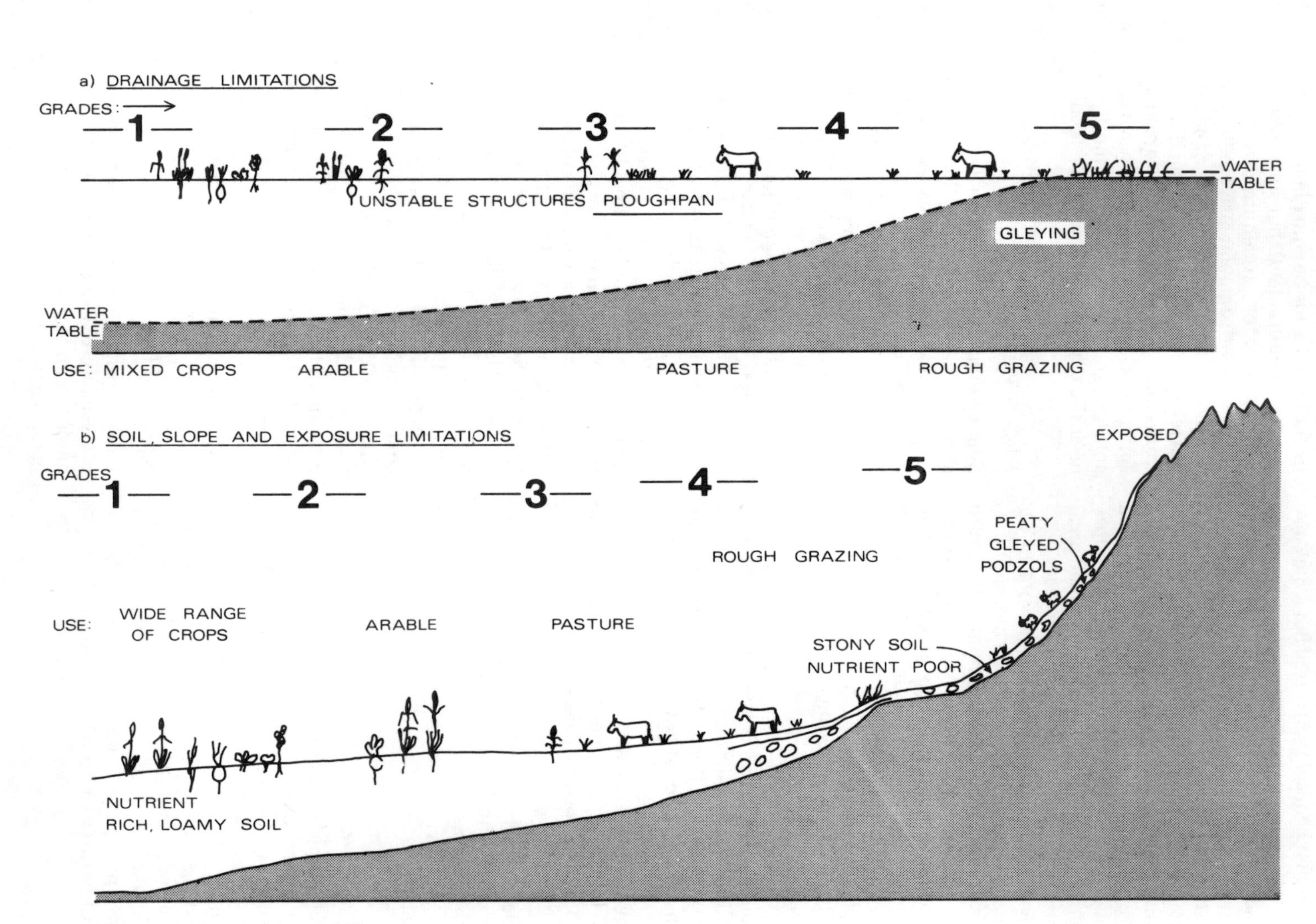

Figure 103 Land capability

drainage, if this is economically desirable. Since the management and economic factors are variable the classification is based on physical factors that limit agriculture under an *average* management scheme.

Land is graded according to the limitations that soil, slope and climate impose on agriculture:

Grade 1: No physical limitations. Deep well-drained loams, sandy loams, silt loams or peats. Level or gentle slopes. Nutrient rich. Wide range of crops with high yields. Any arable or horticultural crop can be grown.

Grade 2: Some minor limitations. Factors make the land slightly less flexible in the choice of crops that can be grown economically, e.g. texture, drainage, depth and climate. Most arable and horticultural crops can be grown.

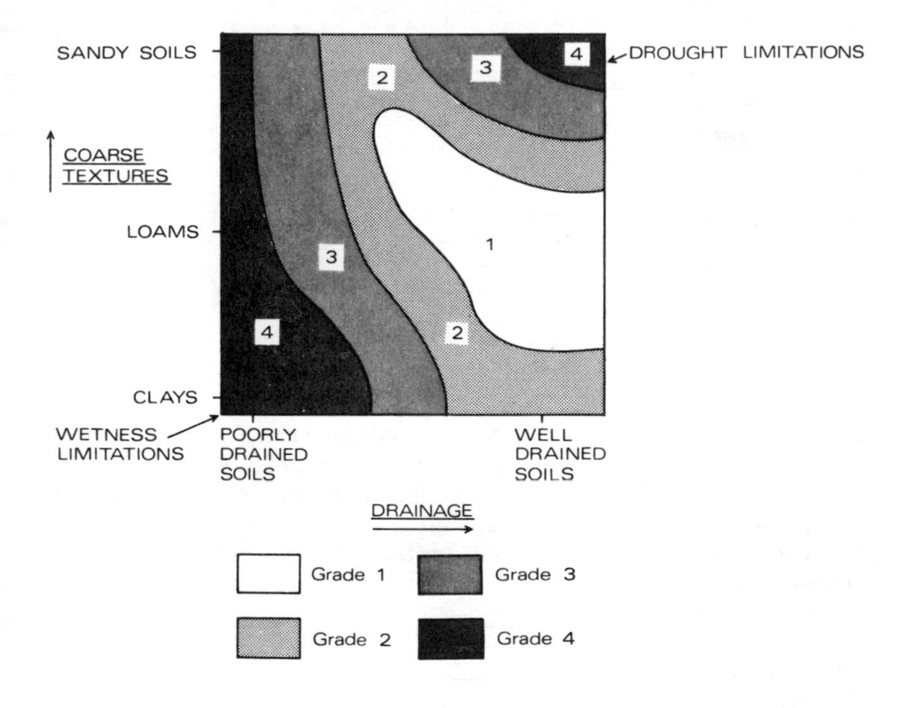

Figure 104 Land capability, drainage and soil texture (for a constant, moderate rainfall input)

Grade 3: Land with moderate limitations due to soil, relief or climate. Factors such as stoniness or altitude may limit productivity. Grass and cereals can be grown. Grade 3 land includes some of the best grasslands but is marginal arable land.

Grade 4: Severe limitations, e.g. unsuitable texture and structures, surface wetness, shallowness and stoniness, steep slopes, high rainfall or exposure. Mostly used as grassland, but some poorer crops of oats and barley can be grown.

Grade 5: Little agricultural value. Very steep slopes, very poor drainage, excessive rainfall and exposure, severe plant nutrient problems. Forestry and rough grazing are the usual land uses.

These grades are illustrated in Figure 103. The first case is where drainage imposes the main limitations, while texture and depth stay the same. The second shows a case where stoniness, soil depth, altitude and exposure all impose limitations.

In much of Britain, where factors such as stoniness and depth are not limiting, texture and drainage impose most of the restrictions on agriculture. Clearly the significance of drainage depends upon the amount of rainfall, but assuming a constant rainfall in one region the grades would be distributed in that region according to texture and drainage as shown in Figure 104. Grade 4 land

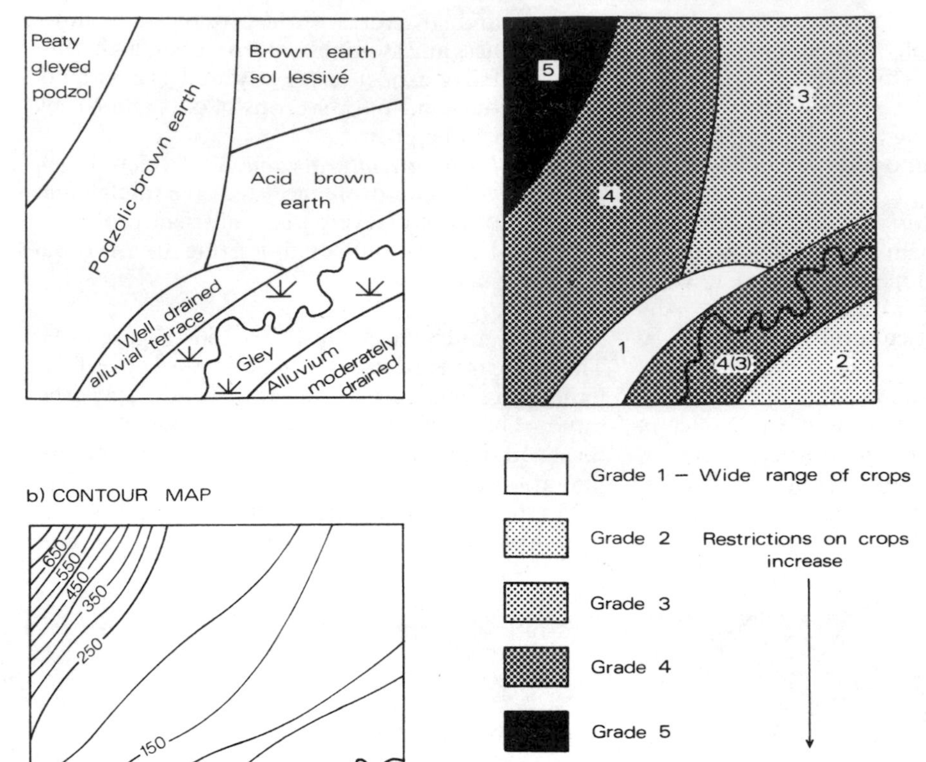

Figure 105 Land capability map

occurs where drainage is excessive or very poor. Grade 5 occurs where the rainfall is higher or the slope steeper or the soil is in a flat marshy area near sea level. In the case of the heavier soils (clay textures), as drainage improves so the capability of the land improves, but Grade 1 is rarely achieved. With the lighter, sandier textures moderate drainage is the optimum but excessive drainage leads to drought susceptibility. The optimum soil is clearly a moderately to well drained loam.

With irrigation a sandy soil prone to drought may be raised from Grade 4 to Grade 3 and some sandy soils in East Anglia are in this category. With drainage a clay soil may similarly be raised from Grade 4 to Grade 3. For this reason a Land Capability Class may be quoted as 4(3), indicating that under better than average management the land capability can be increased.

Land capability maps are produced and they can be compared with soil maps. In Figure 105 a soil map and a topographic map can be compared with a land capability map. The capability units may be larger than the soil units as two soil types may have the same capability.

Crop yields on the different grades can vary as follows:

Grade	*Wheat yield**	*Sugar-beet yield**
1	5000–5500	6000–8000
2	4000–5500	5000–6000
3	3000–4000	4500–5000
4	Uneconomic	c. 4000
5	Uneconomic	Uneconomic

*Yields in kilograms/hectare of grain for wheat, roots or beet

It is difficult to propose general rules about crop yields, soil type and land capability, and crop trials are the main way of experimenting with crop pro-

a) EXPERIMENTAL PLOTS

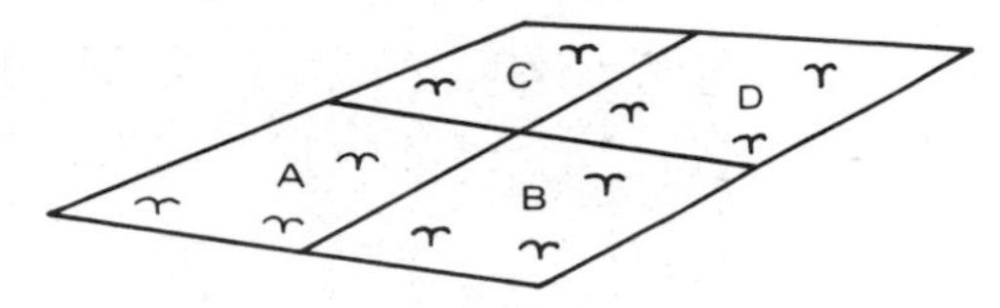

b) YIELD MEASUREMENT

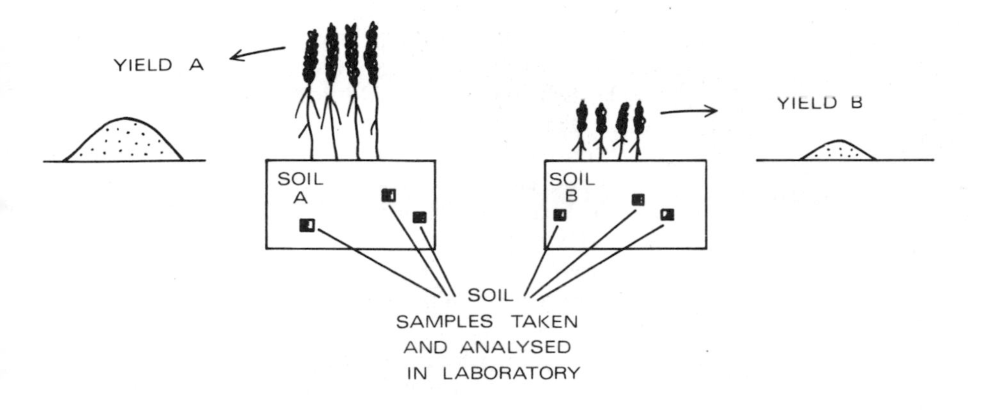

c) DATA ANALYSIS

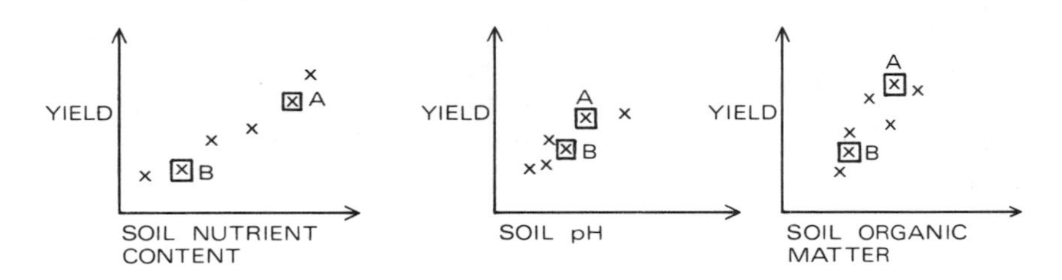

PLOTTING SAMPLE DATA ON GRAPHS DEMONSTRATES ANY RELATIONSHIPS BETWEEN CROP YIELD AND SOIL CHARACTERISTICS

Figure 106 Crop trials

ductivity on various soil types (Figure 106). Experimental plots can be laid out and sown with identical seed, the yield from each crop measured and the soil on which the crop was grown can then be analysed. Information about the relevant soil attributes can be plotted against yield and the reasons for different yields examined. By careful soil study the productivity of a piece of land can be estimated by assessments of soil type and of the overall land capability (including factors such as slope and climate).

In this way the agricultural scientist and soil scientist can combine their skills to work out the factors which influence the productivity of a crop. By using land capability assessments, advice can be given as to the best crop for a particular area.

Summary

Soil resources can be examined in terms of capability for agricultural productivity. This involves studying soil conditions together with climatic and relief factors to evaluate the limits that are imposed on agricultural production. It is then possible to cultivate the most suitable crop which will 1) produce an economic return and 2) not cause soil damage. Soil management – drainage, structure control, fertilizer application and irrigation – together with crop trials, can all improve soil productivity and so help to make the fullest use of a soil.

6

Soil description and classification

6.1 Introduction

This chapter explains methods of soil description, classification and mapping. Since this book is concerned with soil in Britain we have generally described the methods currently employed by the Soil Survey of Great Britain. The descriptive methods owe a great deal to the work of G R Clarke, to whom reference is made in the reading list at the end of the book.

Detailed soil work usually involves making a *soil profile description*. This involves describing as many visible and tactile ('feelable') characteristics of the various horizons as is reasonably possible. Methods of making detailed soil profile descriptions are therefore given first in this chapter (sections 6.2, 6.3, 6.4 and 6.5). Having described a soil profile in detail it is possible to *classify* the soil according to various classification schemes (section 6.6). It is then possible to map the distribution of the resulting soil groups (section 6.7).

It is important to stress that it is not always necessary to make full detailed profile descriptions. Some problems require that only one or a few characteristics are noted or measured. Section 6.8 is a discussion of the ways in which one decides precisely how much field work is really necessary.

6.2 Digging the pit

Examination of a soil is not just a matter of digging a hole and looking at it. First, it is important to stress that permission must be sought and gained before starting to dig a pit. *Every piece of land is owned by someone.* (Common land, paradoxically, is often the most difficult for which to get permission, there being several individuals who have the right to veto your permission!) To find out who owns land ask at the nearest farm. If they don't own the land they'll probably be able to say who does. If you explain to the owner what you want to do, why you want to do it, and take full responsibility to be tidy and careful, most owners are very cooperative. It is also worth remembering that if one individual soil observer causes difficulties he will make problems for any who follow.

In the field you will need a strong sharp spade. (A small light spade is often easier to use than a large heavy one.) In addition a short-handled pick, a geological hammer and two plastic sheets (approximately 2 m^2) are useful. A penknife or small trowel can also be used for clearing up the soil face for final examination.

Having obtained permission, choose your site carefully (providing you are not following a predetermined sampling pattern – see section 6.8). Look critically at any bumps and hollows on the landscape and avoid them. Otherwise you will probably find that you are making an archaeological investigation of an old road or old barn foundations. It is also best to try and get at least twenty metres or so away from paths or field boundaries (hedges, walls, fences): these too may be the cause of soil disturbance. Remember that as far as possible one is concerned to find natural soils, undisturbed except for normal agricultural cultivations. Very often woodlands (especially small woodlands) tend to be more disturbed than agricultural land. Small woodlands are often in position because they cover old quarries or land either too bad or too difficult to farm.

Pits should be about 1 × 0·7 m. Place the plastic sheets adjacent to the chosen site, carefully remove the turves or, if in arable land, the topsoil, and place on one of the sheets. Drag the sheet away from the edge of the pit and leave on one side. Next dig out the pit to a depth of one metre, or until you reach solid rock, whichever is the shallower, placing the soil on the other sheet. When the pit is complete drag this sheet away also. It is not necessary to dig the pit to a uniform depth but only to excavate to full depth the side which one wishes to examine. When choosing which side of the pit to describe it is best to choose the side that gets most light. This will both aid description and make for easier photography, should the latter be desired.

When examination of the pit is complete drag the two plastic sheets back to the pit side. Tip in the subsoil and tamp it well down. Then replace the topsoil and turves. If the soil is very clayey and strongly structured it may be difficult to get all the soil back into the pit. In this case it is best to leave a small grave-like mound over the pit because the soil will settle in a few days. If the pit is in a field frequented by farm animals (for example, dairy pasture), it is best to cover the replaced turves with cowpats or other animal dung. Cows are very inquisitive animals and will otherwise virtually re-excavate the pit! If reasonable care is taken during excavation and examination there should be very little evidence of your efforts after work has been completed.

6.3 Site description

Before examining the soil profile it is important to make a description of the site. Table 1 lists the features noted by members of the Soil Survey of Great Britain when making detailed profile descriptions, but the number and nature of observations depends on the exact nature of the study being carried out. C Mitchell (*Terrain Evaluation*, Longman, 1973) gives a longer discussion of the different methods of site description.

6.4 Soil horizons

How to recognize a horizon

As far as the soil observer is concerned a horizon is a layer of soil which differs in some way from the soil immediately above or below. It may differ simply in one property or it may differ in several; it may differ a great deal or it may differ only slightly; the difference may be only gradual and ill-defined or it may be sharp and distinct.

Nevertheless soil does not usually lie on the

Table 1 Items to be recorded in site description

Item	*Source of information*	*Example*
National grid reference (six figure)	Ordnance Survey map	SP 185547
County	Ordnance Survey map	Warwickshire
Name of observer		W Shakespeare
Locality	Ordnance Survey map	Dunsinane Farm, Shottery
Name of owner or tenant	Inquiry	J Macbeth
Elevation (metres)	Ordnance Survey map	35 m
Slope and aspect	Use of level or by eye	2° SE
Relief	Use of the following terms: convex concave plateau with a verbal description (see Figure 107)	Very slightly convex slope
Drainage of site	Use of the following terms: receiving (i.e. net inflow) normal shedding (i.e. net out-flow) fresh-water flooding (e.g. streamside site) (see Figure 108)	Normal
Parent material	Examination at site and reference to geological literature	Shelly oolitic limestone (Chipping Norton limestone)
Vegetation or system of agriculture	Examination (in the case of natural vegetation the dominant plant species should be named)	Short ley (rye grass)
Weather		Fine; sunny after short heavy shower

underlying rock in a random way, and the processes going on in the soil serve to sort it into distinct layers. Even on tip-heaps that have been recently reclaimed by artificial means incipient horizons can usually be seen. In these cases the processes have already begun to separate the individual horizons and produce a characteristic *soil profile* or combination of horizons (see section 1.1).

Soil horizons may be distinguished, for example, on the basis of colour, texture, stoniness, number of roots, soil structure and presence (or absence) of particular chemicals. Horizons are usually distinguished on the basis of features that are distinguishable *in the field*, rather than in the laboratory. (These features are sometimes known as *field characteristics*.) Thus, whilst the presence or absence of carbonates might be a useful criterion, quantitative amounts (for example, above or below 0·2 ppm) would not, the calculation of quantity of carbonates requiring laboratory equipment. Equally, whilst presence of carbonates might be a useful criterion, distinction between calcium or magnesium carbonate would not. The former can be distinguished using a bottle of dilute hydrochloric acid, but the latter can best be distinguished by laboratory technique.

Fortunately for the observer, visible (or tactile, in the case of soil texture) changes usually correspond with particular chemical variations. The reason for this is quite clear. Features like colour depend on the chemical constitution, whilst features like texture and structure often control the chemical processes.

Horizon terminology

As we shall see later (section 6.6) recent classifications of soil types depend very considerably on horizon classification. Some soil classifications use *diagnostic horizons* to determine ('diagnose') to which group a soil belongs. Thus it is very important that horizons themselves are correctly classified. It might seem that this would be a relatively easy task. However, over the years systems of horizon classification have grown up which not only describe the horizon but also contain an implicit statement about its properties: defining horizons in such a way clearly calls for some skill. (Details of horizon formation have been given in section 3.1.)

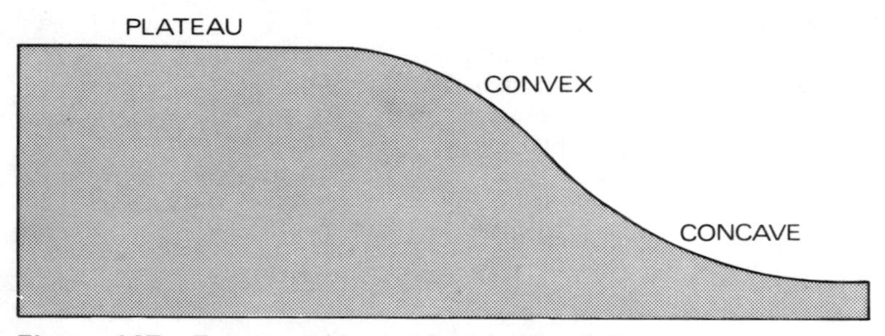

Figure 107 Terms used in the description of slope

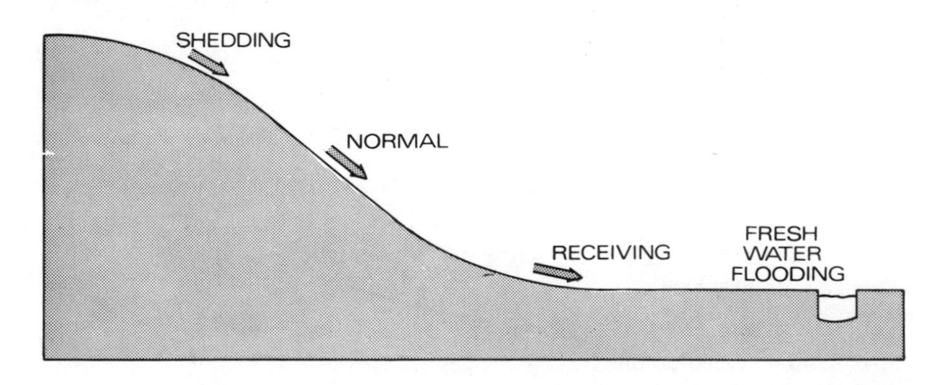

Figure 108 Terms used in the description of site drainage

The most commonly used horizon nomenclature system has three basic units, A, B and C, which are used thus:

A Mixed mineral–organic horizon at or near the surface. This is generally a horizon of eluviation which may lose both soluble salts by drainage and also fine particles of insoluble material by mechanical downwash.
The A horizon contains a high proportion of organic material, but this is fully incorporated in the soil. Where there are thick organic accumulations above the A horizon these are classified separately.

B This is a subsurface horizon of accumulation. Essentially it is a horizon of illuviation, accumulating both mechanically downwashed material from the A horizon and also chemical depositions.

C This is the mineral matter of geological origin in which the soil horizons are forming, otherwise known as a 'parent material' horizon.

Although soils with simply these three horizons are not uncommon, there are a wide range of other possibilities. One situation is where the soil is so shallow that no distinguishable horizons are recognizable and the soil profile is simply one 'A/C horizon'. This situation is typical of calcareous rendzinas and other skeletal soils (see section 6.6).

Other letters which are used to denote particular horizons are as follows:

SURFACE HORIZONS

L undecomposed litter

F partially decomposed litter (equivalent to fermentation)

H well-decomposed humus with little or no mineral matter

A see above
- Ah very dark coloured (humic) horizon often, but not necessarily, associated with high organic matter content
- Ap ploughed layer of cultivated soils

SUBSURFACE HORIZONS

Note: Where two possible alternatives are in common use the recommended method is given first and the other method second.

E eluvial mineral horizon from which clay and/or sesquioxides have been removed
- Ea bleached (albic or ash-like) eluvial horizon in podzolized soils
- Eb brown (paler when dry), friable, weak-structured eluvial horizon depleted of clay

B see above; illuvial concentrations of the following materials may be denoted by suffixes thus:
- Bt illuvial clay (textural B horizon)
- Bh illuviated humus, characteristic of podzols
- Bf, Bfe or Bi illuviated iron (chemical abbreviation: Fe), characteristic of podzols
- Bs brightly coloured horizon containing sufficient accumulations of sesquioxide (iron and aluminium oxides in amorphous forms [extractable by alkaline pyrophosphate]) arising from downward translocation, and/or alteration of silicate minerals *in situ*

C see above

R bedrock, which can be massive, stratified or fragmented *in situ*, with much displacement of the fragments

The following qualifying suffixes are also used:

- c or ca horizons with significant amounts of calcium carbonate; c' is used where common or abundant secondary carbonate concretions are present
- cs horizon with significant amount of calcium sulphate
- g gleyed horizon with greyish and/or ochreous mottling due to periodic waterlogging
- (g) slightly gleyed horizon

A/C or AC } horizons of transitional or inter-
B/C or BC } mediate character

Lithological discontinuities in stratified parent materials (for example, glacial drift) are given by Roman numeral prefixes (e.g. IIB, IIC, IIIC, etc.). I is always understood and not stated. Any minor changes which can be observed and recorded are denoted by arabic numeral suffixes (e.g. Bg1, Bg2 or, alternatively, B1g, B2g, etc.).

Examples of horizon nomenclature in use

Four examples follow in which the standard horizon nomenclature has been applied to actual soil profiles. Full profile descriptions are not given yet, but follow the explanation of classification in section 6.6.

EXAMPLE ONE

Gleyed calcareous brown earth from Gloucestershire, developed in calcareous drift lying over Lias clay. *Podimore* series, calcareous variant described by Cope, 1973.

Horizon	*Explanation of nomenclature*	
Ac	Calcareous 'A'	All in parent material 'I' understood
Bc	Calcareous 'B'	
Bcg	Gleyed calcareous 'B'	
IIBcg/Ccg	Gleyed calcareous 'B' intermediate to 'C'	In parent material 'II'
IIICc′g	Gleyed 'C' with secondary carbonate	In parent material 'III'

EXAMPLE TWO

Gley soil from Gloucestershire developed in drift over grey clay. *Rowsham* series described by Cope, 1973.

Horizon	*Explanation of nomenclature*	
A(g)	Slightly gleyed 'A'	In parent material 'I' (understood)
Bg	Gleyed 'B'	
IIBg	Gleyed 'B'	In parent material 'II'
IIIBc′g/Cc′g	Gleyed 'B' with secondary carbonate concretions intermediate to 'C'	In parent material 'III'
IIICc′g	Gleyed 'C' with secondary carbonate concretions	

EXAMPLE THREE

Podzol developed in drift over carboniferous limestone, from Mendip Hills, Somerset. *Priddy* series described by Findlay, 1965.

Horizon	*Explanation of nomenclature*
A	
Ea	Bleached eluvial horizon
Bfe	Illuviated iron horizon ('iron-pan')
Eb	Brown eluvial horizon
Bt	Illuviated clayey horizon ('textural "B"')
B/C	'B' intermediate to 'C' (in this particular case the horizon is clay over limestone)

EXAMPLE FOUR

Humic gley developed in rivurine alluvium from Devon. *Laployd* series described by Clayden, 1971.

Horizon	*Explanation of nomenclature*	
L	Undecomposed litter	
F	Partially decomposed litter	
H	Well-decomposed litter (in this particular case, black amorphous peat)	
Ag	Gleyed 'A'	
C1g	Upper gleyed parent material horizon	Distinguished by amount of grit, colour and other features
C2g	Middle gleyed parent material horizon	
C3g	Lower gleyed parent material horizon	

6.5 The soil profile

The *soil profile* is the section of soil exposed either in a soil pit or in a roadside or quarry section. It extends from the ground surface down to the parent material in which the soil is developed. As has already been observed, not every soil characteristic need be noted at every soil profile examined: the number and nature of recordings depend on the project being carried out. (More discussion on this point is given in section 6.8.) If a full profile description is to be made the following features need to be noted:

depth of each horizon;

for each horizon:

- colour
- texture
- carbonates, proportion
- stoniness
- structure
- consistence
- porosity
- secondary minerals
- roots
- soil animals
- nature of lower boundary of horizon.

The description of these various characteristics will now be discussed in more detail.

Horizon depth is measured to the nearest whole centimetre. Zero is given as the top of the A horizon (or Ap, Ah or Ag horizon if appropriate). Horizons *above* the A horizon are measured using the surface of the A horizon as a baseline, whilst horizons *below* also use the A horizon surface as zero. The following example should clarify the procedure:

Depth (cm)	*Horizon*
10–7	L
7–5	F
5–0	H
- - - - - - - - zero - - - - - - - -	
0–8	A
8–35	B
35+	C

Colour is noted using the Munsell soil colour system, in the form of both a written description and a numeric code. The Munsell colour system was developed in the USA and is designed to enable the coding of all possible colours (not only those found in soils). It is based on a three-dimensional colour solid (Figure 109), the axes of which represent hue, value and chroma respectively. *Hue* is the dominant spectral colour (e.g. red, yellow-red, etc.), and is related to the wavelength of light. Each dominant colour is subdivided four times. So, for example, 2·5Y, 5Y, 7·5Y, 10Y are increasing hues of yellow. *Value* is the apparent lightness or darkness and ranges from 0 (white) to 10 (black). *Chroma* refers to the purity or saturation of the colour or, in other words, the departure from the neutral greys and whites.

Soil colour is checked by reference to a soil-colour book. These contain carefully matched colour chips, the best known being the Munsell Soil Colour Charts and the Fujihira Soil Colour Charts, both of which use the Munsell system. To check the colour the soil is first moistened to saturation (dry soil often shows different colours from wet soil), and the soil matched as nearly as possible to a colour in the book. The name and number of the colour is then noted, as in the following examples:

light reddish brown (5YR 6/4)
very dark greyish brown (2·5Y 3/2)
dark greenish grey (5G 4/1).

Both the colour of the general soil *matrix* (background) and of any *mottles* need to be noted separately, as do any additional colours in the horizon, such as those in old worm-channels. Mottles are described according to the following three scales:

Amount	*Contrast*	*Size*
abundant	prominent	coarse
common	distinct	medium
few	faint	fine
		very fine

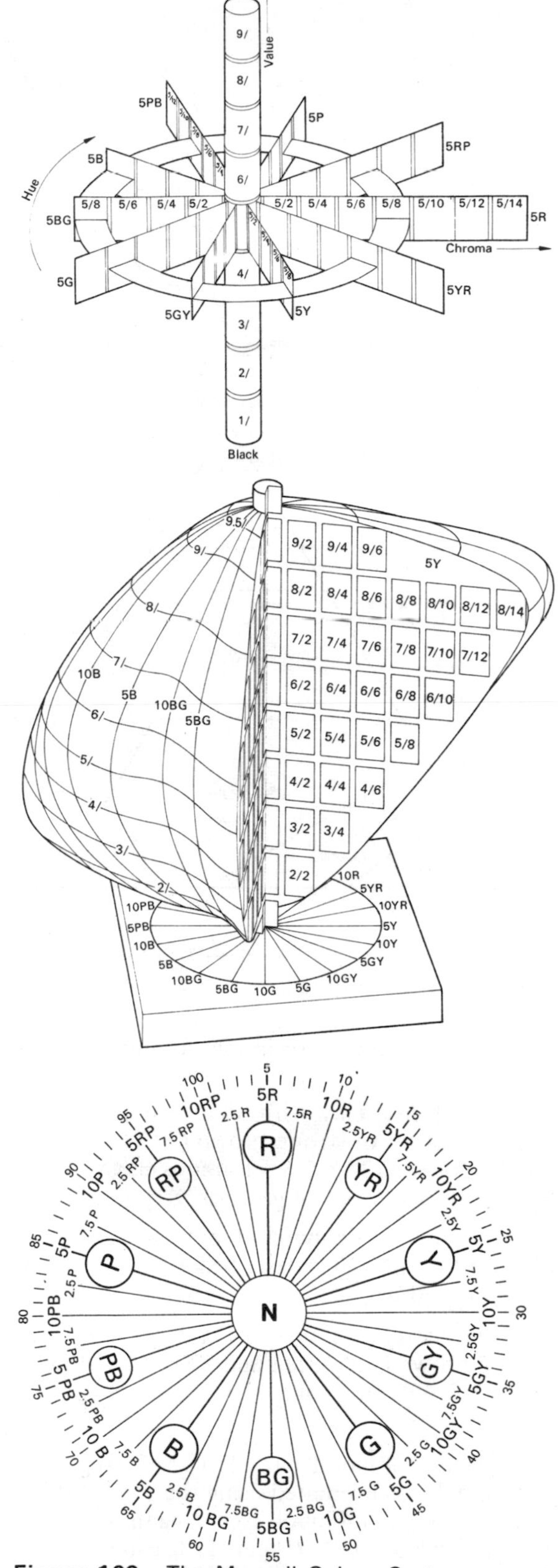

Figure 109 The Munsell Colour System

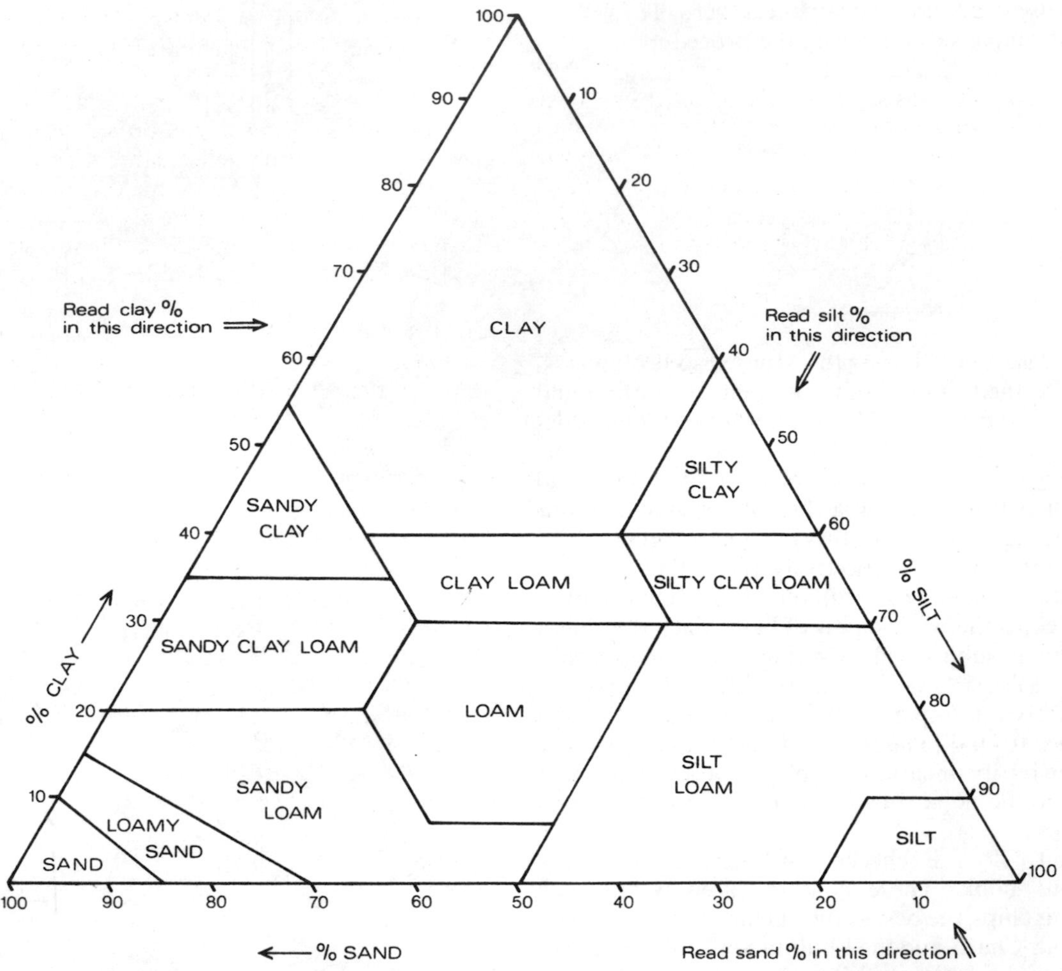

Figure 110 Soil texture

Examples of colour descriptions for mottled soils are:

grey (5Y 6/1–5/1) clay with abundant prominent coarse mottles of strong brown (7·5YR 5/6)

light olive brown (2·5Y 5/6 calcareous clay with few distinct fine mottles of greenish grey (5GY 6/1).

Where there is no definable dominant colour but many mottles the soil is described as *variegated* or *varicoloured*, thus:

Varicoloured reddish brown (5YR 5/4 and 5/3), strong brown (7·5YR 5/6 and 5/8) and yellowish red (5YR 4/6 and 5/6).

Soil texture (mineral particle size distribution: see section 2.3) is estimated using the procedure of 'hand-texturing'. This involves placing the soil in its correct particle size class (Figure 110), using the tactile characteristics of the three main particle size components – sand, silt and clay. (Full details of the size ranges and laboratory techniques are given in section 2.3.). The properties of the main grades are as follows:

'COARSE SAND consists of grains large enough to grate against each other, and can be detected individually both by feel and sight.

FINE SAND consists of grains which are far less obvious, but can still be detected, although individual grains are not easily distinguished by either feel or sight.

SILT. Individual grains cannot be detected but silt feels characteristically smooth and soapy, and only very slightly sticky.

CLAY is characteristically sticky, although some dry clays require a great deal of moistening and working between the fingers before they develop their maximum stickiness.'

(after G R Clarke, 1971)

Individual soil texture classes have the following characteristics:

SAND. Soil consisting mostly of coarse and fine sand, and containing so little clay that it is loose when dry and not sticky at all when wet. When rubbed it leaves no film on the fingers.

LOAMY SAND. Consisting mostly of sand but with sufficient clay to give slight plasticity and cohesion when very moist. Leaves a slight film of fine materials on the fingers when rubbed.

SANDY LOAM. Soil in which the sand fraction is still quite obvious, which moulds readily when sufficiently moist but in most cases does not stick appreciably to the fingers. Threads do not form easily.

LOAM. Soil in which the fractions are so blended that it moulds readily when sufficiently moist, and sticks to the fingers to some extent. It can, with difficulty, be moulded into threads but will not bend into a small ring.

SILT LOAM. Soil that is moderately plastic without being very sticky, and in which the smooth soapy feel of the silt is the main feature.

SANDY CLAY LOAM. Soils containing sufficient clay to be distinctly sticky when moist, but in which the sand fraction is still an obvious feature.

CLAY LOAM. The soil is distinctly sticky when sufficiently moist and the presence of sand fractions can only be detected with care.

SILTY CLAY LOAM. This contains quite subordinate amounts of sand, but sufficient silt to confer something of a smooth soapy feel. It is less sticky than silty clay or clay loam.

SILT. Soil in which the smooth, soapy feel of silt is dominant.

SANDY CLAY. The soil is plastic and sticky when moistened sufficiently, but the sand fraction is still an obvious feature. Clay and sand are dominant, and the intermediate grades of silt and very fine sand are less apparent.

SILTY CLAY. Soil that is composed almost entirely of very fine material but in which the smooth soapy feel of the silt fraction modifies to some extent the stickiness of the clay.

MEDIUM CLAY. The soil is plastic and sticky when moistened sufficiently and gives a polished surface on rubbing. When moist the soil can be rolled into threads. With care a small proportion of sand can be detected.

HEAVY CLAY. Extremely sticky and plastic soil, capable of being moulded when moist into any shape and taking clear fingerprints.'

(From Soil Survey Staff, 1960)

Note: The last two classes (medium clay and heavy clay) are now usually grouped together as 'clay'.

The texture is assessed by rubbing and working a moist sample between thumb and forefinger. It is usually best to use the left hand, thus leaving the right hand free and clean for making notes (or vice versa if left-handed). Accurate texture assessment is inevitably a matter of experience, but it is surprising how quickly one can master the basic groups and knowledge of the others soon follows. It is important to remember that one is only trying to place the soil in a broad group and not accurately determine the proportions of clay, sand and silt.

Organic matter and chalk (or other forms of finely divided calcium carbonate) tend to make the soil feel more silty and it is worth taking account of this when estimating the texture of soils containing high proportions of either.

If there is more than 15 per cent organic carbon the horizon is *humose* and the mineral texture scale is not employed. A separate scale is used as shown in the following table. (Although it is in fact percentage organic matter that is estimated in the field the table is arranged to show equivalent loss on ignition per cent.)

Descriptive term	*Organic carbon* (%)	*Organic matter* (%)	*Equivalent loss on ignition* (%)
Peat	>25	>43	>50
Peaty soil	15–25	26–43	30–50
Humose mineral soil	7·5–15	13–26	15–30

After Cope, 1973

Carbonates are estimated in the field by adding 10 per cent (dilute) hydrochloric acid – which can be carried in a small plastic bottle – and observing the degree of effervescence. If the soil visibly effervesces it contains more than 1% carbonates.

Stoniness is estimated visually by volume. The following *class* descriptions are used:

Percentage by volume	*Descriptive term*
<1	Stoneless
1–5	Slightly stony
5–20	Stony
20–50	Very stony
50–75	Extremely stony
>75	Stone dominant

SIZE of stones is estimated as follows:

Size (diameter, cm)	Descriptive term
0·2–1	Gravel
1–5	Small stones
5–10	Medium stones
10–20	Large stones
>20	Boulders

SHAPE of stones is noted using the following terms:

angular (including cubic and flat varieties)
sub-angular (including cubic and flat varieties)
rounded (including nodular formations)
shaly (cleaved structure)
tabular.

KIND of stones is also noted.

An example of a stoniness description is: 'stony with small to medium angular and subangular cherts'.

Structure DEGREE of structure development is noted using the following scale:

STRUCTURELESS. No observable peds; massive if coherent and single-grain if non-coherent.
WEAK. Indistinct peds; when disturbed the soil breaks into much aggregated material.
MODERATE. Well-formed peds; little unaggregated soil when disturbed.
STRONG. Peds distinct in place; soil remains aggregated when disturbed.

SHAPE of peds is noted as follows:

PLATY. Vertical axis much shorter than horizontal.
PRISMATIC. Vertical axis longer than horizontal, vertical faces well defined, vertices usually angular.
COLUMNAR. Prisms with rounded tops.
BLOCKY. Peds roughly equidimensional, and enclosed by plane or curved surfaces that are casts or moulds formed by faces of adjacent peds; subdivided into angular (sharp edges) and subangular.
GRANULAR. Small, subrounded or irregular peds without distinct edges or faces, usually hard and relatively non-porous.
CRUMB. Soft, porous, granular aggregates like bread crumbs.

Each shape of peds is divided into *size* classes as shown in the following table:

Ped type	*Size description (ped diameter, mm)*				
	Very fine	*Fine*	*Medium*	*Coarse*	*Very coarse*
Platy	<1	1–2	2–5	5–10	>10
Prismatic/ columnar	<10	10–20	20–50	50–100	>100
Blocky	<5	5–10	10–20	20–50	>50
Granular	<1	1–2	2–5	5–10	>10
Crumb	<1	1–2	2–5	n.a.	n.a.

n.a. = not applicable.

Consistence is the term used to describe the degree and kind of cohesion of soil material or fragments and varies with moisture content. The descriptive term applied refers to the moistness when sampled:

LOOSE. Non-coherent when moist or dry.
FRIABLE. Moist soil crumbles under gentle pressure to fairly uniform aggregates which cohere slightly when pressed together; the term *'labile'* indicates stronger cohesion after crumbling.
FIRM. Moist soil crushes under moderate to strong pressure but offers distinct resistance; *very firm* soil material is difficult to crush between finger and thumb.
HARD. Dry soil is moderately resistant to pressure; it can be broken in the hands, but is barely breakable between thumb and fingers; *very hard* soil material can be broken in the hand only with difficulty.
BRITTLE. Material resists deformation and breaks suddenly under pressure.
PLASTIC. Very moist or very wet soil can be moulded and rolled into shapes without breaking.
STICKY. Very moist or wet soil adheres to the hands.
SOFT. Soil yields easily under pressure.

Porosity gives an approximate indication of the quantity and size of fissures and pores in the soil. The following terms are usually used to describe porosity, although it should be remembered that porosity is to some extent a reflection of moisture conditions at the time of description:

FREQUENCY	SIZE
abundant	coarse
common	medium
few	fine

Secondary minerals are described by abundance, form and size. They may include secondary carbonate, manganiferous and gypseous deposits. Examples of descriptions are:

few medium and coarse secondary manganiferous deposits
common fine secondary carbonate concretions.

Roots. The following method of describing roots is in common usage. QUANTITY is noted per 30 cm^2 of profile face:

Abundant	more than 100
Frequent	100–20
Few	20–4
Rare	3–1

SIZE is noted as follows:

Large	more than 1 cm diameter
Medium	1 cm–3 mm
Small	3–1 mm
Fine	less than 1 mm

TYPE of root is noted as either:

woody,
fleshy,
fibrous or
rhizomatous

An example of root description is: 'abundant small fibrous roots'.

Soil animals are recorded in a brief verbal description, e.g. type and abundance.

Horizon boundaries. At the end of each horizon description a note is made of the nature of the lower horizon boundary. The following terms are usually used in the description. SHAPE of the boundary is described as:

EVEN: when the horizon is at the same depth across the whole of the visible profile.
UNDULATING: when upward and downward projections are wider than their depth.
IRREGULAR: when projections are deeper than their width.

CLARITY of the boundary is described as:

SHARP: when the transition zone is less than 2 cm thick.
NARROW: when the transition zone is 2–5 cm thick.
MERGING: when the transition zone is greater than 5 cm thick.

Examples of soil profile descriptions are given at the end of the following section (6.6). One final point is worth remembering when making a soil profile description. It is far better to write down exactly what you see than to avoid doing so simply because you have no appropriate term or phrase available – even if this means a lengthy and wordy description. Once the pit has been filled in you cannot have another look!

6.6 Soil classification

The essential aim of this section is to enable the reader to understand British soil maps and to help him classify British soils. It is not an exhaustive résumé of classification problems and schemes and the reader is referred to the section on further reading (p. 116) for more advanced texts on the subject.

Why classify?

One purpose of soil classification is to enable generalized information to be easily transmitted from one individual to another. If there were no form of classification at all it would be necessary to give a full profile description every time one referred to a particular profile. For many (indeed most) purposes this is more than is required and a simple summary word or phrase is all that is needed. This gives the individual receiving the information a basic idea of the main features and saves a great deal of time.

A second purpose of soil classification is in *theory construction.* The advancement of any science depends upon being able to make generalizations and predictive statements about what will happen in particular circumstances. To enable this development to take place in soil science it is clearly necessary to combine soils in groups (or classify soils) in such a way that all members of any particular group respond in similar ways and are thus able to have predictive statements made about them. For example, for agricultural purposes it is obviously helpful if one can make predictive statements about how a group of soils respond to particular fertilizers. (We will refer to some of the problems of theory constructions with soils in section 6.8.)

To enable these two long-term aims to be achieved, the short-term objective of soil classification is that soils must be grouped in such a way that each group has a minimum of diversity *within* it and a maximum of diversity *between* groups.

The basic unit of classification

In plant and animal classifications, organisms fall into easily recognizable classificatory units – the species. Each species can be fairly clearly defined and it is only in unusual circumstances that species interbreed to produce intergrades. In the case of soil, however, the choice of a suitable basic classificatory unit is more difficult. One has only to look at a soil profile to see that there is a considerable range of variation laterally across the profile (ignoring the obvious vertical horizon variation). It is therefore necessary to state some artificial limits to the chosen basic unit. In Great Britain the basic unit of classification is the *soil profile*, which is:

> 'considered for the purpose as three-dimensional, with lateral dimensions large enough to evaluate diagnostic properties of horizons at a particular place.'
>
> B W Avery, 1973, p. 325

Effectively one is defining the soil profile as having no lateral variation. In the United States a different unit – the *soil pedon* – is employed. This is an artificially defined three-dimensional cuboid having lateral dimensions which depend on the lateral variability of the properties of the defined class.

Development of soil classification

The first significant attempts at soil classification were those of the Russian school – led by Dokuchaiev – which were published in the 1880s. The Russian classification was based on the assumption that soil type was largely determined by climate, and this is, of course, generally true for the large-scale soil belts with which the Russians were concerned. This kind of classification is known as a *typological* classification. It should be remembered that many of our commonly used soil group names, like 'podzol' and 'rendzina', derive from this Russian classification.

Development of typological classifications took place throughout the first fifty years of this century, different versions being produced for various parts of the world. However, whilst the general premise that climate is the main determinant of soil distribution holds true on a large continental scale, it becomes increasingly difficult to apply at smaller scales. Other factors begin to over-ride climate in their importance (see section 3.1), and it becomes more and more difficult to interrelate the factors and therefore to complete the soil classification. Much more research is needed before we are fully aware of the relative importance of the factors responsible for soil formation and variation.

The realization of this complexity has led to the replacement of typological classifications by *definitional* classifications, the first of these which entered common use being that of the United States Department of Agriculture (U S D A). (This is often known as the '7th Approximation' classification because the published classification was indeed the 7th revision produced by American soil surveyors.) Definitional classifications depend on a description of the soil. The soils most alike are grouped together, trying to make little or no implicit statement about soil genesis.

The 1940 British classification

Detailed British soil mapping commenced in the nineteen-twenties, but it was not until 1940 that enough information had been assembled to enable a detailed classification system to be generally adopted. The scheme is based on six main soil groups:

1 brown earth;
2 podzols;
3 gley soils;
4 calcareous soils;
5 organic soils and
6 undifferentiated alluvium.

These groups are basically derived from the Dokuchaiev system and relate closely to those in continental European classification produced at about the same time. Details of the subgroups in the classification are given in Table 2.

With some minor modifications the classification has been used until 1974 in publications of the Soil Survey of Great Britain. However, a great deal of information about British soils has been amassed since 1940 and it has recently been thought wise to develop a new definitional classification, some of the old groupings having been shown to be inappropriate in the light of new discoveries.

The 1973 British classification

The 1973 classification is largely the work of B W Avery and is to be used in all publications of the Soil Survey of England and Wales published after 1974 (apart from one or two items already in press at that time). It is anticipated that the Soil Survey of Scotland will follow in due course, although, at the time of writing, no final decision had been published.

Table 2 1940 soil classification

DEFINITIONS

I Soils of the brown-earth group

Three characteristics form the basic definition of the normal brown earths:

1 The soil has free drainage throughout the profile.
2 There is no vertical differentiation of silica and sesquioxides in the clay fraction.
3 There is no natural free $CaCO_3$ in the soil horizon.

Other morphological and chemical features may vary. Thus the soil may be of any colour, but this colour is more or less uniform throughout the profile; the degree of acidity may vary widely.

The virgin soils are usually characterized by an accumulation of leaf litter on the surface, which is underlain by mull humus. Under cultivation the surface is altered, added bases may be present, and to this extent arable soils will differ from the normal brown earths.

The brown-earth group is divided into soils of low base status and soils of high base status. Soils with a high base status are only slightly acid, and become neutral with depth; they are derived from base-rich parent materials. Those of low base status have a tendency to acidity throughout the profile.

Subtypes of the brown-earth group

A *Creep or colluvial soils*–This group is dependent on topography for its development. In morphological and chemical characteristics it is the same as the normal brown earth soils.

B *Brown earths with gleyed B and C horizons*–The soils of this subtype are the same as the normal brown earths except for a suggestion of gleying in the lower horizons. This gleying is no more than an occasional bluish or rusty mottling. The effect may be due to rare rises in ground water or to a slight impedance in drainage.

C *Leached soils from calcareous parent materials*–These soils are characterized by a red-brown colour and a condition of base unsaturation. They may be quite acid, and if $CaCO_3$ is present it is in the form of hard lumps. Organic matter is light in colour, but is not necessarily low. Secondary $CaCO_3$ may occur at the base of the B horizon, or in the parent material.

II Soils of the podzol group

The chief morphological characteristics of normal podzolized soils are:

1 The presence of a bleached (grey) layer under the surface raw humus.
2 The yellow to rusty coloured accumulation layer which follows.

The chemical characteristics are found in the differentiation of the silica and sesquioxides of the clay fraction. Under cultivation the surface raw humus is absent. Arable soils may show the typical grey and rusty layers, or these may be almost entirely obliterated. All transitions occur, but so long as the clay fraction shows differentiation of the silica and sesquioxides, such soils are included in the podzol group.

Subtypes of the podzol group

A *Slightly to strongly podzolized soils*–These depend on the thickness of the bleached horizon.

B *Concealed podzols*–The soil has a raw humus surface layer but no bleached layer. The translocation of sesquioxide is proved by the changes in the silica–sesquioxide ratio.

C *Peaty podzolized soils*–In these the raw humus has developed into peat. They may vary from a slightly to strongly podzolized condition (for definition of peat see later).

D *Podzolized soils with gleying*–These are essentially podzolized soils in the upper layers, but exhibit signs of impedance by gleying in the B or C horizons.

E *Truncated podzols*–Here the surface soil has the characters of a B horizon. Under grass vegetation the iron colours are washed by humus.

III Soils of the gley group

The characteristic of gleying is the presence of greenish, bluish-grey, rusty, or yellowish spots or mottling:

1 *Surface-water gley soils*–In these the excessive water is on the surface and produces gleying in the surface horizons. In the lower horizons gleying progressively decreases or may be absent altogether.
2 *Groundwater gley soils*–The surface of such soils may be dry, at least seasonally and often permanently, with little or no gleying. Gleying is essentially present in the lower layers. This group includes soils with slow percolation, not necessarily occurring only in depressions.

Subtypes of groundwater gley soils

A *Gley podzolized soils*–These soils have a raw humus surface and a bleached A horizon. The B horizon is thin or absent. Gleying occurs below this level.

B *Peaty gley podzolized soils*–Essentially similar to A but peat replaces raw humus.

C *Peaty gley soils*–These soils are completely gleyed and carry a peaty surface.

D *Gley calcareous soils*–These are characterized by a grey colour and a moderately high organic matter content. Calcium carbonate occurs throughout the profile and increases with depth; the soil is base saturated. There is little change in the silica–sesquioxide ratio down the profile. Secondary calcium carbonate often occurs in the form of concretions, especially in the lower layers. Gleying is shown by the presence of bluish, greenish, rusty, or yellow spots and mottling.

IV Soils of the calcareous group

These soils are developed from calcareous parent materials, contain primary calcium carbonate in the soil horizons, and are base saturated.

Subtypes of the calcareous group

A *Grey calcareous soils* (rendzina type) – Under natural vegetation these soils show a very dark surface horizon, a high content of organic matter, and a well-developed crumb structure. There is no differentiation of the silica–sesquioxide ratio down the profile. Calcium carbonate increases in amount with depth until the parent material is reached. Secondary deposition of calcium carbonate may occur. Under arable cultivation organic matter is lower, calcium carbonate is higher, and the soils may be pale grey or almost white in colour. Crumb structure is less pronounced.

B *Red and brown calcareous soils* – These are formed on hard limestone and do not occur on the chalk. They are shallow, being characterized by a red or brown colour and by the presence of fragmentary calcareous rock. The organic matter content is usually low and the silica–sesquioxide ratio constant throughout the profile. Secondary deposition of calcium carbonate may occur in the parent material.

C *Calcareous soils with gleyed B and C horizons* – These soils, in the upper layers, are similar to either subtypes **A** or **B**, but show slight gleying in the lower horizons.

V Soils of the organic group

The soil character is determined by the presence of twenty or more centimetres of waterlogged organic matter, termed peat. There are two groups:

1 *Basin peat* – Soligenous in origin, i.e. formed under the influence of excessive or stagnant groundwater.
2 *Moss peat* – Ombrogenous in origin, i.e. formed under the influence of heavy rainfall and low summer temperature.

Basin peat. The main development forms of this group are as follows:

A1 *Fen* (including Carr) – This is formed under the influence of calcareous or base-rich groundwater. Transition phase is grass-moor, etc.

A2 *Raised moss* – This is ombrogenous as a result of accumulation of **A1** above groundwater level.

B1 *Acid low moor* – This is formed under the influence of drainage from acid or base-poor rock and soils, e.g. podzolized surface or raised moss.

B2 *Raised moss* – As in **A2**.

Note – Moss peats are predominantly ombrogenous since they develop under conditions of high rainfall on a substratum lying above groundwater level. The ultimate form of these is 'raised moss' and may develop over any organic soil when it grows above groundwater influence. Moss peat covering a region is termed 'blanket moss' and is to be regarded as climatic in the pedological sense.

*Subtype of **B2** – hill peat*

This is a variety of blanket moss formed on hilltops and slopes which varies from the main type in distribution and character and is therefore to be mapped separately as 'hill peat'.

VI Undifferentiated alluvium group

Owing to the great variety of soils which may be encountered in a comparatively small area of alluvium, some surveyors do not attempt to differentiate them. In such cases the soils are allocated to this group. Where alluvial flats are extensive careful survey will be worth while. In this case the different series identified will be allocated to one of the other five groups.

Source: G R Clarke, *Soil Survey of England and Wales, Field Handbook*, Oxford University Press, 1940.

As already stated the basic unit of classification is the soil profile. The classification is, in general, based on observable *field properties* of profiles and, as far as possible, depends on those that are permanent. For example, thin surface layers easily destroyed by cultivation do not qualify as distinguishing horizons.

Table 3 lists the major groups, groups and subgroups. The higher categories are distinguished by combinations of the two main factors:

1 the composition of the soil within specified depth limits, and
2 presence or absence of diagnostic horizons generally reflecting degree or kind of alteration of the original material.

Table 3 New soil classification in England and Wales (source: Avery, 1973)

Major group	*Group*	*Subgroup*
Lithomorphic (A/C) soils Normally well-drained soils with distinct, humose or organic topsoil and bedrock or little altered unconsolidated material at 30 cm or less	*Rankers* With non-calcareous topsoil over bedrock (including massive limestone) or non-calcareous unconsolidated material (excluding sand)	Humic ranker Grey ranker Brown ranker Podzolic ranker Stagnogleyic (fragic) ranker
	Sand-rankers In non-calcareous, sandy material	Typical sand-ranker Podzolic sand-ranker Gleyic sand-ranker
	Ranker-like alluvial soils In non-calcareous recent alluvium (usually coarse textured)	Typical ranker-like alluvial soil Gleyic ranker-like alluvial soil
	Rendzinas Over extremely calcareous non-alluvial material, fragmentary limestone or chalk	Humic rendzina Grey rendzina Brown rendzina Colluvial rendzina Gleyic rendzina Humic gleyic rendzina
	Pararendzinas Over moderately calcareous non-alluvial (excluding sand) material	Typical pararendzina Humic pararendzina Colluvial pararendzina Stagnogleyic pararendzina Gleyic pararendzina
	Sand-pararendzinas In calcareous sandy material	Typical sand-pararendzina
	Rendzina-like alluvial soils In calcareous recent alluvium	Typical rendzina-like alluvial soil Gleyic rendzina-like alluvial soil
Brown soils Well drained to imperfectly drained soils (excluding pelosols) with an altered subsurface (B) horizon, usually brownish, that has soil structure rather than rock structure and extends below 30 cm depth	*Brown calcareous earths* Non-alluvial, loamy or clayey, with friable moderately calcareous subsurface horizon	Typical brown calcareous earth Gleyic brown calcareous earth Stagnogleyic brown calcareous earth
	Brown calcareous sands Non-alluvial, sandy, with moderately calcareous subsurface horizon	Typical brown calcareous sand Gleyic brown calcareous sand
	Brown calcareous alluvial soils In calcareous recent alluvium	Typical brown calcareous alluvial soil Gleyic brown calcareous alluvial soil

Table 3 (*contd*)

Major group	*Group*	*Subgroup*
	Brown earths (*sensu stricto*) Non-alluvial, non calcareous, loamy, with brown or reddish friable subsurface horizon	Typical brown earth Stagnogleyic brown earth Gleyic brown earth Ferritic brown earth Stagnogleyic ferritic brown earth
	Brown sands Non-alluvial, sandy or sandy gravelly	Typical brown sand Gleyic brown sand Stagnogleyic brown sand Argillic brown sand Gleyic argillic brown sand
	Brown alluvial soils Non-calcareous in recent alluvium	Typical brown alluvial soil Gleyic brown alluvial soil
	Argillic brown earths Loamy or loamy over clayey, with subsurface horizon of clay accumulation, normally brown or reddish	Typical argillic brown earth Stagnogleyic argillic brown earth Gleyic argillic brown earth
	Paleo-argillic brown earths Loamy or clayey, with strong brown to red subsurface horizon of clay accumulation, attributable to pedogenic alteration before the last glacial period	Typical paleo-argillic brown earth Stagnogleyic paleo-argillic brown earth
Podzolic soils Well-drained to poorly drained soils with black, dark brown or ochreous subsurface (B) horizon in which aluminium and/or iron have accumulated in amorphous forms associated with organic matter. An overlying bleached horizon, a peaty topsoil or both, may or may not be present.	*Brown podzolic soils* Loamy or sandy, normally well drained, with a dark brown or ochreous friable subsurface horizon and no overlying bleached horizon or peaty topsoil	Typical brown podzolic soil Humic brown podzolic soil Paleo-argillic brown podzolic soil Stagnogleyic brown podzolic soil Gleyic brown podzolic soil
	Gley-podzols With dark brown or black subsurface horizon over a grey or mottled (gleyed) horizon affected by fluctuating groundwater or impeded drainage. A bleached horizon, a peaty topsoil, or both may be present	Typical (humus) gley-podzol Humo-ferric gley-podzol Stagnogley-podzol Humic (peaty) gley-podzol
	Podzols (*sensu stricto*) Sandy or coarse loamy, normally well drained, with a bleached horizon and/or dark brown or black subsurface horizon enriched in humus and no immediately underlying grey or mottled (gleyed) horizon or peaty topsoil	Typical (humo-ferric) podzol Humus podzol Ferric podzol Paleo-argillic podzol Ferri-humic podzol

Table 3 (*contd*)

Major groups	*Group*	*Subgroup*
	Stagnopodzols With peaty topsoil, periodically wet (gleyed) bleached horizon, or both, over a thin iron-pan and/or a brown or ochreous relatively friable subsurface horizon	Iron-pan stagnopodzol Humus-iron-pan stagnopodzol Hard-pan stagnopodzol Ferric stagnopodzol
Pelosols Slowly permeable non-alluvial clayey soils that crack deeply in dry seasons with brown, greyish or reddish blocky or prismatic subsurface horizon, usually slightly mottled	*Calcareous pelosols* With calcareous subsurface horizon	Typical calcareous pelosol
	Argillic pelosols With subsurface horizon of clay accumulation, normally non-calcareous	Typical argillic pelosol
	Non-calcareous pelosols Without argillic horizon	Typical non-calcareous pelosol
Gley soils With distinct, humose or peaty topsoil and grey or grey-and-brown mottled (gleyed) subsurface horizon altered by reduction, or reduction and segregation, of iron caused by periodic or permanent saturation by water in the presence of organic matter. Horizons characteristic of podzolic soils are absent	1 Gley soils without a humose or peaty topsoil, seasonally wet in the absence of effective artificial drainage	
	Alluvial gley soils In loamy or clayey recent alluvium affected by fluctuating groundwater	Typical (non-calcareous) alluvial gley soil Calcareous alluvial gley soil Pelo-alluvial gley soil Pelo-calcareous alluvial gley soil Sulphuric alluvial gley soil
	Sandy gley soils Sandy, permeable, affected by fluctuating groundwater	Typical (non-calcareous) sandy gley soil Calcareous sandy gley soil
	Cambic gley soils Loamy or clayey, non-alluvial, with a relatively permeable substratum affected by fluctuating groundwater	Typical (non-calcareous) cambic gley soil Calcaro-cambic gley soil Pelo-cambic gley soil
	Argillic gley soils Loamy or loamy over clayey, with a subsurface horizon of clay accumulation and a relatively permeable substratum affected by fluctuating groundwater	Typical argillic gley soil Sandy-argillic gley soil
	Stagnogley soils Non-calcareous, non-alluvial, with loamy or clayey, relatively impermeable subsurface horizon or substratum that impedes drainage	Typical stagnogley soil Pelo-stagnogley soil Cambic stagnogley soil Paleo-argillic stagnogley soil Sandy stagnogley soil

Table 3 (*contd*)

Major group	*Group*	*Subgroup*
	2 Gley soils with a humose or peaty topsoil, normally wet for most of the year in the absence of effective artificial drainage	
	Humic-alluvial gley soils In loamy or clayey recent alluvium	Typical (non-calcareous) humic-alluvial gley soil Calcareous humic-alluvial gley soil Sulphuric humic-alluvial gley soil
	Humic-sandy gley soils Sandy, permeable, affected by high groundwater	Typical humic-sandy gley soil
	Humic gley soils (*sensu stricto*) Loamy or clayey, non-alluvial, affected by high groundwater	Typical (non-calcareous) humic gley soil Calcareous humic gley soil Argillic humic gley soil
	Stagnohumic gley soils Non-calcareous, with loamy or clayey, relatively impermeable subsurface horizon or substratum that impedes drainage	Cambic stagnohumic gley soil Argillic stagnohumic gley soil Paleo-argillic stagnohumic gley soil Sandy stagnohumic gley soil
Manmade soils With thick manmade topsoil or disturbed soil (including material recognizably derived from pedogenic horizons) more than 40 cm thick	*Manmade humic soils* With thick manmade topsoil	Sandy manmade humus soil Earthy manmade humus soil
	Disturbed soils Without thick manmade topsoil	
Peat soils With a dominantly organic layer at least 40 cm thick, formed under wet conditions and starting at the surface or within 30 cm depth	*Raw peat soils* Permanently waterlogged and/or contain more than 15 per cent recognizable plant remains within the upper 20 cm	Raw oligo-fibrous peat soil Raw eu-fibrous peat soil Raw oligo-amorphous peat soil Raw eutro-amorphous peat soil
	Earthy peat soils With relatively firm (drained) topsoil, normally black, containing few recognizable plant remains	Earthy oligo-fibrous peat soil Earthy eu-fibrous peat soil Earthy oligo-amorphous peat soil Earthy eutro-amorphous peat soil Earthy sulphuric peat soil

It can be seen that the second factor contains an implicit statement about soil genesis and to this limited extent the classification is not entirely definitional. It can also be seen from the table that the major groups bear names generally similar to those in previous use. However, these are now precisely defined and it is apparent that a number of other new names have been introduced to distinguish the lower order of groups and subgroups. This was necessary because many of the old names were being applied to different types of soil by different workers in various countries.

Explanation of some of the terms used in the new classification will now be given. Details of where to find more information are given in Further reading (p. 116).

Humose mineral materials are those containing more than 4·5–7·0 per cent organic carbon.

Ferritic soils are those in which the Fe_2O_3 content is more than 4 per cent and more than half the measured clay percentage.

Distinct topsoil. A cultivated soil with an appreciably darkened Ap horizon containing at least 0·6 per cent organic carbon in the upper 15 cm, or an uncultivated soil with as much or more organic matter and continuous O or H or Ah horizons (whichever is present) together more than 7·5 cm thick.

Humose topsoil. An A horizon that is humose over at least 15 cm depth or 10–15 cm if directly over bedrock or fragmental material.

Peaty topsoil. An O horizon 7·5–40 cm thick, over a mineral subsurface horizon.

Thick manmade A horizon. A dark A horizon more than 40 cm thick resulting from addition of manure containing earth or otherwise attributable to human occupation.

Podzolic B horizon. B horizon or horizons in which organic matter and aluminium and/or iron have accumulated. It usually underlies a bleached (albic) E horizon or a dark Ah, H or O horizon and is required to extend to at least 15 cm depth, excluding surface litter. The following horizons can form all or part of a podzolic B:

- Bh normally dark coloured with little iron
- Bhs dark coloured, more than 1 cm thick, with proportionately more iron
- Bs brown or ochreous
- Bf (or Bfe) thin iron pan

(More details of these horizons are given in section 6.4, p. 87.)

Humic and *Ferri-humic* podzols have Bhs horizons that are humose over at least 10 cm and generally lack an E horizon.

Argillic (Bt) horizon. Textural B horizon containing translocated silicate clay. It is required to contain significantly more clay than all overlying horizons.

Palaeo-argillic soils show features (e.g. reddish colours due to prolonged weathering) that were derived before the last (Weichselian) glacial period.

Weathered B (Bw) horizon. Non-podzolic, non-argillic B horizon is usually brownish and differentiated by colour and/or structural features. As a diagnostic horizon (Brown soils major group) it is required to extend at least 10 cm below an Ap horizon or to more than 30 cm depth, and extremely calcareous material is excluded. A shallower brownish weathered horizon (A/Bw) characterizes brown rankers and brown rendzinas.

Sandy soils. For inclusion in sandy groups or subgroups, at least half the upper 80 cm of mineral soil must be sand or loamy sand.

Vertic features. The soils classed as pelosols or in pelo- subgroups are identified by the following characteristics:

1. More than 35 per cent clay over at least 30 cm, starting at the surface, directly below the Ap, or at less than 25 cm depth.
2. Blocky, prismatic or wedge-shaped peds with glazed faces, often inclined.
3. Cracks more than 5 mm wide between 25 and 50 cm depth in most years.

Calcareous soils are those lacking an argillic horizon and have a Bw, Bg or C horizon containing at least 1 per cent $CaCO_3$ at less than 40 cm depth.

Alluvial soils comprise those in fluviatile, marine or lacustrine deposits at least 30 cm thick.

Profile descriptions

Four example soil-profile descriptions are now given which are taken from Soil Survey Memoirs. In general the descriptions accord with the terminology outlined in sections 6.4 and 6.5, although minor deviations will be noticed.

Profile 1 Lithomorphic soil: rendzina
(after Jarvis, 1973)

GRID REFERENCE SU 285827
LOCATION Ashdown Farm, Ashbury
RELIEF steep convex valley side
SLOPE AND ELEVATION 10°WSW, 220 m OD
LAND USE permanent downland grassland

HORIZONS

cm

0–13 A	Very dark greyish-brown (10YR 3/2), very friable, very calcareous, slightly stony (medium angular flints) humose silty clay loam; strong fine and very fine crumb structure; extremely abundant fine fibrous roots; many clusters of faecal pellets; narrow even boundary.
13–44 AC	Dark greyish-brown (10YR 4/2), very friable, very calcareous silt loam with stones dominant, gravel to medium brown-stained subangular chalk and common medium to large angular and broken flints; in larger patches of matrix, strong fine crumb or subangular blocky structure; clusters of faecal pellets; abundant fine fibrous roots between stones; merging boundary.
44+ C	Broken brown-stained Upper Chalk (Chalk Rock) *in situ*.

Profile 2 Brown soil: brown earth (sensu stricto)
(after Clayden, 1971)

GRID REFERENCE SX 803699
LOCATION Wotten Cross, Denbury
ELEVATION 100 m OD SLOPE AND ASPECT 2°S
LAND USE Ley grass
HORIZONS

cm

0–20 Ap	Dark brown (7·5YR–10YR 4/2) clay loam; stony throughout with predominantly fine slate fragments and occasional small stones of limestone, vesicular lava and dolerite; moderate, very fine subangular blocky structure; friable; moderate organic matter; abundant fine roots; merging boundary.
20–46 B1	Brown (10YR–7·5YR 4/3) loam; moderate to fine strong crumb structure; extremely friable; moderate organic matter; roots common; merging boundary.
46–58 B2	Brown to yellowish-brown (10YR 4/5–3/5) loam; weak fine crumb structure; friable; few roots; narrow boundary.
58–81+ C	*In situ* slate with very fine laminae, easily broken.

Profile 3 Podzolic soil
(after Clayden, 1971)

GRID REFERENCE SX 797749
LOCATION Rora Wood, Liverton
ELEVATION 120 m OD SLOPE AND ASPECT 20°NW
LAND USE oak coppice with heathy ground vegetation
HORIZONS

cm

10–0 H	Dark brown (5YR 2/2) amorphous organic matter; very fine crumb structure; abundant woody and fibrous roots, particularly concentrated above boundary to Ea horizon; narrow boundary.
0–56 Ea	Light grey (10YR 6/1, dry) sandy loam, with greyish-brown (10YR 5/2) patches in lower part of horizon; stone dominant, mainly gravel and small angular chert; structureless; loose; low organic matter; common fine fibrous roots; narrow irregular boundary.
56–58 B1h	Dark reddish-brown (5YR 3/2) loam with some areas of 5YR 2/2 and variegated with greyish brown and strong brown; extremely stony; structureless, massive; firm, weakly indurated; high organic matter; narrow irregular boundary.
58–96 B2s	Strong brown (7·5YR 5/8) loam, becoming of yellower hue with depth, with occasional coarse pockets of reddish-brown humus staining; extremely stony, small to medium angular cherts; weak fine crumb structure; friable; moderate organic matter; few roots; merging boundary.
96–124 BC	Yellowish-brown (10YR 5/6) loam with some humus staining; otherwise as above; merging boundary.
124–183+ C	Very pale brown (10YR 7/3) loam; extremely stony, mainly gravel to small stones, some soft and partially weathered (but not cheesy) chert; weak fine angular blocky structure; crumbles readily; few roots.

Profile 4 Gley soil
(after Jarvis, 1973)

GRID REFERENCE SP 318003
LOCATION Carswell Marsh, Buckland
RELIEF Thames floodplain
SLOPE AND ELEVATION level; 65 m OD
LAND USE permanent grass
HORIZONS

cm	
0–12 Ag	Very dark grey (10 YR 3/1) humose, friable, stoneless calcareous clay with ochreous staining along 50 per cent of roots; moderate fine crumb structure; extremely fine abundant fibrous roots; many small gastropod shells; narrow boundary.
12–20 Bg	Greyish-brown (10YR 5/2), reddish-yellow (7·5YR 6/8) mottled, slightly humose, friable, very calcareous clay; moderate, breaking to fine, subangular blocky structure; common fine pores; abundant fine fibrous roots; a few small concretions at centre of yellowish-red (5YR 4/8) mottles; many small gastropod shells; merging boundary.
20–37 Cgca	Light brownish-grey (2·5Y 6/2), yellowish-red (7·5YR 6/6) mottled, sticky massive stoneless, very calcareous clay; common fine fibrous roots; a few small gastropod shells; a few secondary $CaCO_3$ concretions and secondary $CaCO_3$ also infilling many pores; narrow boundary.
37–50 IICg1	Grey (N5/0), with large reddish-yellow (7·5YR 6/6) mottles, sticky massive stoneless clay; common fine pores; common dead fine fibrous roots; merging boundary.
50–90 IICg2	Grey to light grey (N6/0), with large reddish-yellow (7·5YR 6/8) mottles, sticky massive calcareous clay; common fine pores and dead, fine fibrous roots; small ferruginous concretions at centre of some mottles.

6.7 Soil mapping

Published soil maps of Great Britain are available at three main scales:

1 : 63 360 or one inch to one mile – For the past twenty years this has been the traditional scale, about thirty maps having been published to date. These maps are published on the 3rd Edition Ordnance Survey sheet lines, so as to provide direct comparability with Geological Survey maps.

1 : 25 000 – Because progress with the one-inch maps was proving to be rather slow, the system of soil mapping was changed in the late 1960s. The one-inch mapping programme was terminated and a new series of 1 : 25 000 maps was started. The idea was not to provide a complete coverage of the country, but to map areas of particular geomorphological or agricultural interest. To date (late 1975) about twenty maps on this scale have been published covering parts of England and Wales. In Scotland the one-inch programme is continuing, at least for the time being.

1 : 250 000 – These maps are intended to complete coverage of the country on a county basis. The maps show *soil associations* (groups of soils of different soil groups occurring in similar landscape situations). At the time of writing only one of these maps has been published – that for Lancashire.

Figures 111 and 112 show the extent of published soil maps in Great Britain.

We have seen already (section 6.6) that soils may be classified into major groups, groups and subgroups. A further subdivision is into *soil series.* These latter are differentiated by profile characteristics, chiefly lithological, that are not differentiated at subgroup level. They are named after places near which they are common or were first described (e.g. Evesham series).

It is very important to appreciate that until now we have been concerned with *conceptual soil classification,* which may differ from *soil mapping classification* or spatial soil classification. To understand the difference between these two it is perhaps best to visualize a hypothetical area of landscape of a size, let us say, of 100 m². Imagine that within this area 100 soil profiles have been described and the profile description notes taken away to the laboratory. If the profile description notes are sorted into piles according to soil series we might end up with a conceptual classification like this:

Pile 1	Pile 2	Pile 3
(series A)	(series B)	(series C)
35 profiles	30 profiles	30 profiles
Pile 4	Pile 5	
(series D)	(series E)	
4 profiles	1 profile	

If we then went back to the original field area we might find that the soil series were distributed spatially as shown in Figure 113 (p. 108). It can be seen that some series overlap in their distribution. To produce a reasonable soil map with more or less

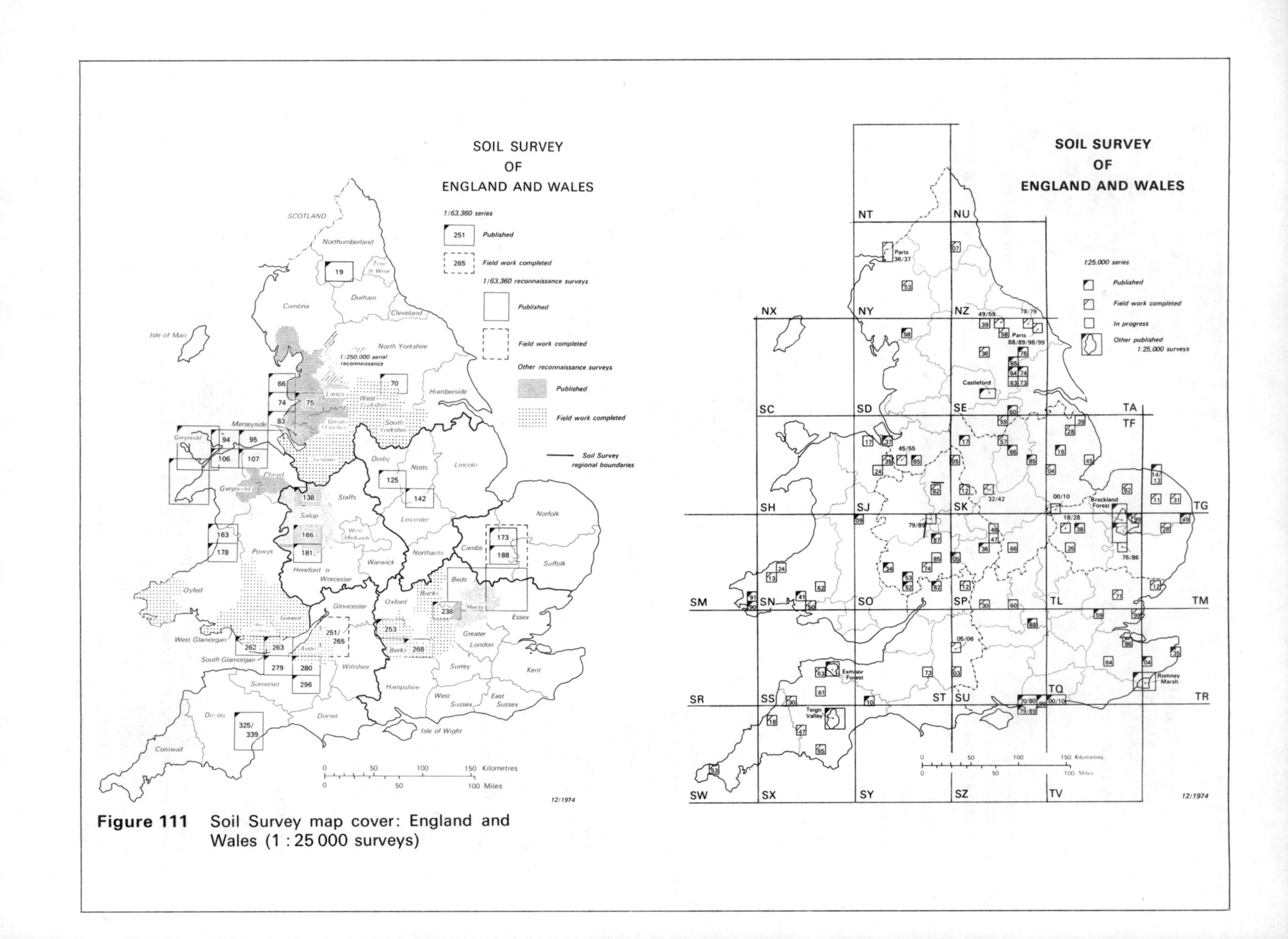

Figure 111 Soil Survey map cover: England and Wales (1 : 25 000 surveys)

discrete areas of the same sort of soil it is necessary to devise a new classification – a soil mapping classification. The lines between the mapping classification units (or *mapping units*) are shown in Figure 114 (p. 108). Each mapping unit will contain certain amounts of the particular series as follows:

mapping unit A: series A–100%, no other series
mapping unit B: series B–85%, C – 11%, E – 4%
mapping unit C: series A – 7%, B – 17%, C – 66%, D – 10%

The mapping units are named after the dominant included series. Thus, on a real soil map, Evesham soil mapping unit is dominantly Evesham series but may (and probably will) contain inclusions of other series. Not all soil memoirs and records state the precise amount of inclusions in percentage terms – most merely give an indication such as 'dominant', 'common', 'rare', etc. This is because to give a precise indication would involve very detailed sampling and checking, which is not always possible.

If a mapping unit contains more or less equal amounts of different series, or if no series contributes more than about 40 per cent, then the mapping unit is known as a *soil complex*. Soil complexes are commonly mapped, for example, on river floodplains where old river courses confuse the soil pattern.

Fortunately for those making maps, boundaries between soil mapping units often coincide with distinct boundaries in the landscape, like the edges of valleys and the crests of hills. This fact – which is not surprising when one considers the factors involved in soil formation – makes the drawing of the lines on the map a great deal easier. Soil mapping takes place in distinct stages, which will now be outlined.

Stage 1: Preparation – Before starting to work in the field the soil surveyor must assemble all the information available about the area in which he is to work. This includes geological maps, aerial photographs (if available) and papers on the geology, geomorphology and agriculture. Contact is made with local farmers and landowners and with officials of the Ministry of Agriculture, Fisheries and Food.

Stage 2: Field reconnaissance – The object of this is to provide background information on the entire area together with detailed studies in small sample areas. These small samples are chosen so that as far as possible they will give an impression of the main types of soil that are likely to be encountered when making the map. Usually small blocks of land or, alternatively, transects across the area are examined. Correlations with other similar soils elsewhere are made at this stage.

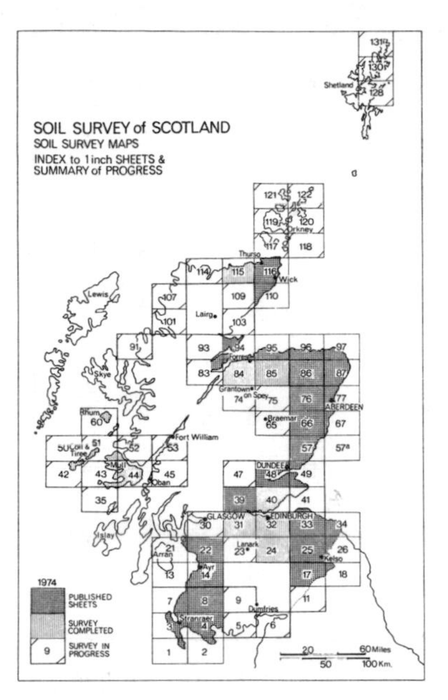

Figure 112 Soil Survey map cover: Scotland

Stage 3: Legend construction – Following reconnaissance it should be possible to decide on soil mapping units for later detailed mapping. This stage is the most crucial to the final quality of the survey. The legend will include not only information on the different series within each mapping unit, but also information on the legend relationships of the mapping units. Each main series is described in a profile pit and samples taken for chemical analysis.

Stage 4: Detailed mapping – At this stage lines are drawn on a field map to correspond with mapping unit boundaries. To aid in this the surveyor will probably use a soil auger. Alternatively (or sometimes additionally) he may use a spade to make a small pit ('keyhole pit'). The scale of field map used depends on the ultimate scale of the published survey: usually a 1 : 10 560 (six inch to one mile) field map is used for a published map of 1 : 25 000, whilst a 1 : 25 000 field map is used for a published map of 1 : 63 360 (one inch). Some surveyors plot boundaries directly on aerial photographs which they carry with them in the field. Detail of recording depends on a number of factors, including the experience of the surveyor and the nature of the mapping

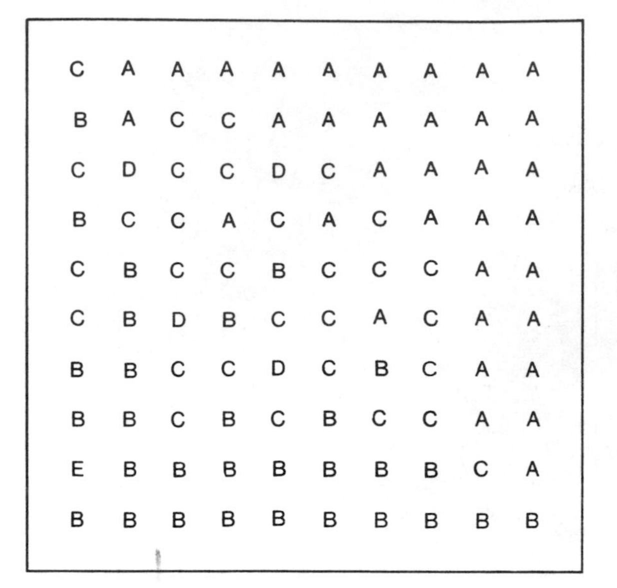

Figure 113 Hypothetical distribution of five soil series (A, B, C, D, E) in a 100m² area

unit variability. Some maps are made by detailed sampling schemes, whilst in others the sampling is more or less *ad hoc*. For a published map at 1 : 25 000 borings are usually made at a density of between thirty and sixty to the square kilometre.

Stage 5: Writing report – A detailed report of the soil map usually accompanies the published map. Reports on the 1 : 63 360 are called 'soil memoirs', those on the 1 : 25 000 maps 'soil records', and those on other scales 'soil bulletins'. There is clearly a great deal of work also in the fair drawing of the completed map.

6.8 Frameworks for research

For a long time soil study has consisted largely of observations of soil in profile pits. Although this type of observation is valuable in many cases, before starting field work it is important to establish exactly what sort of information is required and, perhaps even more important, exactly why the information is required. To enable one to think more clearly it is best to work within a framework of scientific enquiry. This involves setting up an *hypothesis* and formulating a field-work programme which *tests* the hypothesis. Following the acceptance or rejection of the initial hypothesis one is able either to formulate a *theory* or, alternatively, if the initial hypothesis has not been substantiated, set up a new hypothesis for subsequent testing.

Once a general outline of the research has been established the detailed arrangements must be worked out. In order to understand some of the problems that may be encountered we will now consider a hypothetical example.

The subject of the study is trying to relate soil to vegetation in a particular small field area. The steps in the research programme might be as follows.

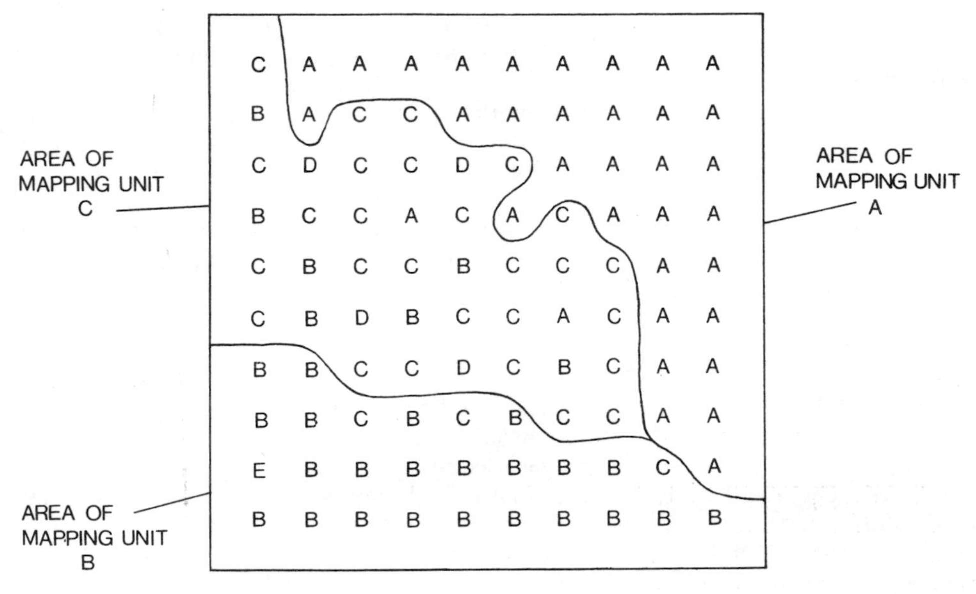

Figure 114 Soil map of the area in Figure 113

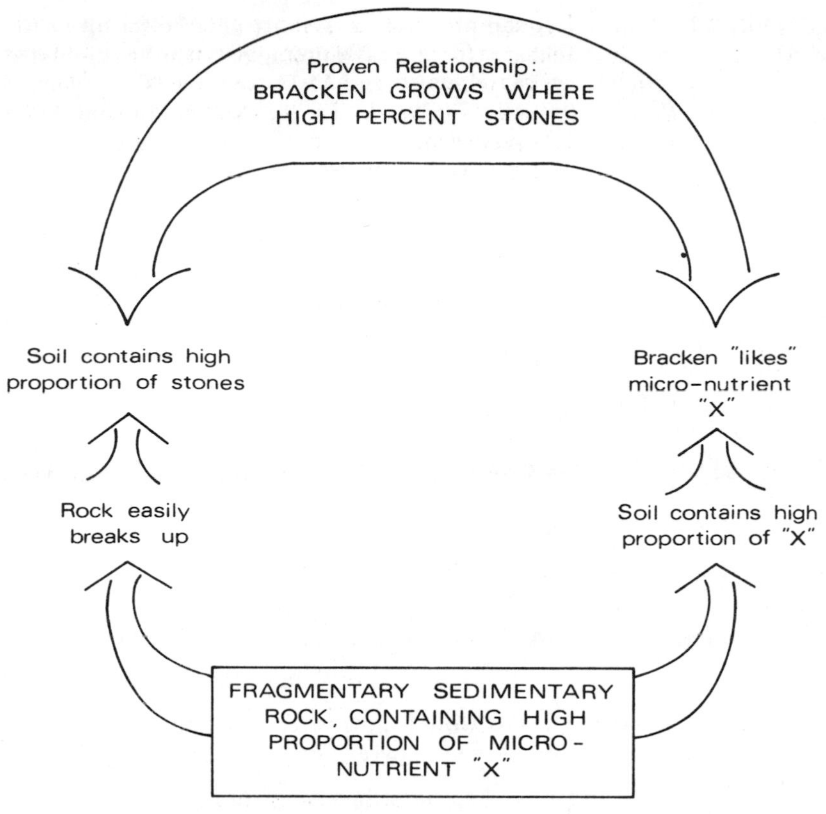

Figure 115 First possible network of relationships linking high proportion of bracken and soil stoniness

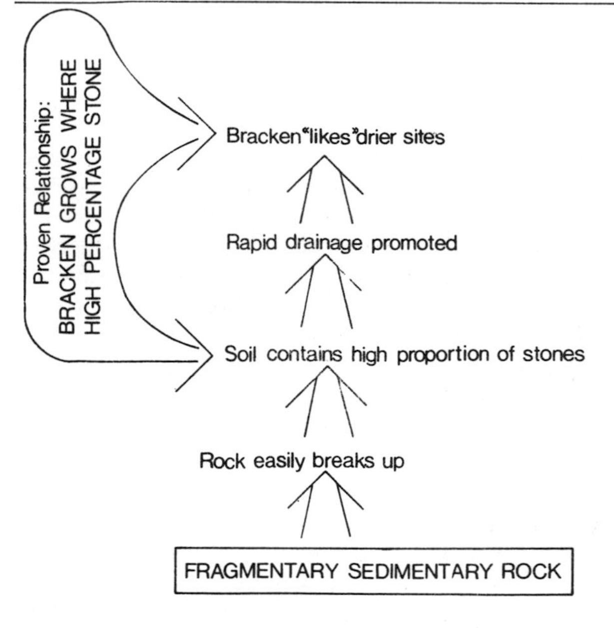

Figure 116 Second possible network of relationships linking high proportion of bracken and soil stoniness

1 *Hypothesis* – A general hypothesis might be 'that vegetation is related to soil type'. However, this is too general a hypothesis for it to be easily tested and one needs to formulate a *working hypothesis.* In formulating this hypothesis one must consider, in this case, two main questions:

a) what is meant by 'vegetation'?
b) what particular features of the soil are we referring to?

Obviously the most straightforward approach is to look at certain aspects of soil and vegetation. In the case of soil one could examine, for example, soil depth and soil stoniness. With vegetation one could, perhaps, look at the percentage cover of bracken. Clearly these sorts of decisions will depend to a great extent on the area being studied, and on the resources and skills of the research worker involved. In the case of the example we have been looking at the working hypothesis might be stated as 'the proportion of bracken in the vegetation community is related to the proportion of stones and/or the soil depth'.

2 *Testing* – The statistical test or other method that is going to be used to test the relationship must be decided upon at an early stage. The detailed requirements of the test will determine the *size* of the sample. (It may be that the worker decides that a simple graphical representation of the results is all that is going to be produced, and worked towards. Whilst in the case of simple relationships these are sometimes very valuable, using such a technique does mean that less validity can be given to the conclusions drawn from such a graph or table – a very subjective element having entered the work.)

In this case an appropriate statistical technique might be the Mann-Whitney 'U' Test, which tests for a significant relationship between two sets of data (more details are given, with examples of use, in R Hammond and P S McCullagh, *Quantitative Techniques in Geography: An Introduction* (1974), Clarendon Press). Decisions must also be made about *coding* the data. In other words, how are soil depth, soil stoniness and proportion of bracken to be estimated, recorded and presented for arithmetic analysis? Possible methods of doing this are:

soil depth: measured by auger to the nearest centimetre,
soil stoniness: percentage by volume at approximately 15 cm depth estimated by eye (this will involve excavating small 'keyhole pits'),
proportion of bracken: percentage cover over 2 m^2 estimated by eye.

A sampling network must next be set up in the field area (details of sampling patterns are discussed by J D Hanwell and M D Newson, *Techniques in Physical Geography* (1973), Macmillan), and the data recorded for each of the sample sites. The final results may be noted in a table, thus:

Site	*Soil depth* (cm)	*Soil stoniness* (%)	*Bracken* (%)
1	35	15	90
2	20	10	70
3	46	20	60
etc.			

On completion of field work the results are taken back to the laboratory and the Mann–Whitney test applied twice, to test in turn:

a) the relationship between soil depth and percentage bracken, and
b) the relationship between soil stoniness and percentage bracken.

The test would show whether or not any relationship existed between the variables.

3 *Formulation of theory* – From the results of the tests one can construct a theory about the relationships in the particular area studied. To see how far this was a substantial general theory one would have to retest the hypotheses elsewhere. It is important to understand that although there may be a relationship between two of these variables (say stoniness and percentage bracken) one cannot necessarily make a *causal connection.* One is not saying that soil stoniness *causes* less (or more) bracken to grow, but merely that there appears to be some link between the two variables. There may, in fact, be a long causal 'chain' connecting the two variables, but further tests would be needed to establish the individual links. Two ways in which they *might* be linked are shown in Figures 115 and 116, although it must be stressed that this example is entirely hypothetical.

Summary
Methods of describing sites, soil horizons and profiles are given. British soil classification and soil map construction are explained. Soil research should be formulated according to scientific method and it is not always necessary to describe detailed profile sections, but selected information can be gathered according to the purpose of the investigation.

Appendix 1

The analysis of soils for grain size (see section 2.3)

Method one – sieving

(The apparatus required may be purchased from A Gallenkamp and Co Ltd, Technico House, Christopher Street, London EC2B 2NA.)

APPARATUS NEEDED

Sieve stack appropriate to desired analysis, e.g. pan; 0·002 mm, 0·02 mm and 0·2 mm sieves (appropriate to differentiating clay, silt and sand on the International Scale). Additional sieves can be used for more detailed analysis.
Sieve shaker
Wire and nylon sieve brushes
Selection of beakers
Pieces of paper larger than the diameter of the sieves
Weighing balance, capable of weighing to an accuracy of 0·01 g
Oven, capable of drying at 105°C
Desiccator
Electrical stirrer and flask
Calgon (sodium hexametaphosphate)

METHOD

1 Take a sample of roughly 150 g. Add 1 litre of distilled water.
2 Stir vigorously with the electrical stirrer, adding Calgon at the approximate rate of 2 g/litre. Stir for between 5 and 10 minutes. (Use less than 2 g rather than more in order to minimize foaming.)
3 To dry the sample, *either* leave to settle and decant and then place in oven at 105°C for 24 hours *or* put the sample direct into the oven at 105°C until the sample has evaporated to dryness. When in the oven stir occasionally to prevent caking. The first method of drying takes rather less time than the second. After drying in the oven cool in desiccator. A third method of drying is to air-dry the sample, which is slower but prevents caking. This is especially useful for soils with a high clay content which easily harden on drying and may otherwise have to be broken up again with a pestle and mortar.
4 Weigh out 100.00 g of the dispersed dried sample. This is Weight 1 (W1).
5 Select and stack sieves, ensuring that the pan is at the bottom of the stack. Place on the shaker, insert the soil at the top and shake for 15 minutes.
6 Turn out the soil retained by each sieve on to separate sheets of paper. Brush each sieve upside down over the appropriate piece of paper using nylon brushes for the finer nylon sieves and wire brushes for the coarse wire sieves.
7 Pour each sample into an appropriate sized beaker and weigh to two decimal places (W2).
8 Weigh the beaker empty (W3).
9 Calculate the weight of soil in each size class (W4), thus:

$$W4 = W3 - W2.$$

10 The proportion of soil in each size class is usually expressed as a percentage by weight of the total sample and is therefore calculated thus:

$$\text{Proportion of soil in given size range} = \frac{\text{W4 for particular size range}}{\text{W1}} \times 100.$$

Method two – Sedimentation

APPARATUS NEEDED

Beakers, 250 ml, 500 ml, tall 600 ml
Evaporating basins for drying soil
Oven
Calgon (sodium hexametaphosphate)
Electrical stirrer and flask
Cylinders, tall glass 1 litre
Soil hydrometers, calibrated from 0–60 g/l
Thermometer

METHOD

1 Weigh out 60 g dry soil (W1). Sandy soils may be oven-dried at 105°C for 24 hours. Clay soils will have to be air-dried for 2–5 days (spread out on a tray to dry) or oven-dried at 30°C for 2–3 days. Otherwise they may set rock hard if baked at 105°C. Dry until constant weight is achieved (±1 g on successive weighings).
2 Place in a tall 600 ml beaker with 500 ml distilled water and a small amount of Calgon (up to but not exceeding 1 g) and stir to dissolve Calgon. Transfer to stirrer flask and stir for 10 minutes, moving the flask round ensuring that all the structures are broken down into individual grains.
3 Choose a period when 6–8 hours is available (for details of time scale, see chart below). Pour stirred suspension into tall glass 1 litre cylinder. Add a further 500 ml distilled water, using some of this to rinse out the stirring flask. 60 g of soil are now mixed in 1 litre of water. Place hand over end of glass cylinder and invert approximately 30 times in a minute. Leave to stand.
4 After approximately 5 minutes take the first hydrometer reading (R1), using the chart below to determine the precise time. R1 is in g/l and as all the sand will have settled the reading is one of silt + clay in suspension.
5 After 8 hours take the hydrometer reading (R2). All the silt will have now settled and thus R2 represents the amount of clay in suspension.
6 Calculation:

W1 = sand + silt + clay = 60 g.
R1 = silt + clay, g still in suspension.
W1 − R1 = sand, g settled out in first five minutes.
R2 = clay, g still in suspension after 6–8 hours.
R1 − R2 = silt, g settled out after 5 minutes but before 6–8 hours.

Express g of sand, silt and clay as a percentage of W1.

TEMPERATURE CHART

As the temperature increases the water becomes less viscous and the particles settle faster. Allowance has to be made for this in accordance with the chart below and care should be taken that the temperature is kept as constant as possible during the measurement. The settling cylinders should be kept away from sunlight or radiators and should preferably be placed in a constant temperature bath if one is available.

	Reading time	*Temperature*
R1	10 minutes	10°C
	5 minutes	18°C
	$4\frac{3}{4}$ minutes	20°C
	$4\frac{1}{4}$ minutes	25°C
R2	$8\frac{1}{2}$ hours	18°C
	8 hours	20°C
	$6\frac{1}{2}$ hours	25°C

Note: Taking the hydrometer reading may require practice. The hydrometer should be gently inserted into the liquid about half a minute before the required time and allowed to settle. The hydrometer should be kept as clean as possible and handled extremely gently as it is fragile.

Appendix 2

The estimation of organic matter content by loss on ignition

(see section 2.6)

Method one uses an oven capable of heating to 500°C and Method two uses a bunsen burner as the source of heat.

Method one

APPARATUS NEEDED

Small crucible for each sample
Oven capable of heating to 500°C (e.g. Gallenkamp Muffle Furnace)
Long tongs

METHOD

1 Oven-dry the soil at 105°C for 24 hours to remove moisture. Cool in a desiccator.
2 Weigh out 10 g of the cooled, dry soil (W1).
3 Weigh the crucible + soil (W2).
4 Place the crucible and soil in the oven at 500°C for 2 hours.
5 Remove the crucible from the oven with the tongs. Carefully inspect the sample. If unignited organic matter remains return sample to oven for a further half hour.
6 Cool the fully ignited sample in a desiccator.
7 Weigh the crucible and ash (W3).
8 Calculate the loss of weight of the soil sample:

W2 − W1 = weight of crucible (W4),
W3 − W4 = weight of ash (W5),
W1 − W5 = loss on ignition (W6).

W6 represents the loss in weight caused by the oxidation of organic matter at 500°C. This is usually expressed as a percentage of the original sample thus:

$$\text{Percentage of organic matter} = \frac{W6}{W1} \times 100.$$

Method two

The procedure is identical except that step 4 is replaced by setting up the crucible in a clay triangle over a bunsen burner, partially covering the crucible with a crucible lid. The crucible is heated until fuming ceases and then the weighing and calculations are carried out as above.

Appendix 3

The measurement of soil moisture content
(see section 2.7)

APPARATUS NEEDED

Container for soil (tin foil or specimen tube)
Oven capable of maintaining 105°C ($\pm 2°$)

METHOD

1 Collect a small sample (about 10 g) of soil in the field and carefully seal it immediately in the field to prevent moisture losses in transit. Suitable methods of sealing include wrapping in tin foil or placing in specimen tube sealed with Sellotape.
2 Weigh the soil and container complete (W1) (removing any Sellotape first).
3 Open up container and place container and soil in oven at 105°C for 24 hours.
4 Remove sample and container from oven and cool in a desiccator.
5 Weigh cooled container and sample (W2).
6 Discard soil and weigh clean container (W3).
7 Calculation:

W3 − W2 = weight of dry soil (W4),
W1 − W2 = moisture loss after heating in oven (W5).

Express the moisture loss as a percentage of the dry soil thus:

Percentage water that the mineral soil was holding $= \frac{W5}{W4} \times 100$.

Further reading

The following is a selection taken from the large amount of available literature.

A General texts
B Reading allied to specific topics

A General texts

Brady, N C (1974), *The Nature and Properties of Soils*, Macmillan. A standard, comprehensive American text for detailed and advanced study.

Bridges, E M (1970), *World Soils*, Cambridge University Press. A lucid introduction to soil properties and soil types of the world. The colour illustrations of soil profiles are especially useful (and are available as a set of slides).

Bunting, B T (1972), *The Geography of Soil*, Hutchinson. A broadly useful text.

Cruickshank, J G (1972), *Soil Geography*, David & Charles. A broad treatment of soil types and properties.

Donahue, R, *et al.* (1971), *Soils: an introduction to soils and plant growth*, Prentice-Hall. A detailed American book dwelling on ecological and agricultural viewpoints.

Eyre, S R (1968), *Vegetation and Soils*, Edward Arnold. Of especial interest to geographers and ecologists. A very useful text dealing with processes and interrelationships on a world scale.

Fitzpatrick, E A (1971), *Pedology*, Oliver & Boyd. Intended for undergraduates. An individualistic soil classification is used which is confusing for the general student of soils.

Fitzpatrick, E A (1974), *An Introduction to Soil Science*, Oliver & Boyd. A useful introduction.

Russell, E W (1973), *Soil Conditions and Plant Growth*, Longman. A comprehensive, detailed advanced book. The standard British work on soil, dealing especially with the agricultural viewpoint. A necessary book for the serious advanced student of soils.

Townsend, W N (1973), *An Introduction to the Scientific Study of the Soil*, Arnold. Dealing with soil formation and especially with soil fertility, it is of especial value to the agricultural student.

B Reading allied to specific topics

Soil texture and structure

Brewer, R (1964), *Fabric and Mineral Analysis of Soils*, Wiley.

Griffiths, J C (1967), *Scientific Method in the Analysis of Sediments*, McGraw–Hill.

Organic matter

Kononova, M M (1966), *Soil Organic Matter*, Pergamon.

Soil water

Childs, E C (1969), *An Introduction to the Physical Basis of Soil Water Phenomena*, Wiley.

Hillel, D (1971), *Soil and Water*, Academic Press.

Soil biology

Alexander, M (1961), *An Introduction to Soil Microbiology*, Wiley.

Burgess, A, and Raw, F (1967), *Soil Biology*, Academic Press.

Jackson, R M, and Raw, F (1966), *Life in the Soil*, Edward Arnold.

Philipson, J (1971), *Methods of Study in Quantitative Soil Ecology*, Blackwell.

Wallwork, J A (1970), *Ecology of Soil Animals*, McGraw–Hill.

Soil chemistry

Bear, F E (1964), *Chemistry of the Soil*, Van Nostrand/Reinhold.

Jackson, M L (1962), *Soil Chemical Analysis*, Constable.

Hesse, P R (1971), *A Textbook of Soil Chemical Analysis*, Murray.

Ecosystems, water flow and plant ecology

Van Dyne, G M (Ed.) (1969), *The Ecosystem Concept in Natural Resource Management*, Academic Press.

Chorley, R J (1969), *Water, Earth and Man*, Methuen.

Gimingham, C. H. (1972), *The Ecology of Heathlands*, Chapman & Hall.

Rorison, I. (1969), *Ecological Aspects of Mineral Nutrition in Plants*, Blackwell.

Soil management

Cooke, G W (1967), *The Control of Soil Fertility*, Crosby Lockwood.

Hudson, N (1971), *Soil Conservation*, Batsford.

Lambe, T W, and Whitman, R V (1969), *Soil Mechanics*, Wiley.

Smith, M J (1970), *Soil Mechanics*, Macdonald & Evans.

O'Riordan, T (1971), *Perspectives on Resource Management*, Pion.

Practical soil study

Andrews, W A (Ed.) (1973), *Soil Ecology*, Prentice–Hall.

Clarke, G R (1971), *The Study of Soil in the Field*, Oxford University Press.

Hanwell, J D, and Newson, M (1973), *Techniques in Physical Geography*, Macmillan.

Hodgson, J M (1974), *Soil Survey Field Handbook*. Soil Survey Technical Monograph No. 5. Harpenden: Soil Survey.

Chapter six – additional references

In Chapter six the following Soil Survey Memoirs and Record were referred to:

Cope, D W (1973), *The Soils of Gloucestershire I*. Harpenden: Soil Survey Record.

Clayden, B (1971), *The Soils of the Exeter District*. Harpenden: Soil Survey Memoir.

Findlay, D C (1965), *The Soils of the Mendip District of Somerset*. Harpenden: Soil Survey Memoir.

Jarvis, M G (1973), *The Soils of the Wantage District*. Harpenden: Soil Survey Memoir.

Sources

Figure 4 – Carson, M A, and Kirkby, M J (1972), *Hillslope Form and Process*, Table 9.4, p. 248.

Figure 7 – Adapted from data given in Deju, R A, and Bhappu, R B (1965), *Surface properties of silicate minerals*, N. Mex. Inst. Min. and Tech., State Bur. Mines and Min. Res. Circ. 82.

Figure 9 – As for Figure 4 plus data from Huang, W H, and Keller, W D (1972), *Organic acids as agents of chemical weathering of silicate minerals*, Nature, Physical Science, vol. 239, pp. 149–151.

Figure 11 – Adapted from Buckman, H O, and Brady, N C (1969), *The Nature and Properties of Soils*, Macmillan, Figure 1.4, p. 10.

Figure 71 – After Hubbard, C E (1968), *Grasses*, Pelican.

Figure 80 – After H M S O (1970), *Modern Farming and the Soil*.

Figure 85 – After Smith, M J (1970), *Soil Mechanics*, Macdonald & Evans, Figure 38.

Figures 86 and 87 – Both after Lambe, T W, and Whitman, R V (1969), Wiley, *Soil Mechanics*, Figures 9.4, 9.5 and 9.6, pp. 118–119.

Figure 109 – The Munsell Soil Color Company Ltd. The charts may be purchased from Munsell Color or from Tintometer Ltd (sole distributor in the U.K.).

Figure 111a and b – Soil Survey of England and Wales Annual Report 1973 (Rothamsted Experimental Station, Harpenden, Herts).

Figure 112 – Soil Survey of Scotland, Macauley Soil Research Institute, Craigiebuckler, Aberdeen (from information booklet *The Soil Survey*).

page 90 – Horizon nomenclature: Cope, D W (1973), *Soils of Gloucestershire*, Harpenden: Soil Survey of England and Wales Record.

page 90 – Horizon nomenclature: Findlay, D C (1965), *Soils of the Mendip District of Somerset*, Harpenden: Soil Survey of England and Wales Memoir.

page 90 – Horizon nomenclature: Clayden, B (1971), *Soils of the Exeter District*, Harpenden: Soil Survey of England and Wales Memoir.

page 92 – Properties of main sand grades taken from Clarke, G R (1971), *Study of Soils in the Field*, O U P.

page 93 – Texture descriptions taken from *Soil Survey Staff* (1960), *Field Handbook*, Harpenden: Soil Survey of England and Wales.

page 93 – Table after Cope (1973), see under p. 90 above.

page 96 – Comment from Avery, B W (1973), in *Journal of Soil Science*, O U P, vol. 24, p. 325.

pages 97–98 – Table 2 from Clarke, G R (1940), *Soil Survey of England and Wales, Field Handbook*, O U P.

pages 99–102 – Table 3 from Soil Survey of England and Wales, *Annual Report 1973* (see under Figure 111).

page 103 – Profile 1 from Jarvis, M (1973), *Soils of the Wantage District*, Harpenden: Soil Survey of England and Wales Memoir.

page 104 – Profile 2 from Clayden, B (1971), see under p. 90 above.

page 104 – Profile 3 from Clayden, B (1971), see under p. 90 above.

page 105 – Profile 4 from Jarvis, M (1973), see under p. 103 above.

Index

Figures in bold type thus, **22,** indicate the major reference of a group of references.

S0-BSR-990

the WILD HUNT

Also by Jill Tattersall

MIDSUMMER MASQUE
LADY INGRAM'S ROOM
LYONESSE ABBEY

the WILD HUNT

by *Jill Tattersall*

WILLIAM MORROW & COMPANY, INC.
NEW YORK 1974

Design by Helen Roberts

Printed in the United States of America.

1 2 3 4 5 78 77 76 75 74

Library of Congress Cataloging in Publication Data

Tattersall, Jill.
The wild hunt.

I. Title.
PZ4.T222Wi [PR6070.A68] 823'.9'14 73-13639
ISBN 0-688-00227-7

the WILD HUNT

1

In my dream the light was golden on the lake except where the boy was swimming, spreading black water like an inky stain behind him. He dived, and the ripples drew coins of gold and bronze together across the place where he had been. I waited but he did not surface and I knew that no one would help me to find him. The pale foxlike face that had been peering between the reeds was laughing silently at my distress, and disappeared as the scream I had been trying to force through my lips wrenched free at last.

Hands were on me then, shaking me out of high summer and into sharp spring, dragging me back from that golden afternoon into the cold confusion of the present night.

Elizabeth's calm voice reached me like a lifeline. "Hush, hush, Chantal," she soothed me and I felt my trembling beginning to die away, and my racing heart stumbled and slowed its beat.

"Poor Miss Fabian," I heard Mary Bowers solemnly remark. "I am sure it is no wonder that she suffers from the nightmare, even though it must be six years since the tragedy."

I turned my head protestingly. It seemed important to convince her that my nightmare was born of a memory much older than that. It was never of my guardian's murder that I dreamed, only of Rowley who I had thought was drowning one summer afternoon long before—Rowley, and that other boy who turned the dream to nightmare, the boy who used to follow us, slipping like a fox through the leafy woods, the pale-eyed boy who laughed when he thought Rowley was dying. . . .

"Come, Miss Fabian," said Mrs. Mason irritably. "This will not do. You are disturbing the whole school."

My eyes flew open. The darkness had been banished to the corners of the dormitory by the oil lamp Mrs. Mason was holding aloft in her large brown-freckled hand. She was an awesome sight as she loomed over me, a match for any nightmare with her lace cap pulled on askew over a head full of curlpapers, her square mannish face set with annoyance above the tobacco-stained front of her voluminous wrapper. I could smell the Otto of Roses which she used to scent her snuff, as a compliment to the Prince of Wales who had once had the good taste to bow to Mrs. Mason.

"What is all this, Miss Fabian?" she demanded, her dignity unimpaired by her inelegant appearance. "I thought you were cured of these nightmares at last. If you are not, I shall be obliged to mention the matter in my Report, you know."

There was a faint murmur of protest from the other girls, carefully unidentifiable. Mrs. Mason set down her lamp on the nearest table and clapped her hands sharply together. "That will do, young ladies. Go back to your beds at once. Good night, Miss Fabian. Miss Bowers, be so good as to precede me with the lamp."

She swept magnificently from the room, leaving me to press my burning face into the pillow, grateful for the darkness that hid my deep embarrassment. I was even grateful to Mary Bowers, too, for once, when she came in presently and put a stop to the whispers that had broken out, thus enabling the dormitory to regain its customary nocturnal calm.

Experience had taught me that if I went back to sleep too soon after waking from a nightmare it might return. I employed my thoughts, therefore, in wondering for a few moments why this particular one, though it originated so far in the past, had begun to haunt me only after that night of violence six years ago. Finding no answer to this problem, I turned to the question of why it should have visited me now, after leaving me in peace for many months. It was possible, I suppose, that if I went carefully over the events of the preceding day, I might find some hint of a cause which I could then in future take pains to avoid.

But as far as I could remember, it had been a perfectly ordinary day.

We had been roused by the bell and dressed by candlelight, complaining as usual of the shortage of basins and of the unseasonable cold. We then went down to prayers, read by Mrs. Mason, and proceeded to a breakfast of porridge and coffee while the grey drizzling day reluctantly lightened outside the windows. After the meal, Mrs. Simmonds instructed us in the use of the globes, and I was reproved for passing a note to Elizabeth. She, of course, had been more adept in passing hers to me, and had not been caught. History followed in a soporific murmur of repeated dates, and then the day improved with French and Singing, two of my favourite lessons. In the afternoon, the rain having stopped, Miss Hood took the whole school for a walk across the Heath. We set off two and two, according to the rule, and I walked with little Emily Collingwood because she complained that she had seen nothing of me for an age, while Elizabeth, after a mild protest, walked ahead of us with Sarah Gibson. I had been obliged to reprove Emily, I recalled, for giggling when a horseman, a dark, neat, nondescript sort of a man on a fine black stallion, who had reined in to allow us to cross the road, seemed to be staring at us. Once on the Heath, we were allowed to walk as we pleased and Emily took my hand as we passed near the gibbet where some wretched tarred remains were hanging in irons, which groaned in an eerie way with every wind that blew.

"Is that Black Dick?" Emily asked, between delight and awe; but Mary Bowers overheard her and informed her crushingly that though Black Dick had indeed been hanged in chains on that very gibbet, it had been quite fifteen years before. There had ensued a rather morbid conversation, in the course of which Mary explained her special knowledge as being due to the fact that she was a second cousin twice removed of the Lord Mortmain who had shot him and thus enabled the law-officers to bring Black Dick to justice.

"Lord Mortmain?" I repeated unwarily. "What a singularly odious name!"

"The Earl of Mortmain was my father's second cousin," Mary frigidly reminded me.

"But even so, you must own that the meaning of the name is curiously sinister. No doubt you are proud of the connection with a gentleman so aptly titled, however."

"Proud!" Mary exclaimed, while Elizabeth nudged me. "Certainly I am. The Quentins must be one of the most powerful families in Sussex, I suppose. Quentin, Emily, is the family name of the Earls of Mortmain," she added informatively, "and their family seat is Lindenfold. The present Earl—"

"Was he dead when they hanged him?" Emily demanded irrepressibly. "Black Dick, I mean? Did he die of his wound?"

"Certainly not," declared Miss Bowers. "My kinsman only shot at him in order to protect his property and to ensure that the highwayman could not escape—his intention was not to assist the felon to avoid justice by an easy death."

"Odious indeed," murmured Elizabeth, "and entirely in accordance with the Quentin reputation."

At this point Miss Hood called us to order and we had been obliged to march on obediently.

Macabre though the episode might seem, I did not think that the seeds of my nightmare lay within it. What of the rest of the day? We had returned to eat a plain dinner of stewed beef and potatoes, boiled fowls, and some of last year's withered apples. Then the day continued with a lesson in deportment during which I was condemned to half an hour strapped to the hideous back-board, after which sewing had seemed quite pleasant, and a writing lesson positively delightful by contrast. All went on very much as usual until the favourite hour when we had eaten our supper of bread and soup and were free to converse before the fire. I remembered well this evening's subject of conversation: it was a lively discussion as to which of us was the most likely to become the toast of the town on our emergence into Society.

"Miss Bowers is so well-connected and has already the air of a great lady," Sarah Gibson declared judicially; "but Elizabeth Tremaine is elegant and witty—and she has such narrow hands and feet."

"Charming of you," Elizabeth murmured. "But how in the world does one know what the gentlemen are going to admire the most? Your ample curves, dear Sarah; or a pocket-Venus like Emily? Or someone like Chantal who is not classically beautiful but whose face is very interesting and engaging, and whose figure—"

"Pray stop," I begged her. "You are putting me to the blush."

"No, but Elizabeth is right, Chantal," Sarah insisted. "Your hair is neither brown nor blonde, but it is soft and shiny, and your eyes and mouth are lovely—"

"Different gentlemen prefer different styles of beauty, I dare say," Mary Bowers decreed weightily. "To be the toast of the town would argue, I suppose, an obvious appeal that I doubt if any one of us can pretend to—or indeed would want to possess. Miss Fabian is certainly very pretty. You are to make your curtsey soon, no doubt," she added, condescending to speak to me directly. "May I ask who is to present you?"

"Yes, Chantal," Sarah cried. "I have been meaning to ask if you have been told when you are to leave. For you are past eighteen now, the oldest of us all, and I should have supposed that you would have gone to London this spring."

Her words sent a shiver down my arms, despite the fire, and I pulled my shawl closer about me. Yes, on reflection it must have been then that the nightmare began to lie in wait for me. I turned my back on the room, the firelight flickering on the polished furniture, the sagging chairs, the worn carpet, and the faces of my friends which displayed too much interest, too much concern. I looked instead over the tall fender directly into the glowing heart of the fire.

"No," I said, rather more loudly than I intended, "I haven't the least notion what is to become of me." I had meant it to sound adventurous, or amusing. That I had failed I knew when Elizabeth took my hand in a gesture of unspoken sympathy. I added quickly, "I dare say Mrs. Mason and the lawyers will settle it all between them."

Emily had then been inspired to ask what was the difference between an executor and a trustee and in answering her the remaining moments passed until it was time for prayers and bed.

That was the explanation then, I thought, turning over and clutching at a slipping blanket. It had, as usual, been fear which had caused my nightmare; but on this occasion it must have been fear of the unknown future and not, as always before, fear of the violent past. Now I knew where my fears were centered I must be on my guard against arousing them. From now on, I determined, I must do my best to live each day as it came, thinking of neither past nor future—though I somehow doubted my ability to control those delightful daydreams of my reunion with Rowley with which I was wont to beguile many an otherwise tedious moment. But those, I told myself, I knew to be foolish and they could therefore have no significance.

The very next day, as it chanced, was to lift a corner of the veil behind which I had decided not to desire to look.

I had climbed the cedar tree to rescue Emily's new ball and had just knocked it down into her waiting hands when I heard a sash window being raised in the house behind me. My heart sank correspondingly as Mrs. Mason's angry tones rang across the garden.

"Come down at once, Miss Fabian, and present yourself to me. What do you mean by such wicked, hoydenish behaviour, miss?"

Between alarm and my consciousness that I had no head for heights, I descended rather too quickly, ripping my muslin gown in the process and adding to the tracery of angry-looking scratches on my arms. Emily stood aghast while I crossed the short new grass as reluctantly as I had forced myself to climb the tree, and faced Mrs. Mason through the open window. She still held the quill with which she had been writing at the desk in the window-bay but as I waited for her to speak she sank down again in her chair and laid the pen upon its silver tray.

"Well?" she demanded, with less heat but an even more intimidating effect. "What have you to say for yourself, miss?"

Before I could begin to form an answer, Emily had rushed to my side and begun the case for the defense. "Oh, please, Mrs. Mason—it is my new ball, and Miss Fabian knew how sorry I should be to lose it. It was for my sake that she fetched it down and—and I beg you will not punish her, ma'am."

"And I wish I need not, Miss Collingwood, but I fear I cannot condone such vulgar behaviour in the most senior young lady in the school. What would Miss Fabian's trustees say if they were to hear of it, pray?"

"Oh, Mrs. Mason, you would not mention it to them, I hope?" cried Emily, quite as alarmed as I at the notion.

"I see no reason why I should not, Miss Collingwood," she declared in a manner which proclaimed her implacable decision to do so, "particularly as I am presently engaged in writing Miss Fabian's Monthly Report—"

So that was what those large sheets of hot-pressed paper were, covered with neat copperplate script and scattered across the writing-table before her, one of them barely a foot away from me. I had had no notion Mrs.

Mason could have thought of so much to say about me, and found myself consumed with an insatiable curiosity to know what manner of conduct, or lack of it, she found worthy to report on at so much length. The first sheet was turned half towards me and I began to read it swiftly without giving myself time to consider that what I did was wrong—far worse than climbing any cedar tree.

I was vaguely conscious of the two voices continuing their theme, while I skipped over a description of a cold I had suffered at the end of March, and a brief mention of last night's disturbance, and found my eyes riveted to the following lines:

"With reference to your letter of 31 March 1809, in which you mention the possibility of a post being found for Miss Fabian shortly—" I nearly exclaimed aloud at this, but luckily Miss Hood, who had been supposed to supervise our playtime, had now joined us and her indignant voice drowned the slight gasp which I was unable to prevent escaping me. Hastily I read on—"and your expressed desire that she should accept the first suitable such offer, I feel it my duty to point out that in my opinion Miss Fabian is as yet unsuited to have younger persons in her charge. As far as general responsibility for her juniors is concerned, no doubt she would do well enough, and she is certainly popular with all the girls; but she has not yet acquired the authority to instill discipline in others. Also, industrious as is Miss Fabian in many subjects, I fear that in Arithmetic and the Casting of Accounts, she continues to show no aptitude. As the executors of Arthur Medlicott, decd., and the trustees of Miss Fabian, yours are unquestionably the hands in which her future disposition lies. It is my duty, however, to bring to your notice what is my firm conviction, that Miss Fabian would greatly benefit from a further twelve months in my seminary in order to enable her to correct the faults of a somewhat passionate, if outwardly gentle and retiring character, before she goes out into the world: for as I have had occasion to mention many times, she is prone to quick temper, impetuous decisions and impulsive action—"

There the page ended and though I could not read the others from where I stood they no doubt contained ample illustrations of these faults. It was decidedly galling to realize I had just conveniently provided Mrs. Mason with another, for of one thing I needed no reflection to be certain: that the prospect of a further year at school held no attractions for me.

Mrs. Mason then startled me by addressing me directly, ordering me to my room, and condemning me to a dinner of bread and water. There was no disputing the finality of her tone and I turned to go upstairs while Emily was still protesting on my behalf.

The hours of punishment for once did not go slowly, for it was not to be expected that I could continue to hold to my decision to avoid looking into the future after being given such a hint of it, and I had much with which to occupy my mind. The thought of earning my own living did not trouble me so much as the certain knowledge that I was not, after all, to go into Society as did the other girls when they left Mrs. Mason's, and as I had sometimes dared to hope I might. This was no mere matter of relinquishing the idea of dancing at balls, flirting at Ranelagh, attending concerts and card parties in amusing company, and attired in the latest fashions. It involved my final renunciation of any hope of meeting and enchanting Rowley, my childhood hero who had once rescued me from the Horsefield Woods, in which, though they were so close to my home, I had contrived to lose myself after deliberately giving my long-suffering guardian the slip. Though I did not know his real name or anything of his circumstances but that he too had once lived in these parts, Rowley was, I felt certain, only to be encountered in that level of society to which I could no longer aspire. Slender though that hope had ever been, my farewell to it was not accomplished without pain, and I think until that moment I had not fully appreciated how much the thought of my reunion with one who was little more than a dream-memory had meant to me.

I must put all the past behind me now, I determined with more resolution than before, and prepare for a very different and more practical future—and after all, I told myself irrepressibly, it was always possible that something quite unexpected and exciting lay before me.

2

My solitary confinement was eventually interrupted by Mrs. Higgins, the housekeeper, who brought the cheerful news that the Reverend Shorncliffe was come to call on me, and that I might go down to Mrs. Mason's parlour. I hastened to obey the summons and was delighted to find that the pressure of Mrs. Mason's duties had obliged her to leave him there alone.

"How charming to see you, Godfather," I exclaimed, gazing lovingly at his thin stooped frame, his silvery hair and knotted hands, the worn leather volume which always protruded from his pocket. "It seems an age since you came to Horsefield House."

He took my hand and patted it vaguely. "Yes, yes, my dear, I am afraid it is. I have been away, you know—to London, no less. Such fine churches—so many books! And yet, after all, it is good to be back, despite the sundry annoyances of this ill-omened week. . . . How have you been, my dear?"

"Oh, very well, sir—that is, I had a trifling cold, and last month I had the measles—such a fuss there was—letters of enquiry every day from the solicitors, fruit and flowers and trinkets streaming from their chambers, the door knocker never still. I believe Mrs. Mason would have been glad if I had succumbed, to get a little peace. And then old Mr. Underwood did me the signal honour of calling on me just afterwards, to see for himself that I was recovered, I suppose, and that Mrs. Mason had not buried me in the garden with the intention of continuing to receive the fees for my education."

"Very proper in Mr. Underwood," he said dryly. "You are in good spirits, I observe, despite the fact that you are in disgrace."

"Oh, did Mrs. Mason tell you that? Well, it does me no great harm to starve for a little, and the others usually contrive to smuggle me up something later. But you are thinner, Godfather. Does that man of yours give you enough to eat? I wish I could come and look after you."

"An admirable notion, my dear, but since I am a bachelor I fear it will not do."

"What a pity, for it would suit us both so well. Oh dear, I have forgotten to ask you to sit down. What will you be thinking of Mrs. Mason's education of me? Pray be seated, Mr. Shorncliffe." I gestured him towards a handsome chair of carved walnut and myself took a plainer one of modern mahogany. "Did you have some particular reason for riding over to Horsefield House, sir?"

He looked puzzled and ran a veined hand through his thinning silver hair. "I wonder if I did? I think it was only to give myself the pleasure of seeing you, my dear. But while I am here I should value your opinion on a snuff the Bishop was so kind as to give me, when I was in London."

"Should you indeed, sir? I am immensely flattered. I suspect, however, that it is Mrs. Mason's opinion you really desire."

"I won't deny I asked the good lady to pronounce on it," he owned, pulling out a small enamelled box and taking a pinch of the tobacco it contained. "But I have a greater respect for your judgment, I think." He inhaled thoughtfully. "Mrs. Mason's opinion coincided with my own, that there is a little too much Brazil in it. I believe I prefer my Spanish Bran, after all. What do you say?"

I peered into the box before taking a cautious sniff. "Those large grains are Brazil, I suppose? Yes, it is certainly a little too strong for my taste."

"I am glad we are in agreement. Well, my dear—" He began to rise.

I put out a detaining hand. "Please, Godfather. There is something I must tell you. . . ."

"A confession, my dear?" he said gently, subsiding.

"Not exactly, but I hope you will treat it as such—and I beg you not to mention it to Mrs. Mason—but the fact is that—that it has come to my knowledge that my trustees wish to—to find me a suitable post that I may begin to earn my living—"

He looked startled. "A post, Chantal? But are you really old enough to be leaving school?"

"Godfather! I am eighteen," I assured him indignantly.

"Are you, indeed? But you still seem very young to go out into the world. However, if Mrs. Mason thinks you are ready for it she is certainly a more experienced judge than either of us."

"No, well, as it happens, I believe it is Mrs. Mason's opinion that I should stay at school another year."

"Indeed? And that does not please you, Chantal? Well, if it will comfort you, it is my opinion that the trustees will be a match for Mrs. Mason, if their wills are in opposition to hers."

"Oh, do you think so, Godfather?"

"I do, but are you so anxious to be earning your own living?"

"I—really, I hardly know. The suggestion has come as a surprise to me."

"What, my dear, do you mean that you had envisaged a different future?"

I began to sigh, and checked it. "Perhaps now it sounds absurd but I fear I had somehow hoped—the other girls rattle on about drawing rooms and London Seasons, you know. Horsefield House is so very select, I can hardly recall a girl who did not leave here to go into Society."

He looked at me more kindly than I felt I deserved. "My dear, in view of your circumstances, you are really very fortunate to have had the advantage of these five or six years at what is considered, as you say, to be a very superior boarding school."

"Yes, sir, I am sure you are right, but I must admit that I know nothing of my circumstances, my financial position. You will think me foolish beyond permission, perhaps, but I never thought of asking the extent of my guardian's fortune. I suppose the fact is that it has come to an end?"

Mr. Shorncliffe looked astonished. "But, Chantal—poor Medlicott left nothing."

"Nothing?" I stared at him blankly. A cold gust of wind rattled the window and a puff of smoke eddied out of the fireplace. I said slowly, "But Underwood, Jones and Underwood are my guardian's executors. If he left nothing, how have they been able to pay my bills all these years? I fear I don't understand, sir."

Mr. Shorncliffe cleared his throat. "You were not aware, then, that your guardian had once had some unfortunate experience of a bank's closing unexpectedly, in which he lost a good deal? He was in the habit, therefore, of converting his available funds into diamonds—"

"Diamonds! Yes, I remember now . . ." Diamonds, of course, had been my guardian's greatest interest. He was always buying them, unattractive stones, I had thought them, much like any other pebbles except that they were kept in soft leather bags.

"He would not entrust them to the bank," Mr. Shorncliffe continued, "so he had a safe made for them, a safe in which he also kept his papers. It was concealed, you may remember, behind that singularly ill-painted portrait of his grandmother. When Medlicott was killed, the safe was found empty. One has to assume that the—ah, murderer took everything that had been in it."

I became aware that I felt cold and was clasping my hands uncomfortably tightly. "But in that case, sir, who has been supporting me since then?"

"H'm, well, my dear, this is rather difficult. I wonder if I had better not ask Mr. Underwood to come and explain the matter—"

"Oh, please, Godfather! Surely it cannot be so—inexplicable?"

"Very well, my dear Chantal. I shall do my best, as far as I understand the situation. It seems that Medlicott added a codicil to his will the very evening of his death and delivered it himself to Underwood. In it Medlicott named a new trustee, and left you in his charge. The solicitors merely carry out the wishes of this person in regard to you, and one assumes that it is he who has—ah, also accepted the financial responsibility."

"But—who is he, sir?"

My godfather shook his head. "Mr. Medlicott requested that the trustee should remain anonymous and the solicitors have succeeded in protecting the secret of his identity."

I felt extremely agitated at the thought of this unknown person directing my life. I jumped up. "This is extraordinary! Why was I not told of this before? Why was I always allowed to assume—but who in the world can it be?"

"My dear, even if I knew, I believe it must be his own desire to remain incognito, and as I was saying, it was certainly Medlicott's last wish that this should be so."

I still could hardly grasp the situation. "A secret benefactor! It is very strange. I wonder if it could be some relative of mine, upon my father's side?"

"It could be so, my dear, but even if it is, I should not count upon his choosing to make himself known to you—rather the contrary, I fear, for as you know, your grandfather Fabian disowned your father at the time of his marriage and subsequently refused to acknowledge any responsibility for you, even when poor Medlicott took you to him when you were newly orphaned."

"So I understand; and even if he did relent when my

guardian died, he is now tired of me, it seems, and plainly considers his obligation to be at an end."

"Who knows, dear child? But if it is so, you are not to be thinking yourself abandoned, alone in the world. When this post, whatever it may be, is found for you, I promise to be less neglectful of my goddaughter." He took my hand. "I engage most solemnly to visit you, to correspond, to—ah, watch over you as assiduously as Underwood, Jones and Underwood have done hitherto."

"Godfather!" I exclaimed. "Is it—can it be you?"

"I, my dear? Do you mean, am I your benefactor? No, no, I don't want you thinking that—why, how could I be? I haven't a groat to my name, bless me. Certainly not enough to pay the monstrous fees that woman asks, over a hundred pounds a year, I believe, to teach and board you—"

"And how do you know that, sir?" I demanded triumphantly.

He looked quite distressed. "No, no, my dear—I must have heard it somewhere, but I beg you not to suppose—"

"Very well, I won't tease you, Godfather, for I see you don't like it." But he had not positively denied the charge, I had noticed, and I could not forbear to add that indeed I was most grateful to my dear unknown benefactor and I hoped some day he would give me the chance to tell him so. "And I suppose it was for him that my miniature was painted at Christmas time," I suddenly exclaimed. "I trust you liked it, sir?"

"What miniature? I don't—oh, this is terrible! You really must believe me when I protest—oh dear, I wish I had stayed in London. You have no notion what a week it has been: first the chimney at the rectory caught fire—nothing in itself and easily put out, but smoke and soot everywhere, some of my most valuable books ruined—"

So he might not possess a large bank account, I thought, but he did have a library of valuable books, and it would be just like him to sell some of them to support his old friend's foster daughter.

"I am not sure the water was not worse than the smoke," he was continuing. "Then on Sunday the Offertory Box was broken into and in just the same manner—I hope it may not be the beginning of another outbreak—and now you accuse me of a generosity of which you must, you really must know quite well that I am innocent."

Again I noticed that he did not positively assure me that he was, and I relinquished the fleeting and unlikely image of a kindly long-lost grandfather in favour of the much more probable one of my dear Mr. Shorncliffe in the role of my secret benefactor. But he seemed seriously upset and I hastened to change the subject.

"Very well, Godfather. But what is this about an outbreak of stealing in the church? It sounds very wicked. I think I never heard of such a thing."

He was obviously grateful to be diverted. Helping himself lavishly to his over-scented snuff, he explained that there had been a serious outbreak of such thefts a few years ago. "Several churches were kept locked up in consequence, or the boxes removed; but I did not care for either solution. Supposing some poor wretch needed consolation, and found the church door barred against him, or a rich man felt an impulse to charity and there was no poor box to receive his gift? No, I preferred to make good the damage myself, and to keep a watch to catch the thief."

"And did you ever catch him, sir?"

"I did, my child, as it happens. It was most disturbing, really. I caught him *in flagrante delicto*—"

"And let him go, I suppose?"

"Ah, well, yes—I am afraid now I may have been rather weak—he told me his story, there were extenuating circumstances, poor lad."

"He was starving, no doubt."

"No, no, to do the young man justice, he never pretended that. It was a sort of madness that consumed him, to steal, he said. It was not that he was in want."

"I don't think that is a very good excuse for stealing," I said, considering it.

"Ah, youth! Always so quick to judge," he murmured. "No, if the young fellow was to be believed—and I own that it did occur to me that he might be secretly laughing at me but I felt it proper to give him the benefit of the doubt—he had, he said, this mania to steal, not always money but often quite useless objects. It gave him, he said, a sort of comfort. But allied to his mania was a frightful fear of discovery and exposure and it was that, together with the fact that there were others who would be affected by his disgrace, which persuaded me to release him."

"And did the stealing cease?" I asked with interest.

"Certainly it did," he replied with dignity. "I am not quite a fool, my dear. I obtained a written confession from him, to be used if I had ever the merest suspicion of his having stolen again."

I smiled lovingly at him. "And did you keep this paper, or did you immediately lose it, dear Godfather?"

"Neither, Chantal. I lodged it with my solicitor, your guardian. In any case, the young man did not transgress again."

"Until now, perhaps."

"Oh dear, I hope you may be wrong. It is possible, I suppose—the sides of the boxes were removed in a particular fashion intended to prevent immediate discovery—and of course the fellow may feel himself safe since that paper went with the rest when poor Medlicott was killed. But though there is this similar feature to those other thefts it is so long now since then that I can't believe this latest felony to be the work of Mr. Ex—"

"Mr. X?" I exclaimed. "What a very curious name—or is that merely what you call him to yourself?"

"No, my dear, but I suddenly recollected how very wrong it would be in me to mention his name. I dare say he has quite settled down by now, after all, and is leading a blameless life. Well, Chantal, I have kept my horse waiting long enough. I am so glad I came—so extremely interesting to hear your news, and that reminds me, in case you are—er, removed from this establishment before we meet again, you will be sure to send me your new address, will you not, as soon as you know it yourself?"

"Of course I will, sir. Oh, I do wonder where it will be!"

He stood up. "Try not to wonder, child, nor to worry. Determine, rather, to accept what comes to you."

It was much the same as the advice I had given myself, but I found it quite impossible now to prevent myself from attempting, aided by Elizabeth's fertile imagination, to visualize what my life might be like in a hundred different situations and as many households, from the appalling to the idyllic. During the next ten days I grew rapidly accustomed to the idea of leaving school, and became ever more impatient with the petty tyrannies that Mrs. Mason imposed on us. I began to be careless of my reputation as a pupil and it was only after something of a struggle that I was able to continue to put any effort into my work. The time began to go more slowly as the month advanced and gradually I became convinced, to my distress, that Mrs. Mason had persuaded Messrs Underwood, Jones and Underwood to agree to my being sentenced to another year in prison—as her famous establishment was rapidly beginning to seem to me.

It was with interest, therefore, that I heard the unmistakable sounds of the arrival of a visitor one cool mid-April day, and almost incredulous delight with which I received the subsequent summons to attend Mrs. Mason in her parlour without delay. I brushed my hair, straightened my sash, and ran downstairs immediately. Only as I paused outside Mrs. Mason's door to regain my breath did I notice my heart was beating rather fast, but before I had time to feel any real apprehension I heard Mrs. Mason speak within the room. I took a deep uneven breath and turned the door handle.

Instead of the formidable female whom I had half-expected to be waiting to interview me, there was a gentleman in Mrs. Mason's parlour, standing with his back to the room looking out of the window with his hands clasped behind him. He was a youngish man, I instantly observed, and immaculately dressed in a dark blue coat, white breeches, and gleaming black Hessian boots. He was making no pretense at conversation with my imposing headmistress, which argued an enviable degree of self-possession, and as soon as she spoke it was evident that Mrs. Mason recognized a superior in him.

"This is the—er, young lady, my lord," she announced with rare deference. "This is Miss Fabian."

The nobleman, for I was certain the full dignities of a peerage were his and not merely a courtesy title, began to turn slowly round and I, remembering my manners, dropped hastily into a curtsey. As his stern profile—high forehead, Roman nose, determined chin, not in itself unpleasing—was for a moment silhouetted against the clear April daylight it seemed to strike a chord of recognition in my brain, a disagreeable sensation that almost caused me to lose my balance though I did not recognize him in any precise sense of being able to recall the circumstances of our former meeting, beyond that it had been in some way harrowing. Then he was facing me and I saw with an almost equal shock of relief that he must be a stranger after all, for I could never have forgotten the livid scar that twisted his mouth and rendered saturnine features that might otherwise have been handsome. His dark eyes seemed to burn into mine until, with a faint shrug, he turned away again. Only then did it occur to me that he would no doubt have supposed my initial revulsion to have been caused by the sight of his disfigurement. It was a bad start and I rose from my curtsey blushing with embarrassment and wishing that I could sink through the floor and disappear.

"Miss Fabian," said Mrs. Mason sharply, apparently restored by the sight of my confusion and unaware that she was about to add to it, "this gentleman is the Earl of Mortmain, who has done us the honour of calling at Horsefield House in order to consider your suitability for the position of governess to his lordship's young cousin, Sir Hugo Perowne."

Mortmain, I thought incredulously. Was this the scourge of highwaymen and worthy heir to the Quentin reputation of whom we had been talking on the Heath the other day? He certainly looked devilish enough. But who could ever have been so mad as to commission such a person to interview a young lady for a domestic post?

Mrs. Mason, belatedly recalling that I was presumably ignorant of the sudden necessity for earning my living, said in kinder tones, "I have to inform you that it is the desire of your trustees, Miss Fabian, that you should accept this post if you are offered it."

So Godfather had been right, and Mrs. Mason had lost her battle!

Lord Mortmain, having surely absorbed all that was of interest in his restricted view of an unkempt and muddy garden, turned again to subject me to an arrogant stare, of the kind, indeed, that I had previously supposed to be ill-bred. He said rather harshly, "The news appears to come as a shock to Miss Fabian. She had better sit down."

"Yes, yes, of course," cried Mrs. Mason, quite struck by his consideration. "Sit down, Miss Fabian. I am afraid the fault is mine, my lord. I was hoping to persuade the trustees that Miss Fabian is a little young to be leaving school just yet, so I decided not to unsettle her by even mentioning the possibility of employment to her, until I heard from them again."

"Understandable, ma'am—if somewhat mistaken in the event. But Miss Fabian is eighteen, I believe. Do you imply that she is particularly young for her age?"

"Oh, no, my lord," exclaimed Mrs. Mason unguardedly. "Well, in some ways, perhaps. She is responsible enough when she chooses to be, but as a governess—it is my duty to warn you, sir, that Miss Fabian is not precisely academic."

"Indeed?" The Earl advanced towards me and I stared

angrily down at his glossy boots as they halted before my chair, conscious that there had been something unusual in his manner of walking. "She is not—backward, I hope?"

"Certainly not, sir—but I understood you to have read Miss Fabian's Reports?"

"I believe I did, now that you remind me of it," he said in a bored tone. "It is true, of course, that if Underwood, Jones and Underwood had not recommended her as suitable to be employed by Lady Perowne, who is also their client, I should not now be here. Lady Perowne enjoys delicate health, as I believe Underwood must have explained in his note to you, ma'am, and that is why I have come to look the girl over in her stead."

I wondered if his offensive manner were natural to him, or if he had assumed it in order to punish me for having flinched at the sight of him, and decided it was probably the latter. I raised my head and stared at him deliberately. His scar was disfiguring but not revolting and he had no need to be unduly sensitive about it, I thought. His eyes were almost more distinctive. One might well flinch at the sight of them, for they seemed to emit a sort of power that I, and I suspected even Mrs. Mason, found more than a little frightening. I recalled Mary Bowers describing the Quentin family as one of the most powerful in Sussex. This man had no need of titles and estates to add to his consequence, however. He carried his strength within himself.

At this point the Earl disconcerted me by catching my eye. Fortunately for my reputation as a demure young lady, there was no question of my being able to sustain my gaze and I immediately looked away, annoyed to find I had so little strength of purpose.

"Miss Fabian is very good with the younger girls," Mrs. Mason was saying placatingly, having apparently decided to abandon her final attempt to make another year's fees out of me.

"I hope Miss Fabian may be as good with a spoiled brat of a boy."

Mrs. Mason smiled nervously. "Oh come, I am sure your lordship exaggerates—"

"Well, I believe she will do," declared the Earl, plainly tiring of the whole affair. "The post is yours, Miss Fabian."

3

For a wild moment I wondered whether I could refuse the post; but supposing Lord Mortmain reported unfavourably on me to the solicitors so that they felt they could not recommend me for another? Was my poor anonymous benefactor to be expected to pay my bills for the rest of his life or mine, out of a stipend already inadequate to his needs? I had no option, I decided, but to accept, and I did so in terms of the barest civility.

Mrs. Mason drew an audible breath of relief. She had apparently observed my momentary conflict and also wished no adverse report to come to the ears of the solicitors. She said severely, "Let us hope that you prove worthy of his lordship's trust, Miss Fabian," before turning to the Earl with a very different air. "Now, my lord, how soon do you wish Miss Fabian to take up her post?"

Lord Mortmain pulled a slim gold watch out of his waistcoat pocket and consulted it, plainly distrusting the ornate marble timepiece that stood on Mrs. Mason's mantelshelf.

"I shall take Miss Fabian down with me this afternoon," he declared, to our astonishment. "If she can contrive to pack within half an hour we should be at Lindenfold before dark."

"Really, my lord, I hardly think—"

The Earl turned towards Mrs. Mason with an air of

courteous attention and I was fascinated but not surprised to hear the protest die on her lips.

"Very well, my lord," she murmured, quite cowed. "Miss Fabian, you had better send Higgins up to the attic for your trunk."

"No trunks," declared Lord Mortmain implacably. "If you have one, it can be sent down by waggon. I have room for two small boxes only." He turned back to Mrs. Mason. "Our business is concluded, ma'am, and if you will excuse me I had better be looking to my horses. Wilson is an experienced groom, but they are a mettlesome team."

"But will you not take a dish of tea—a glass of wine? No? Well, I hope your decision with regard to Miss Fabian has not been too sudden, sir. Is there nothing more you wish to know about her?"

"I believe not," he replied carelessly, as if a five-minute acquaintance must have been sufficient for him to have plumbed the depths of my character. He picked up his tall hat from the writing-table, bowed to Mrs. Mason, and walked haltingly to the door before she had time to pull the bell. Rather to my surprise he held it open and waited for me to pass through it before him. I murmured my thanks and he flashed me a strangely intent look before closing the door and limping away down the hall.

I hurried up the stairs, my mind in a turmoil. Mary Bowers waylaid me on the landing, her eyes round and amazed. "Surely that gentleman who just left was Lord Mortmain!" she exclaimed. "I only saw the top of his head but I looked out of the window and certainly recognized Wilson, his groom—but Elizabeth tells me that Mrs. Mason sent for you. I do not understand, Chantal."

"I am far from understanding it myself, Mary," I assured her, "but the fact seems to be that Lord Mortmain has appointed me to be governess to Sir Hugo Perowne, and I must be packed in half an hour—oh, thank Heavens you are there, Elizabeth."

"Sir Hugo Perowne?" Mary exclaimed, while Elizabeth ran to fetch my bandbox and cloak-bag. "Well! You quite astound me, Miss Fabian."

I looked at her curiously. "You do not envy me, I collect?"

"No indeed," she returned with every appearance of sincerity, "I pity you from the bottom of my heart, for quite apart from everything else, Sir Hugo is a dreadful little boy. My sister Charlotte quite detests him."

The sight of Elizabeth returning reminded me of the need for haste. "Miss Bowers, you are related to everybody," I said to Mary, rushing to the chest of drawers and beginning to hurl clothes onto the bed. "Tell me quickly about the family, pray—there is no time to waste for his lordship is waiting to drive me down himself."

"We certainly must hurry," Elizabeth agreed, picking my gowns out of the wardrobe. "You won't want to be driving in the dark with him. I never saw such a wild-looking person—though beautifully dressed, I did observe. It is a pity about his scar, for he probably was handsome once, like Lucifer."

"It was another cousin of mine, Maurice Bowers, who gave him that fearful wound," Mary remarked with a certain complacence, folding my Sunday cambric. "They fought a duel some years ago—it was a dispute over cards, I believe—and Cousin Ivo, who had not come into the title then, though he was soon to do so, was shot in the head. He was out of his senses for a time; indeed, his life was despaired of at one point and Maurice was ready to fly to France—we must have been at peace then, I suppose."

"So it was six years ago," I murmured, for poor at history as I was, I should not soon forget the dates of that short uneasy peace during which my guardian had been killed.

"Or it could have been seven," Mary demurred. "But no, I believe you are right, it was in 1803 and they had both just come down from university in Edinburgh. It was a dreadful time for the Quentins because Ivo was the only son and the heir, of course, and had just come of age—you know that his parents separated soon after they were married, it was the scandal of '82? But though Ivo had lived with his mother for the whole period of his minority so that his father hardly knew him, it seems that old Lord Mortmain never recovered from the shock of having nearly lost his son and in fact he died soon afterwards."

"The old Earl was the fire-eater who dispatched the highwayman, I collect?" said Elizabeth, with the merest flicker of a wink in my direction as she rolled my stockings into a ball and thrust them into the toe of my slippers. As soon as she spoke I knew that it must have been so, for the Lord Mortmain who had interviewed me must have been less than thirty; yet my relief was not so great as one might have supposed. The present Earl, I felt, was likely to be just as dangerous as his father had been, even if he was innocent of that particular crime.

"What did you mean by the Quentin reputation, Elizabeth?" I asked urgently. "You remember, you mentioned it on the Heath that day—"

Elizabeth glanced at Mary. "Oh," she murmured, "I was referring to Quentin's Law—an expression still used in Sussex, I believe, when any member of that family appears to have forced some unpalatable decision on an inferior. The implication is that in those circumstances there is no right of appeal."

Miss Bowers had assumed the haughty look that indicated she was about to take offense, a tedious and time-consuming procedure, so I attempted to forestall it by thrusting my workbag into her hands and begging her to sort it out for me.

"Half of this I shall have to leave to be packed up in my trunk," I said, giving a despairing glance at the pile upon my bed. "But I suppose I had better take my workbag, and these handkerchiefs, and my nightgowns . . . I noticed that Lord Mortmain is lame, Mary. Was he born so?"

Never so happy as when imparting information about

her grander connections, Mary allowed herself to be mollified.

"No, indeed, Miss Fabian. The Earl was badly injured in a carriage accident only a few weeks ago, racing his curricle to Brighton when there was ice on the roads. My brother William alluded to it when he last wrote."

"Poor Chantal," Elizabeth sympathized. "It seems you are to take your life in your hands. I would advise you to take care, only I don't see how you can now you are committed to travelling with his lordship. But perhaps you will do well to bear in mind that he is not only, it appears, a reckless driver and a dangerous man to cross, but a very heartless flirt in addition, so I have been told. At least," she amended, "I have never heard of him escorting the same female for more than a few weeks and he is the despair of ambitious mothers, they say."

"Oh, that is all over now," said Mary, fitting my best bonnet into the bandbox. "It is rumoured that Thelma Curtis has taken him in hand at last."

"And who is Miss Curtis, pray?"

"Why, her father is nothing more than a country squire, master of our local foxhounds, but her mother was a Howard and therefore a cousin of ours. Thelma is a particular friend of Lady Perowne's, though Miss Curtis is a few years the younger, and a notable horsewoman, which Lady Perowne is not."

"I fear I cannot congratulate Miss Curtis on her choice," Elizabeth declared, sweeping my toilet articles into a towel. "I think the gentleman looks extremely fierce and I believe I would as soon marry a Turk as ally myself to him."

"He is a great catch," said Mary with unusual vulgarity, "for at twenty-one he had already inherited his maternal grandfather's fortune, to say nothing of Old Mortmain's, and Miss Curtis is a dutiful daughter. I believe Lady Martha has had Mortmain in her eye for Thelma for some time, only, as Elizabeth said, he has always been reluctant to come to heel."

I would have laughed at such a phrase being applied to the Earl, had not another thought occurred to me. "You said, 'Apart from everything else,'" I reminded Miss Bowers abruptly. "What else strikes you as so particularly unpleasant about my future circumstances, may I ask?"

She looked exceedingly put out and I realized she had thought better of her remark. "There is Mr. Piers Perowne," she said evasively. "I suppose he will be at home just now—the late Sir Hugh's brother, you know."

"And what of him?" Elizabeth enquired with interest. "Is he a Minotaur and does he devour young maidens?"

"You speak lightly, Elizabeth," said Mary stiffly, "but there is a certain justice in what you say. He is the frivolous flirt, rather than Mortmain, in my opinion. I wish he may not take advantage of your innocence, Chantal," she continued, warming to her theme. "I feel it my duty to warn you that I fear Mr. Perowne would never stoop to marrying a person in your position."

"Miss Bowers," exclaimed Elizabeth indignantly, "even from you, this is rather too offensive!"

"That is as may be, Miss Tremaine, but it is better for Miss Fabian to be told these things now than realize too late that Piers Perowne does not mean honourably by her. The Perownes are a proud family and have never, so far as is known, been guilty of a *mésalliance*."

"Thank you, Miss Bowers," I said coldly. "You have now performed your duty in warning me of that. What else do you feel I should know about my new position—for that there is more, I am positive," I added quickly, seeing her about to deny it. "I should not like to accuse a Bowers of either lying or cowardice."

She flushed darkly. "There is no need for either," she said, bending her head, tightening the strap of my bandbox. "It is merely that there is . . . a good deal of talk about Lindenfold."

"Talk?" Elizabeth said quickly. "What manner of talk?"

"I don't know that I should . . . and yet, is it not best

to be prepared?" she asked herself. Elizabeth and I assured her that undoubtedly it was, and she nodded. "Yes, you are very right. Poor Miss Fabian, already so nervous. . . . In a word, then, they say that Lindenfold is one of the last strongholds of witchcraft left in England."

Elizabeth was the first to break the ensuing silence. "But, Mary," she began in rather an awed tone, "are we to assume that you believe such talk?"

"I did not say that, Miss Tremaine," Mary protested. "I am merely reporting what I have heard—what is common knowledge in those parts."

"Witchcraft!" Elizabeth repeated wonderingly. "Well, Chantal, you are fortunate indeed to be driving off with a devilish Earl in order to look after a horrid little baronet in a witch-infested village—I believe you will not soon complain of boredom in your new appointment!"

The housekeeper hurried into the dormitory before I could reply, her keys jingling at her waist. "Are you ready now, Miss Fabian?" she enquired breathlessly. "His lordship sent to ask—and 'tis my belief he means to drive off without you, if you don't go down this very minute."

"So much the better," cried Elizabeth, but I gave her a rather tremulous smile and shook my head. She seized my hands, we kissed and promised to correspond and to make every effort to meet at some time in the undetermined future. Mary then condescended to touch her cheek to mine, and I found myself suddenly filled with impatience to be away from school and out into the world, even in such company as that of the gentleman with the detestable name, who was removing me to a position and a destination of which I had as yet to hear one good word.

I raced through the building making my farewells and arrived in the front hall to find it crowded with pupils and staff. One of the footmen was holding the front door wide, while another carried out my baggage to a gleaming open carriage drawn by four long-tailed chestnuts, who were pawing the drive and snorting impatiently, seeming to be barely held in check by the Earl, already in the driving seat. I gave a last hug to little Emily Collingwood, Mrs. Mason bestowed a cool salute upon my cheek and pronounced a blessing, and then Higgins was assisting me to scramble up beside the Earl, who hardly paused to allow me to become settled in my seat before cracking his whip and sending the horses plunging forward. I had scarcely time to wave at what appeared to be the assembled school before the turn in the drive had snatched them from my sight. By that time I was aware that to take my hand from the brass rail even for a moment was an act of wildest recklessness, for my other hand was occupied in holding my muff and there was nothing else, it seemed, to stop me from being hurled onto the road at the pace we were making. To make matters worse, I began to be concerned with the possibility of losing my second-best bonnet, for as soon as the wind began to tug at it I realized I had not tied the ribbons nearly tightly enough. I could feel it slipping, yet I dared not raise my hand to secure it. Quite soon, the inevitable happened and the bonnet blew away.

"Stop, stop, my lord," I cried. "Oh, I am very sorry, but please stop. I must get my bonnet."

"I don't halt these horses for such a paltry consideration, ma'am," the Earl obligingly informed me.

"Paltry?" I gasped, hardly able to believe my ears.

"A very unbecoming piece of millinery," he explained, hardly to my satisfaction. "Put up your hood and be thankful you have one."

I shot him a look which it was perhaps fortunate he did not see, for his gaze was intent on the road, and continued to cling to the rail. Imperturbably, the Earl slowed his team as we came to the gates, and transferred the reins to his right hand. With his left, he dragged the hood forward over my hair. Behind me I heard the groom clear his throat to cover a snort of amusement or surprise, while I sat stiffly, wondering again if Lord Mortmain treated me in this cavalier fashion because he supposed me to have flinched at the sight of his scar, or because I was his inferior, or whether he would have subjected even Miss Curtis to precisely the same treatment, because it was his nature. I stole a glance at him; he was again preoccupied, which was just as well for he was now overtaking every vehicle in sight. Despite his recent accident, he seemed an expert whip and I could not but admire the precision with which he drove his highly spirited team.

We raced through the village to the accompaniment of boys whistling, curtains twitching, women pausing by the pump to stare at us, the blacksmith hurrying out of the forge to see whose horses were passing at such a thundering pace. Nobody recognized me, of course, but I felt foolish and out of place, perched up high on a curricle for all the world to gaze at, and I was forcibly struck by the Earl's eccentricity in insisting on driving me to Lindenfold himself. Why had he done so? Was he in such a hurry to find a governess for his cousin, and install her? Why could he not have allowed me time to pack my trunk and find my own way to Lindenfold, either by stagecoach or hired chaise? There were a thousand questions I wanted to ask but he looked so stern I dared not interrupt his thoughts, and the longer the silence lasted between us the more difficult I found it to address him.

After some miles we passed through a tollgate and then turned right at a crossroads into a wood. Spring was late this year and though it was already mid-April the leaves were not much more than buds. The tangled branches were starkly etched against the silver sky and reminded me uncomfortably of clutching hands. It grew darker as the trees clustered more thickly. I suddenly felt absurdly afraid, and knew my fear to be inspired by the man at my side. What was this extraordinary power that emanated from Lord Mortmain? It could not be purely my imagination, fed by Mary's reluctant warning, for I had felt it before then, as soon as I had set eyes on him.

"Are you cold, Miss Fabian?" the Earl demanded abruptly, making me start.

"No, sir, my lord," I stammered.

He frowned. I realized he must have seen my involuntary shiver and, misconstruing its cause, believed I lied; or perhaps he realized I was afraid of him and resented it. I felt more ill-at-ease than ever and greeted the sight of the end of the wood a mile or so farther on with what I feared to be an audible sigh of relief. As the road came out into the open, it began to rise up a long hill and the horses were reined back to a trot.

"You are very silent, Miss Fabian," the Earl observed. "Is there nothing you would like to know about your new situation?"

I said in a low voice, "I should certainly be glad to know whatever you can tell me about it, sir."

"But what, in particular? Where shall I begin?"

"With the child, if you please, sir. How old is he?"

"Eight or nine—no, ten, I believe."

"And who spoils him, sir?"

Lord Mortmain said dryly, "You have a good memory, Miss Fabian. Hugo's mother is the worst offender. His grandfather Carlingford showers him with gifts when he is not bullying him. Then there is the terrible old nurse to whom the child is completely in thrall, to the extent of doing anything she asks, whether it be right or wrong, because she, too, alternately threatens and indulges him. You would suppose she might use this dominion she has over him to his advantage, but it does not seem to be the case. He performs mysterious errands at her command, and for the rest he runs as wild as he pleases. Even his Uncle Piers allows him anything he asks for. They are all sorry for the lad because his father was killed at Corunna."

"I am not surprised at it," I heard myself declare. "Poor little boy, I am sorry for him too."

The Earl said coldly, "The tragedy was not exclusively Hugo's, Miss Fabian. Lady Perowne has lost her husband, Piers his brother, Mrs. Cumber her former nurseling, and I my cousin and closest friend. Do you think we should all be overindulged and have our characters ruined as a consequence?"

"I suppose the child must feel it most," I protested.

"You have no right to suppose any such thing." He met my indignant look with a cool glance. "Perhaps your youth and inexperience may be held to excuse you, but you must learn not to leap to conclusions so impulsively. You are no longer a schoolgirl and the sooner you grow up, the better." He turned his head. "Wilson, how much was the Colonel at home in the last few years?"

"Precious little, me lord," replied the groom. "Sir Hugh was only ten days at Spithead afore leaving for Portugal, and that were nine months gone. Afore that he was in Sweden with General Moore, God rest his soul, waiting on the mad king. The Colonel did have three months stationed at Hythe earlier, but they kept sending him to Ireland and Guernsey and such places to recruit volunteers for the Regiment. No, I don't reckon as Sir Hugh spent much more than six weeks at home in the last three years, me lord."

"And before that he was in Egypt and Sicily, Miss Fabian. Wilson's brother was Sir Hugh's groom, by the way, and accompanied him on most of these expeditions. I don't think I exaggerate when I say that Hugo hardly knew his father. The boy feels no more bereaved than he has ever been, and I trust I shall not find you joining the ranks of those who defer to his every whim and give him a shockingly inflated idea of his own importance."

I could find nothing to say to this and stared silently at the fields of springing wheat beyond the greening hedges of blackthorn and brier.

The Earl continued calmly, "I have every hope, however, that your sense is at least equal to your sensibility. It had better be, indeed, for otherwise your new home will tax your nerves severely. Holy Mote is a mouldering old place and Lady Perowne has allowed it to deteriorate shockingly. The moat, of course, makes it very damp."

"Holy Mote?" I repeated. "I had thought you were taking me to Lindenfold, sir."

"Did you think so indeed, Miss Fabian?" He gave a curious smile.

"Mrs. Mason said as much," I informed him rather hastily. "And so did Miss Bowers."

"They were speaking of the village and not my house. No, you are destined for Holy Mote, I fear, where nothing would persuade me to live. Miss Bowers, did you say?" He frowned. "Good Heavens, I had quite forgotten my cousin Mary was at Horsefield. Was she a particular friend of yours?"

"We were members of the same class and dormitory, sir," I informed him, wondering why he should sound quite so disconcerted by the realization.

"That does not precisely answer my question," he pointed out irritably. "You say she spoke to you of Lindenfold?"

"Yes, sir. Miss Bowers was kind enough to help me pack just now."

"Indeed? No wonder you had no information to seek about your post. Miss Bowers had already satisfied your curiosity."

I understood that the Earl was attempting to determine how much I had been told and that he would have been better pleased had Miss Bowers and I not been on speaking terms. The strong impression I received that he would have preferred me to be completely ignorant of Lindenfold and the Perownes would in any case have been disturbing; in view of Mary's extraordinary disclosure it was not perhaps surprising that all my earlier fears now returned in force.

I said defensively, "Miss Bowers did not have time to tell me much. We were too concerned not to delay you, my lord."

"How extremely gratifying." The horses slowed to a walk. I glanced round to see the Earl regarding me with an oddly speculative look. I responded, I was aware, with an expression of surprise, even alarm, and this appeared to reassure him in some way. I saw his long fingers relax on the reins. A moment later we came over the crest of the hill and Lord Mortmain pointed with his whip to a grey smear on the horizon. "There is the sea, Miss Fabian," he remarked quite pleasantly. "It is not far now to Lindenfold. We should be there by dark."

4

It was twilight when the Earl turned the curricle off the the dusty road into the short carriageway to Holy Mote. The stacked chimneys could be dimly seen rising in the smoky air beyond a little copse even as we passed between the gates, and a few moments later a curve in the drive revealed the whole edifice.

My first impression was not reassuring. The house was long and low and appeared to be extremely old, with many gables in the mossy tiled roof. Latticed windows frowned high under the eaves or peered like furtive eyes through the tangle of climbing plants that covered the ends of the stone lower story. These creepers had begun to reach out even above the moat and to straggle up over the half-timbered upper floors as if they sought to fling their choking net over the whole building. The moat seemed to have been partly filled in for at the nearer end of the house I could see a buttress of vivid grass tumbling across the green and murky depths that fronted Holy Mote. Something splashed in the water and I shivered. There was a smell like autumn in the air, rather than spring, I thought; an evocative scent compounded of damp and earth and smoke.

I realized that Wilson was standing by the horses' heads and that the Earl had also descended and was waiting to help me down. I murmured an apology for having kept him waiting, and rather diffidently put my hand in his. He held it firmly but did not immediately move to assist me.

"No need for haste, Miss Fabian," he said in a mocking tone. "It is understandable enough that you should wish to drink your fill of the horror of this ancient place, which in my opinion is fit only to be pulled down. It looks as if it should be haunted, does it not? But so far as I am aware, no ghost has ever been seen to walk here, despite the history of rough handling the house received in the time of Henry the Eighth, when it was a religious foundation, and when at least one of the nuns ejected from it at the Dissolution is reputed to have drowned herself in the moat."

I shivered again. Lord Mortmain said curtly, "I thought I was making it plain that there is nothing to be afraid of, except that the house might fall to pieces."

"I am not afraid," I declared haughtily and not, I fear, altogether truthfully, "but I am now decidedly cold, sir."

He moved back at once, steadying me as I stepped down onto the gravel. He detained me when I would have walked on. I looked up at him enquiringly. His nearness to me in the fading light, the sardonic twist the scar gave to his face, combined to make him seem a sinister figure. "You are not to be afraid, Miss Fabian," he said emphatically. "One of your duties is to teach your charge to be carefree and lighthearted—to let the sunlight into Holy Mote."

He turned away and began to instruct his groom to walk the horses until he returned, and to water them when they were cool, while I reflected that if it was for such singular duties that the Earl had chosen me perhaps it was not after all so curious that he had been able to discount Mrs. Mason's warnings of my unsuitability to be a governess.

While he was still speaking the great brass-studded oak front door swung open, sending out a warm shaft of yellow light across what must once have been a drawbridge, and a bent old man appeared, perfectly in keeping with the general air of decrepitude, holding up a lantern and peering suspiciously into the gloom.

"Who is that?" he quavered, obviously expecting no good to come of a visitation at this hour.

The groom answered him. "'Tis his lordship, Mr. Percival," he announced with a suggestion of reproach, while the Earl took me by the arm and drew me towards the door.

"Is her ladyship at home?" he enquired.

"Oh, what a start you gave me, my lord. Ay, her ladyship is in the parlour, but she is poorly. I couldn't say if she will be at home to your lordship, I am sure."

The impression I received was unmistakably that the Earl was not a welcome visitor in this house and I hesitated on the threshold, disconcerted and apprehensive. Lord Mortmain's fingers closed more firmly about my arm. "Don't trouble to take us up or call the footmen, Percival; I shall announce myself." He led me into a cold stone hall, sparsely furnished with oak settles, its walls gleaming with damp between the threadbare tapestries and ancient weapons that decorated it, his boots ringing unevenly on the worn flagstones. I caught a glimpse of us in a dusty blue-toned mirror, myself pale and shrinking, long amber hair streaming untidily over my shoulders, my eyes seeming rather black than blue, shadowed and questioning in the dim light; and the man at my side, tall and dark, with something almost predatory about his limping walk, the hold he kept on me as if he feared I might escape him. Then we had passed out of the area of reflection and were climbing the broad polished stairs which led us up on to a gallery where portraits stared us down with as little welcome as the butler had offered us, ladies haughty in powder and ruffs and wigs, gentlemen reserved in satin and spaniel-locks.

The Earl paused before a panelled door. He opened it and motioned to me to precede him. I began to do so, but hesitated in surprise at the sight of a lady dressed in black and heavily veiled, reclining on a sofa near the fire, while over her leaned an elderly woman, also in mourning, muttering and moving a lit candle rhythmically before the lady's shrouded face.

Almost before I had taken it in, the scene had changed. The old woman became aware of us, glancing round with an expression of incredible malevolence before swiftly blowing out the candle and unwinding the lady's veil. The lady gave a faint moan of protest, which the old woman ignored, and pressed a hand to her eyes. The woman tucked the veil beneath her arm and shuffled across the room towards us. She passed me as if I did not exist, dropped the sketchiest of curtseys to the Earl and went out on to the gallery.

"Oh, is that you, Mortmain?" said the lady languidly, sitting up and staring in our direction. "I wonder that Percival admitted you. I have a shocking head, I fear, and am in no case to entertain you."

She looked curiously at me. Her dark beauty struck me as quite startling but the Earl did not seem particularly impressed by it.

"I am sure it is no wonder you have the headache, Cousin," he said sternly, advancing into the room. "Such company as you keep—and all the windows tightly closed. A gallop on the Downs would be my prescription for you, ma'am."

"You know I don't care for riding," she murmured. "And you must have forgotten the miasma—to open the windows would be well-nigh fatal at this hour. Are you going to tell me who this young lady is, sir?"

The Earl turned to look at me almost as if he were surprised to find me still standing there.

"Ah, yes," he said carelessly. "Come here, Miss Fabian. Cousin, allow me to present to you Hugo's governess."

Lady Perowne drew in her breath sharply. "Hugo does not have a governess," she objected.

"You are mistaken, ma'am. Here she is."

Lady Perowne raised her fine dark eyebrows and stared hard at me, while I for my part felt ready to leap into the moat. As if, I thought despairingly, I had not had enough to try me this day without now discovering that Lady Perowne had not, after all, commissioned her autocratic cousin to find a governess for Sir Hugo. My face burned, but I could think of nothing to say.

"You surprise me, Mortmain," announced Lady Perowne at last, mildly enough.

"I thought I should do so," he returned calmly, stripping off his gloves and bending over her pale hand to salute it. Then, straightening, he commanded me to come and make my curtsey to her ladyship.

"Good evening, Miss Fabian," she said with understandable reserve, when I had done so. "What is all this, Mortmain? This young lady does not look in the least like a governess."

The Earl turned again to look me up and down. The sardonic smile dragged at his mouth. "I am forced to agree with you," he declared. "However, you must not hold it against her. Miss Fabian has been driving with me and lost her bonnet, and I would not stop for it. She looked quite respectable when I engaged her, I assure you."

"Respectable? I should hope so, indeed, for she scarcely looks more than a schoolgirl. I meant something quite different, as I am sure you are aware, when I doubted Miss Fabian of being a governess."

"Nevertheless, as from this moment she is one."

Lady Perowne allowed a small sound of irritation to escape her. "I suppose you will tell me what is in your mind when it suits you to do so, sir. You had better sit down, Miss Fabian. May I ask where Lord Mortmain found you?"

I spoke with as much dignity as I could manage. "I am afraid there must have been some mistake, ma'am. I understood that you were in need of a governess for your son. Lord Mortmain obtained my name and address from his lawyers, who are also mine and whom my trustee had directed to find employment for me. They sent his lordship to my school, where he interviewed and engaged me."

"There is no mistake," remarked the Earl. "All that you have said is correct. Merely, I omitted to mention to Lady Perowne that I had undertaken this onerous duty on her behalf."

"And that is very like you, Mortmain, if the rest is not. But did you say Miss Fabian had been driving with you? You cannot mean that you brought her from her school in your curricle, I hope?"

"I did, indeed."

"But, Mortmain, that is really too much! No wonder the poor child is bewildered. What in the world persuaded you to such a sudden start?"

Remembering that the Earl, as well as Mrs. Mason, had accused me of impulsiveness as if it were a tiresome attribute, I waited with interest to hear how he would answer this. He did not seem disconcerted by the question, however.

"I am a man of quick decisions," he said austerely. That, he implied, was a very different and superior thing from being merely impulsive. "I formed the impression that Miss Fabian might have changed her mind about taking up the position I offered her, if I allowed her time to consider it."

"I see," said Lady Perowne reflectively. "You thought Miss Fabian so very suitable for the post that you preferred to have her come reluctantly than not at all?"

The Earl bowed slightly.

Lady Perowne turned again to me. "In that case," she said affably, "it should not take long for you to persuade me also of your suitability, Miss Fabian. You will not object to it, I trust, if I ask you a few questions in my turn? Who is your family, for example? Where do you come from?"

I was grateful, though somewhat surprised, when the

Earl answered on my behalf. "Miss Fabian is an orphan," he declared, and limping to the hearth, leaned one arm along the mantelshelf and stared down at the flames. "Her father was a younger son of the distinguished family of that name, who eloped with a French heiress while making the Grand Tour. Both families heartily disliked the match and repudiated the partics to it. It is doubtful if the French side even yet knows that there is a child of the marriage. When Miss Fabian was still quite a baby her father returned to England with his wife and child in order to seek employment in London. They were crossing the Thames on a ferry when it sank. Miss Fabian was saved by another passenger, a Mr. Medlicott, who, on discovering that her parents had drowned, that they had been almost destitute, and finding when he applied to her grandfather that he wanted nothing to do with her, took Miss Fabian to live with him and be his ward. When he died, she was put into Mrs. Mason's boarding school at Horsefield, close by her old home, where she has remained until today."

"Interesting," murmured Lady Perowne. "And did you think of enquiring into Miss Fabian's qualifications when you interviewed her, Cousin?"

"I am quite well-versed in them, I believe. Mrs. Mason was in the habit of reporting on Miss Fabian each month to her trustees and I have had access to those reports," he returned calmly, while I blushed in confusion and annoyance to think that he had probably learned more about me than I knew myself.

Lady Perowne gazed at her cousin with the faintest of frowns. "Really, Mortmain, all this is singularly unlike you," she exclaimed. "I am quite at a loss to account for it."

"I have an obligation to Hugo," he reminded her patiently. "It is my intention to turn the boy out a credit to his father's memory. Miss Fabian is no bluestocking but I am persuaded she is just what Hugo requires at this period of his life."

She sighed. "Very well, we must give her a trial, I

suppose. So much depends on how she can contrive to rub along with Old Nurse."

"On the contrary, her relationship with Mrs. Cumber is completely irrelevant," the Earl said crushingly, "unless Old Nurse bullies her, in which case I shall expect you to inform me of it."

"And then there is Piers," she murmured. "Has it not occurred to you that Piers may prove something of a difficulty?"

The Earl frowned. "Is it his intention to pass the whole vacation at Holy Mote? I thought there was some talk of a reading party. In any case, he will be going up for the Trinity term before long."

Lady Perowne shook her head. Finding a scrap of white cambric in the folds of her black gown, she raised it delicately to her eyes.

"Oh God," the Earl exclaimed wearily. "Is he in trouble again? What is it this time, Adela? Come, don't try to hide it from me."

"Oh dear, I hardly know how to tell you. Will you promise you will not be cross? If you had not been away all this last fortnight—how was dear Lady Mortmain, by the way?"

He stared at her in exasperation, while I looked at him with rather more than surprise. Could he be married? No, of course not, there was Miss Curtis; and besides, there was something untamed in him which seemed to proclaim him a bachelor.

"My mother is no worse, thank you, Adela, though she has given up all hope of a cure," he said grimly. "As for Piers, pray don't distress yourself. I shall soon get the truth out of the young fool."

"No, I shall tell you myself," she declared heroically. "Only I beg you to be lenient with him, for he is only a high-spirited boy, just as Hugh was at his age, I am sure. The truth is that Piers won't—can't—in short, that he will not be going back to Oxford, unfortunately."

"What? Do you mean he has been sent down?"

"Yes, I am afraid he has—but really it has all turned out for the best, for he proposes to spend the summer here at least and I must confess I was just thinking tonight that his presence really obliged me to think of providing myself with some respectable companion. I very much disliked the idea—indeed, I believe that was what brought on my headache—but now Miss Fabian will do instead, at least for the time being."

Lord Mortmain gave a slight groan and relieved his feelings by kicking the smouldering log in the hearth. "Has Piers seen fit to tell you why they sent him down, or wasn't it considered a suitable subject for your ears?"

"He did not tell me exactly—only that it was for something very wicked," she returned vaguely. "Heresy, or some such thing—but of course he did not mean it. It is only that it is all the rage among his set just now."

"And where is he tonight, ma'am?" Lord Mortmain asked, with an air of restraint.

Lady Perowne fluttered her pretty hands. "How should I know, sir? He rode off somewhere after dinner. I am not his keeper, I suppose."

"No, alas. I, if anyone, am that. Ah well, I dare say I shall find a full catalogue of his misdeeds awaiting me at Lindenfold with the rest of my post. You may tell him that I shall be calling on him in the morning."

Lady Perowne sighed. "Don't be too angry with him, pray. You are always so unfair to the poor boy."

"Unfair? How can you say so, ma'am?"

"Easily, Mortmain," she declared with a heightened colour. "Piers says—"

"Well?" he demanded as she paused. "What does Piers say?"

"Why, that you can't forgive him for being alive when poor Hugh is dead—and that is why you thwart him at every turn."

The Earl looked so stunned that I could almost feel sorry for him. Then he said so savagely as to drive out softer emotions, "Piers is a fool."

Lady Perowne looked alarmed. "You will not tell him that I said—"

"No. Well, ma'am, I shall leave you now to instruct Miss Fabian in her duties. Remember that she is quite inexperienced." He regarded me thoughtfully. "Is there anything else about her that I should bring to your attention, Adela? Ah, yes, I understand she suffers from the nightmare, so don't be unduly alarmed if you hear her screaming in the night." He bade us a brief farewell, bowed, and limped out of the room. His boots sounded noisily on the stairs and a few moments later the front door closed heavily behind him.

"Thank Heaven he has gone," exclaimed Lady Perowne with relief. "I cannot think why my late husband was so fond of him, and as for leaving all our affairs in Mortmain's hands! But there, so it is. Now, are you hungry? I am sure you must be. Pull that bell, Miss Fabian, and then come and sit by me and tell me what subjects you learned at Horsefield House."

I obeyed her contentedly. Lady Perowne seemed kind, and if her understanding was not great she had a good deal of charm. I felt that we would agree quite well and that my new position might not be so bad as I had feared.

The footman came and Lady Perowne ordered my soup and told him to ask for a room to be prepared for me. He left, and she returned her attention to me. "So Mortmain thinks Hugo should have a governess," she said reflectively. "Well, there is no going against his decisions, alas, but I wonder why he never mentioned it before. I suppose he intends you to supplant Old Nurse—he has always hated her and maintains that she is bad for Hugo. I dare say he hopes that you will become indispensable to me and that then I shall be more amenable to losing Mrs. Cumber. Well, he shall never succeed in that. I depend upon Old Nurse in a thousand ways; she knows Holy Mote better than anyone else and furthermore she brought up my husband and his brother from the cradle, and Hugo, of course, and was a nurserymaid here when my father-in-law was born. I shall never turn her out."

"I should think not, indeed. But, Lady Perowne, is it not possible that Lord Mortmain was thinking only of your son's good in this?"

Her affability vanished. "When you know my cousin as I do, Miss Fabian—not that your paths will cross, I suppose, for I believe he hates this house since Sir Hugh has gone and visits it as little as he can—but if you did, you would have learned to question his every casual act. Ah, here is your supper," she added more gently as the footman returned followed by another who pulled out a small mahogany table and assisted his colleague to transfer the contents of the tray to it. "It is all so difficult, suddenly having another member of the household," Lady Perowne mused when they had left the room. "There is this question of where you ought to eat—tonight is an exception, of course. I suppose you should properly eat with Hugo in the schoolroom, but Old Nurse is used to having him with her and I am sure she would not tolerate your joining them in the nursery. And then you are not quite like the usual sort of governess, educated as you have been and sponsored, as it were, by Mortmain himself." She sighed. "You had better eat in the dining room with us—unless we have company, of course. Have you finished your soup?"

"Yes, ma'am," I said hastily, laying down my spoon with regret, for it was delicious, I was hungry, and I had not finished, but that same consciousness of her superiority that enabled Lady Perowne to say whatever came into her head without considering the effect of it upon her audience made it difficult to contradict her even in such a minor matter.

"Then be so good as to pull the bell, Miss Fabian. I am ready to retire and no doubt your room has been prepared by now."

When John appeared she sent him for Jackson, her maid, and Randall, a chambermaid. "Randall will show you to your room," she said dismissively. I rose obediently and curtsied. It was very uncomfortable being in this anomalous position in the household, I felt already; and never having had a governess of my own I had really not the least notion of how to go on, or what was expected of me. But in gratitude to Mr. Shorncliffe I must make every effort to repay him by succeeding in this post, I determined as, having bidden my employer good night, I followed the maid down the long rambling corridors.

"It seems a long way to my room," I remarked eventually.

The maid glanced over her shoulder Her thin yellowish face looked apologetic, I thought; a little fearful, also. "Mistress Cumber gave the order, miss. The last time there was a governess here she slept in the Stair Room, it seems."

"In that case I must conform to tradition. How long ago was that, Randall?"

"A good while now, miss. They say she was governess to Sir Hugo's great-aunts, them as is in the portrait with the greyhounds on the gallery. That would be more than fifty years ago, of course."

Fifty years ago! I had told Lord Mortmain that I was not afraid of this old damp house, but I heard my voice tremble in spite of myself as I hurried after the maid before she could turn the corner with her branch of flickering candles and leave me in the dark, calling to her in an attempt at bravado, "I hope the bed is aired then, Randall. Half a century is a long time for a mattress to remain unwarmed." And I tried to console myself by thinking that there was small chance of Lady Perowne's being disturbed by my screams after all, if I did have nightmares here, so far did my room seem to be from the main part of the house.

At last we came to a flight of blackened oak stairs

plunging to darkness below, rising to darkness above. A heavy door faced us across a narrow landing. It groaned in protest as Randall forced it open, and creaked slightly as she stood against it to allow me to pass.

"Oh, but this is charming," I exclaimed with relief, looking down two steps into the room. It was certainly an awkward shape and full of little cupboards and chimney doors, but a small fire burned brightly in the hearth and the bed was turned down invitingly, with the handle of a warming pan protruding from it. The oak floorboards had been polished until they looked like water and the chintz hangings, though obviously old, were clean and crisply ironed. The latticed window gleamed and there was not a cobweb to be seen. My boxes were there awaiting me, and hot water steamed in a pink china bowl on the wash-stand. There was even a modern oil lamp on the table by the bed, and a box of scorched tinder.

"How many did it take to ge this room prepared to such a standard in the time?" I asked, smiling to think of the very different chamber my imagination had prepared me to find, all dust and mould and birds' nests in the hearth.

"Just us four chambermaids, miss," Randall replied, "and one of the housemaids come in to help—there's four of them, too, and two parlourmaids, though it's not too many in a house this size."

"What other staff is kept?" I asked, drawing off my gloves and unstrapping my bandbox.

"Well, there's Mr. Garland, miss, the steward—but we don't see much of him. He keeps to his apartments, mostly. Then there is Mr. Percival, the butler, John Butcher, footman, and Thomas Smithers, second footman. Mrs. Cumber, the old nurse, she acts as housekeeper too; Jackson, her ladyship's maid, and Ashe, Mr. Piers's valet, we don't see much of them, neither, except at meals, for Miss Jackson presides over the housekeeper's table, being as Mrs. Cumber eats in the nursery. There is Mrs. Crabbe, the cook, and she can cook, miss. 'Tis more the custom to have a man cook in an establishment of this size but Mrs. Crabbe is as good as any man—and temper!" cried Randall admiringly, "I'm sure I've never seen her equal for temper, man or woman. Then there are the maids, as I said, ten of us, and a still-room maid and two in the kitchen, and a steward's room boy. Twenty-two in the house, that is, miss, and fourteen outside, not counting the stables, of course."

I pushed aside a chintz curtain and began laying my clothes on the shelves it had concealed. "And where do you all sleep then, Randall? On the floor above?"

"Ay, in the attics, miss, but at the other end of the house. Mr. Garland and Mr. Percival don't sleep in the attics, of course. They has their own apartments."

I looked thoughtfully at her. "Is there anyone in the room above this?"

"Oh no, miss. There is no room above this one—only the stairs and the chimney, miss. Well, if you'll excuse me—"

I suddenly felt very tired and sat down on the wide bed, which no longer seemed wonderfully luxurious but merely too large to be comforting.

"Goodnight, then, Randall."

"Goodnight, miss. I'll be bringing your water in the morning."

She left, closing the heavy door noisily behind her. I sat, listening to the myriad sounds of an old house settling itself for the night and the less alarming ones of the fire spitting in the hearth as the sap ran down the apple logs into the flames, the casement rattling slightly in the breeze. After a while I persuaded myself I was not afraid to be alone and began to prepare for bed. When I was ready I turned the lamp down low and lay watching the curtains rippling gently in the draughts, grateful to the warming pan for having done its business so effectively and wondering drowsily what the morrow would bring.

I awoke reluctantly at a sudden exclamation that sounded as if it were almost in the room.

"The Devil! Oh, it's you—Cumber's Familiar."

I lay still, pressing a hand to my mouth.

Another voice spoke. "I did not mean to disturb you, sir."

Who were they? Could they not see that the room was occupied, with the fire and lamp both lit and my clothes folded on the chair? But perhaps they were outside the door, upon the stair.

"Oh, never mind it," the first voice was saying carelessly. "How goes it at All Saints?"

"We prosper, sir."

"Indeed? That's since Hewett's cow took sick, I'll wager. Or was it the Wild Hunt last Esbat? I'd a notion envious eyes were watching that. Thank Herne the night was cloudy: I'll take Hecate for all your Selenes."

"Artemis for me," said the second man, and I shivered at something in his tone. "I fancy our Master likes them young."

"Pray hush," broke in another voice, whispering urgently. There was a scuffling noise, more whispers, a distant rumble, a click, and then comparative silence. I put out my hand to hold back the bed-curtain but the room looked just as when I had last seen it, with the chintzes gently moving in the draught and the firelight burning low. No voice, I realized, could have penetrated the heavy door so clearly, no person could have passed through it without my knowing it. It was as if I must have dreamed it all, I thought, and yet I knew I had not. How could I have dreamed of voices saying things I did not understand yet which made sense and which I could remember? Those voices belonging to two young men, one arrogant and at first startled, later at ease; the other respectful but furtive; and the whisperer who had chivied them away—I had heard them, I was certain, though I was puzzled to know where they had been standing.

I reminded myself that there were no ghosts at Holy Mote. I yawned and felt the tension slipping out of me. Soon my eyelids were heavy and I knew the welcome sensation of drifting back to sleep. Then, some time between midnight and dawn, I found myself aware that I was really dreaming, weaving a dream out of an old memory of riding with Rowley on his great black horse in the summer woods . . . and then Rowley had gone and I was looking for him everywhere. It is a wild-goose chase, Elizabeth was saying scornfully, but Mary Bowers whispered that my pursuit of Rowley was called a Wild Hunt. . . .

5

I felt confused and stiff when I awoke next morning. The bed had been damp after all, under the deceptive warmth, and also, I remembered, I had dreamed and slept badly. I forced open my heavy eyes and got out of bed, shivering as I stepped into the draughts. I crossed to the window, pulled back the curtains and opened the latticed casement.

At once the clamour of birdsong became almost deafening. It was as if, with the late arrival of spring, the birds had to do all their courting and mating and nesting at the same time, which necessitated three times as much singing as usual. It had rained in the night, the grass was sparkling with it and the water in the moat below my window looked stirred and muddy. I could see two gardeners at work on the lawn at the end of the house, scything before the grass should dry, and a man raking the gravel drive with slow, careless movements. The sky was heavy with more rain to come, there was a light mist in the air and the music of springing water and dripping leaves mingled with the feverish song of the birds.

Drawing in my head, I looked at my bedroom by daylight. I was already attached to it, I decided, and would resent a move to a more stately chamber; only I should have to master the worst of the draughts before winter. I pulled back one of the chintz curtains to find out why it should billow into the room from time to time, but there was nothing to be seen but the cream-painted shelves and row of hooks among which I had disposed my clothes the previous night. The draught was coming from the cracks between the shelves, something to do with the chimney, no doubt. Perhaps the cupboard had been made from an old smoking-oven, which could be a disadvantage on windy days.

Randall interrupted my thoughts, knocking and bringing in a jug of hot water. "I hope you slept well, miss," she said. "It struck me afterwards as I hadn't seen to it the stair was fast. I hope that didn't bother you none, miss."

"The stair?" I repeated, puzzled, thinking she referred to the stairs outside my door, but she walked past me and twitched aside the second curtain, on the far side of the fireplace. I looked over her shoulder. Instead of the shelves I had expected, there was a narrow blackened stair winding upwards to an old warped door, swinging ajar. Immediately I remembered the voices in the night. Was it on this stair that those two men had stood to have their curious conversation?

"What is up there, Randall?" I asked in a low voice.

"That don't lead nowhere, miss, but Old Nurse generally keeps that door locked, even so—Mistress Cumber, that is." She stepped up and began to close the door. "That needn't trouble you, miss. There's nobody can use this stair nowadays."

"One moment, Randall. May I look?"

She swung the door wide towards me. The light was dim up there but I could see that after two more steps the stairs led straight into a wall.

"How extraordinary!"

"Oh, that used to lead somewhere once, I dare say, but now the way is blocked. 'Tis often the same with these old places," she explained, snapping the door shut with a sound that was certainly not the click I had heard in the night. "These big old houses, miss, they've always been altered and built on to and altered again till it's all topsy-turvy, like." She turned the key in the lock and stepped down into the room. "Do you need any help with your dressing, miss? That ain't properly my place to offer it, but seeing as you're only just out of school—"

"Thank you, Randall. I have been dressing myself for years but I would be most grateful if you would brush my hair."

I sat down before the mirror and tried to pretend I was a young lady of the house while she brushed my hair to gleaming silk, but it was useless, for my thoughts were stubbornly determined to occupy themselves with the conversation I had overheard. Why should someone prosper because somebody else's cow had taken sick? What was the Wild Hunt and who was Herne? I had never heard anyone swear by Herne before—it sounded like Horn and might, I supposed, be another way of calling on the Devil—or in this case, thanking him. I remembered Mary's warning: "Lindenfold is one of the last strongholds of witchcraft left in England. . . ."

I shivered slightly and Randall laid down the brush.

"I must go now, miss. I have to take a tray up to Sir Hugo, for he had a bad night, they say, but I'll do the same for you tomorrow, and glad to. That's hair that pays for brushing, miss, that I will say."

She had gone before it occurred to me to ask her the way down, so I hurried into my blue sprigged muslin, tied my sash and pulled on my gloves wishing that like Theseus I had taken a ball of string to guide me when I went into the unknown the night before. In the event it was not so difficult. I went down the stairs that passed my door and came out into a stone-flagged passage that was obviously little used, for ferns were sprouting from damp niches in the walls and green moss grew between the stones. There was another passage branching to the right that led to an outside door with a cracked pane of glass admitting a watery light through the creepers that covered it, but I continued straight on towards the heart of the house and eventually emerged into the Great Hall. Once there, common sense and the smell of roast beef led me to the eastern corner of the house where I discovered the

Breakfast Parlour and, somewhat to my confusion, Lady Perowne and her brother-in-law already well into their meal.

Lady Perowne, still in deepest mourning and even more beautiful by daylight, was toying with some confection of eggs in sauce, while the young gentleman opposite her was ladling oysters over his dish of beef with a generous hand. As I had rather expected, Mr. Perowne was exceedingly good-looking. He had very bright light blue eyes, a healthy colour in his face, and light brown hair that was artistically tumbled about his well-shaped head in a manner which I thought must have taken some time to achieve. His clothes were elegant but rather exaggerated, a very high white stock and a tight dark grey coat with an M-cut collar.

He looked up, catching my eye, stared for a moment, and then leapt to his feet, almost knocking over his chair. "I don't know who you are," he exclaimed frankly, "but you are *charming*."

"Really, Piers," complained Lady Perowne, "must you be so uncouth? This is Miss Fabian, Hugo's governess. I was just about to tell you about her. Good morning, Miss Fabian."

"Hugo's governess? Lucky Hugo," cried Mr. Perowne fervently, while I, reflecting that in this matter at least Mary Bowers had already been proved correct, suppressed a smile and curtsied demurely.

"Sit down, Miss Fabian. John will bring your coffee. Piers, behave yourself. Miss Fabian is Mortmain's protégée and if you upset her he will throw you over the church tower and cut off your allowance. Don't forget that he is coming to see you this morning, if you want a curb for your exuberance."

"Forget it?" groaned Mr. Perowne, subsiding into his seat and loading his fork. "How could I forget it with the note you sent by Ashe last night ensuring that I did not sleep a wink—and why do you suppose I am got up as fine as fivepence when you know I had intended to try out Smithers' mare today?"

"I wish you will not use these vulgar expressions—and is not Smithers the man they call Born-To-Be-Hanged?" she added distastefully.

"Ay, and brother to your second footman, ma'am—though I doubt if Thomas Smithers told you his brother's nickname. I must say for you that you keep your ear dashed close to the ground, for all you never set foot outdoors."

"Oh, Piers, do you have to associate with that man?"

"Why, do you think his future may be catching? To answer your question, ma'am, while Smithers breeds the neatest horses in the county, yes I do. I am already incensed enough at his having sold Midnight to Holloway without having to run the risk of losing the mare as well—confound Mortmain!" He turned to me. "Now, what may I pass you, Miss Fabian? Toast and butter? Here you are, and you must try some of Mrs. Crabbe's marmalade. Sugar for your coffee?"

"Thank you, sir."

He looked at me with flattering attention. "How delightful! Your voice is in keeping with your appearance! Do you sing, by any chance?"

"I trust Miss Fabian has many accomplishments," said Lady Perowne rather petulantly, "since she is now a governess. Ah," she added in a very different tone, "here is darling Hugo—I was afraid he might have to spend the day in bed. Come to Mother, my precious lamb."

I looked with interest at my charge, a thin pale child with limp brown hair and light eyes a less brilliant reflection of his uncle's, who stood on the threshold clinging to the hand of the elderly cross-looking woman whom I had seen with Lady Perowne the previous evening.

"Stand up, Miss Fabian," Lady Perowne directed me in mild surprise that I should require it. "Do you not observe that Sir Hugo has come in?"

I rose slowly to my feet.

"See, Hugo," continued his mother fondly as he languidly crossed the room to her side, "here is your new governess who is to make you mind your books and teach you to be a clever boy."

The child cast me a sulky look. "I don't want a governess, Mama," he declared.

"No, my love—but Cousin Mortmain decided it would be best—"

"I hate Cousin Mortmain," he muttered, pressing his face into her shoulder. Lady Perowne stroked his hair. "I know, my darling, but we must do as he says. How do you find Sir Hugo, Nurse? Did he make a good breakfast?"

"I fear not, my lady. A piece of toast and but half his egg—"

"I threw the rest of it on the floor," Sir Hugo contributed with a gleam of animation and a sideways glance to make sure the rest of us were attending to him.

"Oh, naughty Hugo," purred his mother, attempting to wind a lock of his lank hair into a ringlet about her finger.

"But, Mama, Nurse knows I hate boiled eggs."

"I only gave you what the doctor ordered, my duck," said the crone triumphantly, and it was plain that she and the doctor were old enemies.

"I want beef," cried the baronet, catching sight of his uncle's laden plate. "Why can't I have beef, Mama?"

"Oh, Hugo, you know that with your delicate digestion—"

"Oh, give the boy what he wants," suggested Mr. Perowne easily, taking a slice of red meat off his own plate and offering it to the child, who snatched it up and stuffed it into his mouth with a defi nt look at Mrs. Cumber

"Oh, well, perhaps it will do him good," Lady Perowne said philosophically. "Now, Nurse, this is Miss Fabian, whom Lord Mortmain has appointed to be Sir Hugo's governess; and this is Mrs. Cumber, Miss Fabian."

For the first time I met the old woman's eyes directly. They were very light and cloudy and the malevolence they seemed so clearly to express was due more to the creasing of their lids than to the actual orbs, I decided. I held out my hand, but the old woman chose to ignore it, merely inclining her head slightly and remarking untruthfully that she was pleased to meet me, before turning back to Lady Perowne. "Is Sir Hugo to have lessons this morning, then, my lady?"

"Oh, Nurse, I'm sure I don't know. I suppose his lordship and Miss Fabian have it all decided between them and will inform us of their plans as they think fit."

I could see that though Lady Perowne had been affable enough the night before, she was not prepared to defend me in the face of Mrs. Cumber's disapproval. Accordingly I prepared to defend myself. "Lord Mortmain left you to instruct me in my duties, ma'am," I reminded Lady Perowne.

She pushed her plate aside. "How should I know what your duties ought to be?" she demanded plaintively. "We had a governess at home but we spent all our time evading her, so far as I remember. Surely Lord Mortmain gave you some suggestions?"

"He said that I was to be a companion to Sir Hugo," I recalled. "Perhaps I had better begin by taking him out for a walk while we become acquainted."

"A capital suggestion," approved Mr. Perowne, throwing down his napkin. "May I have the pleasure of escorting you, ma'am?"

"You have a previous engagement, sir."

"Oh Lord, so I have—curse that confounded cousin of ours."

Mrs. Cumber said heavily, "So we are to take our orders from miss now, are we, my lady?"

"No, of course not, Nurse; only she must have charge of Sir Hugo in his study hours, I suppose. Oh dear, I believe I have the headache coming on again."

With a murmur of sympathy the old woman moved

behind her and began to smooth her forehead, crooning some rhythmic incantation as she did so. Sir Hugo, feeling no doubt that he had gone without attention for too long, ejected his mouthful of roast beef with a vomiting motion.

"There," cried his mother in accents of despair, "I knew his digestion was not strong enough to stand it."

"It would not have happened in my nursery," declared Mrs. Cumber, snatching up my napkin to wipe her charge's mouth. "You should be ashamed of yourself, Mr. Piers."

Mr. Perowne rose hastily. "Lord, you are a little beast, Hugo. If anyone is to be ashamed, it should be you, sir. That is the last time you may expect anything off my plate, I promise you."

Sir Hugo burst into tears, which Old Nurse smothered against her apron. Mr. Perowne assured his nephew that he did not mean it and left the room without finishing his breakfast. Lady Perowne addressed herself to me. "I suppose you will not be thinking of taking the poor child for a walk after this, Miss Fabian?"

It was the moment to assert myself, if I was ever to do so, and I found the necessary resolution by picturing Lord Mortmain's scorn if I had to report failure to him. I said firmly, "Certainly I shall take Sir Hugo for an airing, ma'am, with your permission. He appears to me to be in need of exercise."

Lady Perowne gave way immediately. "I am sure you need not ask my permission," she said with a suggestion of a sigh. "I am a mere cipher in my own house, it seems. Do just as you please."

Imperceptibly, I let out the breath I had been holding. "Thank you, ma'am. Perhaps Sir Hugo will show me where he keeps his outdoor clothes?"

Sir Hugo stuck out his lower lip. Old Nurse said angrily, "I have the dressing of Sir Hugo, I'll thank you to remember, miss."

"Very well then, Mrs. Cumber." I turned back to my charge. "I shall meet you in the hall, Sir Hugo, in a quarter of an hour, and shall expect you to be dressed for walking." I bowed to Lady Perowne with an assumption of calmness and left the room with my head held high. In the hall I paused to press my shaking hands together and take some deep breaths before climbing the stairs and beginning the attempt to find my room.

I lost my way twice among those rambling corridors before I recognized the oak door opening off the landing on the back stairs, and by the time I had put on my cloak and boots and retraced my steps more successfully Mrs. Cumber had Hugo ready and waiting by the great front door.

I thanked the old nurse and held out my hand to Sir Hugo, who promptly put his own behind his back. "Good," I said cheerfully, having expected this of him. "I thought you might be too old to hold my hand, but I was not quite sure."

Old Nurse, whose hand he had been holding until that moment, gave me a baleful glance and shuffled away, muttering beneath her breath.

Sir Hugo looked up at me, his narrow eyes bright in his pinched face. "She's ill-wishing you," he said spitefully.

I started. Should I ignore what he had said? But any conversation between us, I felt, was better than none and I asked him what he meant.

He answered in a half-whisper that I was ashamed to discover had a chilling effect on me. "Old Nurse is cursing you. It is a spell, I expect. Perhaps you didn't know that she has the power?"

I found I had bundled him out of the front door almost before either of us had realized it. "Don't talk such nonsense," I said in tones borrowed from Mrs. Mason. "Now, shall we turn right or left?"

He picked a long sliver of mossy wood off the railing of the little bridge and dropped it into the moat. "Does it matter?"

"It doesn't matter to me," I admitted, looking from the huddle of stable buildings on the one hand to the drive disappearing under the wet trees on the other. "I haven't walked in either direction and it was too nearly dark to see much last night. Where would you like to go?"

"The quarry," he said quickly, with a sly glance. "I like the quarry."

"Very well, then." No doubt it was forbidden ground but at least it would get us started on our walk and I fancied I could prevent him from falling into it when we got there. "Which way is it?"

"Past the stables. You don't call me Sir Hugo," he observed as we set off down the drive.

"I was waiting for you to call me by my name first."

He reflected for a moment, kicking a stone along ahead of him. "Miss Fabian . . . I never heard that name before. It's not English."

"Indeed it is—at least I believe it arrived in England with the Romans, some eighteen hundred years ago. As for yours, I thought you would not have got used to people calling you Sir Hugo yet."

He considered that. "I am quite used to it, I think. It is my—my proper style."

"Indeed it is, though it seems rather incongruous somehow for a person of your age to be a baronet."

"My father was killed," he announced with another of his expectant sidelong glances.

"Yes, of course, or you would not be Sir Hugo. It was at the beginning of this year, was it not?"

"The sixteenth of January, 1809," he said precisely. "Didn't you know the date of the battle of Corunna? It will be in all the history books one day, Uncle Piers says."

"I must remember the date, then; but I am not very good at dates," I added with deliberate cunning. "I wonder when it was that William the Conqueror invaded England?"

"Everyone knows that," declared Sir Hugo airily, a match for my wiles. He added in that sinister half-whisper, "Everyone knows about the Cumbers, too, don't they, Miss Fabian?"

"Well, I for one know nothing about the Cumbers," I replied rather tartly.

Sir Hugo stopped, his eyes quite round. "You mean you don't know what they are—both men and women?"

"No, sir, I fear I don't."

"Why, they are witches, Miss Fabian—witches! And you must never, never forget it," he added with another of his sly looks.

6

I stared at Sir Hugo blankly, wishing I had not felt a shiver down my spine.

"Mrs. Cumber was not born a Cumber," was the only protest it occurred to me to make.

"Yes, she was. She never married. She says she only ever loved one man and he was wild and free and made his own laws—and in the end he danced on the stars. She is called 'Mrs.' from respect because people are afraid of her."

"I am sure you are exaggerating."

"I'm not, I'm not," he cried furiously. "She is a witch—she makes potions and sells them to the maids—and she puts spells on people and they work. That is why Cousin Mortmain had his accident."

"Oh, Hugo, it isn't true!"

"It is—and you must call me Sir Hugo or I shall tell Mama."

I pulled him to the side of the drive as a groom rode out of the stables on a brown mare with another horse on an exercising rein. "I will call you Sir Hugo when I feel respect for you but not so long as you behave like a silly little boy."

"I'm not!" he shouted, stamping his foot.

"I knew a boy once," I remarked, "who I am sure was never silly, though he was three or four years older than you when I met him first. He had a horse of his own, not just a pony but a real horse. It was rather like that one the groom was leading, only black, with a long flowing mane and tail—"

"A blood horse," muttered Hugo, kicking his stone into the trees.

"It was partly Arab, he told me."

"All the best horses have Arab blood," he said scornfully. "Didn't you know that?"

"No, but I am glad to have learned it."

He seemed mollified. "We go through the wood here, up this path. Go on about that boy."

"Oh, that boy . . . he could make boats out of wood."

"I can do that."

I stopped, for this was an important admission. "Can you really, Sir Hugo? Would you show me one?"

But that was going too fast for him and he shook his head. We walked on. He had a curious mode of progression, skipping every third step or so and touching certain trees as if it were a ritual. Sometimes he muttered, as if he were talking to an invisible companion. We came to a stream and he peered silently into it for a few moments before leaping up again. The path wound out of the trees and we began to cross a chalky down. I could see a village nestling in the valley—Lindenfold, I assumed.

Hugo came close and holding out his finger quite straight, jabbed it into my arm. "Go on," he repeated imperiously.

"You should say please . . . well, that boy had a sharp knife he used to carve his boats. Sometimes it seemed to me he spent more time sharpening the knife than he did in using it—"

"Did he keep it in a white silk cloth with herbs?"

"N-no," I said, startled as much by the whisper as by the unexpectedness of the words. "Why do you say that?"

"I found one once like that. It was a long time ago." He shook his head as if to dispel the ancient memory. "Go on."

I considered and rejected another attempt to teach

Sir Hugo manners. For the moment, his interest in my story must suffice me as a reminder that we had made some progress this morning. "That boy, then, took a long time to make each boat. Each one had paper sails, mains and mizzens, staysails, top-gallants and royals, all the proper shapes and sizes. He used to copy them from books. On two occasions—I think it was only two but it is hard to be sure, it was all so long ago and my life changed so much afterwards—he took me with him to a lake to watch him sail them. The last time we were there one of his boats tipped over sideways and filled up with water and sank. The boy—I never knew his name, I used to call him Rowley, I don't know why—took off his coat and kicked off his shoes—very shiny, they were, with silver buckles—and jumped into the lake—" I reflected that this was not a very improving tale to be relating to my charge but it was too late to leave it there.

"I was only a very little girl," I went on. "Only four or five, I suppose, and I was very frightened. I had nearly been drowned myself not long before when only my skirts had kept me afloat . . . I called to Rowley to come back, but he could swim and took no notice of me. He swam to the place where his boat had sunk and he dived down until I couldn't see him. I cried and started to go into the water—"

"You couldn't swim, could you?"

"No, but I didn't pause to reflect. I was very willful and impulsive then. . . . I saw the boy come up at last and he had a piece of root in his hand. He went down again and this time he stayed down even longer. I started to struggle towards the place just as he came up with the boat—and then he saw me and dropped the boat after all his trouble and came flying through the water and picked me up and scolded me till I cried, and wrapped me in his coat and carried me all the way home. I don't know why I remember it so well. It is one of my earliest memories."

After a moment Hugo said, "What happened to the other one?"

"He—he was in the reeds. I suppose he went home."

"I meant the other boat," said Hugo with a puzzled look.

"Oh! The other boats had sailed right across the lake and were safe against the bank. Rowley must have gone back for them later, but he never let me go again to the lake to watch them sail."

"You sound sad, Miss Fabian."

Such a remark, from him, was encouraging, I felt. "As you know, Sir Hugo, it is always sad when you lose people you love. I cannot remember which was the last time I saw that boy, but after a while he went away and I never saw him again."

"Why don't you look for him?"

"I used to believe I would do so, when I was grown, or that he would look for me, but now I realize these things are not so simple—"

"Ooh, look, there are cows over there by the stile. I don't like them, do you? I don't like things with horns."

"Why do you keep whispering Hugo?"

"Sometimes . . . it is safer t whisper."

"But why?"

"So—they—don't hear you."

I was afraid of losing his confidence and did not pursue the subject. Presently we came up to a broken railing that edged the quarry. There was a path down into it but the chalk was smooth and slippery after the rain. There were thickets of brambles, half a broken waggon and something that looked like the carcase of a sheep.

"It is my secret place," Hugo declared proudly. "There are blackberries down there."

"Not at this time of the year, alas. But perhaps we could have a picnic here when the weather is warmer." I could imagine how unpleasant such a picnic would be, the airless dell, the flies buzzing, wasps fighting for our apple cores.

"I think we had better go back, Sir Hugo. It looks as if it is going to rain again."

He looked up and screwed his eyes shut, sniffing like a dog. "The wind is off the sea. Uncle Piers says he can smell it on a southeast wind—the sea, and France behind that."

"I was born in France."

His face went red. "You said you were English," he cried accusingly.

"My mother was French."

"I hate the French," he shouted. "They killed my father!"

"And the Thames killed both my parents," I retorted quickly. "That doesn't mean I can't bear the sight of water, I hope, or I should be very thirsty."

"You wouldn't be able to wash your face," he agreed, won back to good humour. "Yes, let's go, before the cows come. Look, there's an egg. If I collected birds' eggs—"

"Don't you?" I said in surprise. "I thought all boys did so."

"Uncle Piers has a collection—he said he had to climb trees to get some of them and come down with the eggs in his mouth." He shuddered. "I even hate cooked eggs. I'm not allowed to climb trees anyway."

"I can climb trees but I don't enjoy it much. Look at that cloud. I think we had better run." I spread my arms like a bird and ran up the path before the wind. Sir Hugo followed more sedately. He was not pretending to be a bird, I realized, glancing back at him trotting solemnly along. He was pretending to be a pony. "Do you like horses?" I asked.

"Some of them. I don't like Snapdragon."

"Who is that? Your pony?"

"Yes, but I'm not really strong enough to ride, Mama says."

"If we go for walks every day you will get hungry and then you will eat and grow strong. Would you like to ride then?"

He frowned. "Grandfather Carlingford says I must

learn to gallop and jump big fences like a man—the sooner the better."

"Well, I have no ambition to gallop and jump big fences but I should certainly like to be able to ride, to walk and trot and canter."

"Oh, I can walk and trot already. I can't canter yet."

"Perhaps we can arrange to be taught—or you can, at least," I added, belatedly remembering my position.

"I shan't learn unless you do it too."

"Why, thank you, sir. If you would like to collect eggs, I will climb the trees for you," I offered, thinking that he was turning out to be more agreeable than I had feared at first.

He studied me gravely. "I don't think your mouth is large enough for an egg to fit in."

"It certainly is," I cried indignantly, and then the absurdity of our argument struck us and we both laughed. The next instant a light rain began to fall and we raced through the wood, up the drive, and into the house.

As we stood panting in the Great Hall, Hugo noticed a coat and stick laid across a carved oak chair. "Mr. Pole-Carter is here," he announced sulkily, his face reverting to its original peevish lines. "I will go up to Old Nurse. She hates him, too—he is forever asking questions."

I put out a hand to restrain him but he twisted away and ran up the stairs. Percival, the elderly butler, tottered towards me. There was a curious mouldy smell about him. It was not his ancient greenish coat, for that emitted a strong odour of camphor. It was his boots, perhaps.

"My lady would like you to go up to the parlour, miss."

"Very well, Percival." I took off my cloak and neatened my hair before the mirror. The glass had been dusted this morning and it might have been reflecting a different girl, I thought, so much confidence had I derived from having made Sir Hugo laugh.

I went upstairs to the parlour where a cheerful middle-aged clergyman was in conversation with Lady Perowne. He had the face of a Roman emperor on an old coin, I thought as Lady Perowne introduced us: one of those profiles that looks blown by a high wind, hair flowing back off a receding brow, nose jutting aggressively, and chin sloping away like the forehead.

"And how is my darling little Hugo?" asked Lady Perowne when Mr. Pole-Carter had acknowledged the introduction.

"Very well indeed, ma'am. We had a most enjoyable walk."

"I am so glad." But she was frowning. "Those flowers," she exclaimed. "Just look at that arrangement! Only the veriest clod could have thought of putting those blooms in that vase." She rose gracefully and crossed the room to do what she could to mitigate the floral disaster.

The vicar, after a tolerant smile at this evidence of Lady Perowne's sensibility, turned back to me. "You went walking, ma'am? I hope you did not get a wetting. Where did you go?"

"We—we walked through the meadows, sir. There was a very pretty view of the village and the sea beyond."

"Ah, you must have seen the church, then," he exclaimed. "It is mostly Norman, as you no doubt observed, and we have a famous sixteenth-century bell named Clap Hammer, which you must be sure to listen for. However, my particular delight is in the gargoyles, the symbols of the Old Religion in the form of goats and devils, carved so cunningly among the saints. You must look out for them on Sunday, Miss—er, Fabian. Of course, this house is closely linked with the church and with that older faith also, I sometimes feel."

"Do you mean an older faith than Roman Catholicism, sir?" I said, staring a little.

"Ay, the Old Religion—worship of the old horned god of hunting, Herne, and the moon goddess, his bride, in her various guises."

I caught my breath. Instantly a distant memory of my

guardian reading from old Roman mythology merged with the more recent one of the conversation I had overheard. "Hecate, sir?" I hazarded.

He smiled widely, full lips flattening over the large teeth. "Ah, she is only the goddess of the dark moon. Selene is the one I had in mind, Queen of the Esbat and of the full moon. Artemis rules the new moon, of course. But it is delightful to find a young person interested in such things."

"I have—heard a little," I admitted truthfully enough. "What is an Esbat, sir?"

"Why, they are Selene's feasts, celebrated at full moon when it falls between the ritual Sabbats—you have heard of witches' Sabbats of course, Miss Fabian, Hallow E'en, Candlemas, Beltane?"

"I have, sir, I believe. Witchcraft is much associated with Lindenfold, is it not?"

"Ay, but more particularly with the Parish of All Saints, some way to the east of us here. There is a deserted churchyard there, the very place for such secret rites. However, you do not need to travel so far in order to appreciate the full flavour of the thing. The old chalk quarry, which you can see from this very window, Miss Fabian, has magical associations . . . thorns, a pit, and water . . . I dare say it has seen many supernatural ceremonies in its time. Then the very nearest of the Downs is known as Holly Hill, which must have derived from Hollow Hill, the traditional dwelling place of Herne. Holly Wood, on the other hand, I am certain must come from Holy Wood, for it contains three great stones, the center one of which was surely a pagan altar, the scene, no doubt, of many a human sacrifice. As for this wonderfully antique house," he continued in a lower tone, with an anxious glance in the direction of Lady Perowne as if to ensure that her sensibilities were not shocked, "who knows but that Holy Mote was actually the very center once—"

He broke off as the door burst open and Hugo launched himself upon his mother, sobbing incoherently.

"There, there, my pet," she cried, dropping her flowers and putting her arms about him. "What is it, darling?"

"They said he wasn't there," Hugo gasped. "But he was there, he was—the foxy man!"

"What man, Hugo? Calm yourself, darling."

Angrily he dashed the tears from his eyes. "Cumber's man," he muttered, finding he had a larger audience than he had expected.

Lady Perowne looked up helplessly. "But if Mrs. Cumber said he wasn't there—"

A board creaked. I turned round. Old Nurse was standing just within the door, her hands clasped beneath her apron, her lined face calm and patient.

"What is all this, Nurse? Sir Hugo seems sadly upset."

"He has been a naughty boy, my lady, and has frightened himself, telling fibs."

Hugo sucked in a long breath and discharged it in a ringing yell. "I wasn't telling fibs, he was there, you know he was."

"Was he, Nurse—whoever this person is?"

"Of course not, my lady. Sir Hugo came upon me where he is not supposed to be, in the little sewing room I have upstairs. I spoke sharply to him, for you know the floors are rotten in many parts of the attics and he is not allowed up there; and that was when he pretended he saw a man."

"You see, Hugo, you must have been mistaken."

"Don't believe her," he whispered. "Uncle Piers was there, he knows."

"There was no one there, my lady," Mrs. Cumber said with finality. "Mr. Piers was running after Sir Hugo, to stop him, and was in the room almost as soon as he screamed. He can vouch for it that there was no one else in there with me."

"Nurse is quite right," said Piers Perowne from the doorway behind her. He stepped past her into the room. The only signs of his meeting with Lord Mortmain, I thought, were that his hair was slightly more disordered and his eyes restless and bright.

"I was in the sewing room almost as soon as Hugo," he said. "And there was no one there—no one but Nurse, of course."

There was a moment's silence filled only with the soft crackling of the fire, the hissing of the hot sap; and the raindrops blowing against the leaded windows.

Sir Hugo leaned forward. His small hand pointed at his nurse with a dreadful intensity. His whisper when it came must have been clearly audible to us all. "You conjured him," he said.

Into the shocked pause that followed came the distinctive sound of Lord Mortmain's limp.

He stood in the doorway surveying the scene with a raised eyebrow and mouth slightly twisted as if in scorn. "Good morning, Cousin Adela; Miss Fabian; Pole-Carter. What ails you all? You remind me of nothing so much as a tableau of Mrs. Salmon's Waxworks."

Hugo whimpered and ran back to his mother's side. Mr. Perowne shrugged and went over to the fire. It was left to the vicar to make some sort of explanation.

"Most interesting," he cried, rubbing his hands together with an unpleasing rustling sound. "Most interesting. The boy claims he saw a—an apparition, sir, and the two adult witnesses deny it. Furthermore, by way of accounting for the phenomenon, he accuses Mrs. Cumber—"

"Sir Hugo is overwrought," the old nurse interrupted sharply. "He vomited at breakfast and then was made to walk I don't know how far in the rain. 'Tis no wonder he's delirious, if you ask me, my lady. If the doctor has to be called in I shall have nothing to reproach myself with. Orders is orders, and if inexperienced persons are set above us, then those that set them there must abide by the consequences."

"That will do, Nurse," said Lord Mortmain quietly. "I am sure her ladyship will excuse Sir Hugo if you wish to remove him to bed with a hot brick, or whatever you think best." He looked steadily at Mr. Pole-Carter, who started and hastened to make his farewells.

"Oh dear," exclaimed Lady Perowne, when the three of them had gone. "I am sure this kind of upset cannot be good for my nerves, which were already upon the stretch knowing you were taking Piers to task, Cousin. Poor little Hugo. Nurse seems to think Miss Fabian is to blame, Mortmain."

"Indeed?" said the Earl coolly. "And what has Miss Fabian to say to that?"

I replied in a low voice that I had merely taken Sir Hugo walking. "We were making each other's acquaintance, sir, and—and went some way towards friendship, I hope. Surely it could not have been that which upset him?"

"He is so sensitive," said Lady Perowne rather uncertainly, as if she were not sure whether this should be occasion for pride or regret. "You did not realize, perhaps. Did you tell him fairy stories, work on his imagination—"

I gave a guilty start. I had remembered telling Sir Hugo about the boy Rowley and the summer lake, the foxy face among the reeds. The Earl appeared to have no difficulty in reading my burning face. He said curtly, "Perhaps you will confine yourself to more factual matters in the future."

I stammered some such assurance, relieved that there was to be a future, and receiving Lady Perowne's dismissal, hurried to leave the room.

7

Once I was alone, walking down the gallery between the uncaring well-bred painted stares of all those long-dead Perownes, my self-confidence began to return. It was unfair that I should be so blamed for Hugo's experience, I thought; foolish that I should have blamed myself. Only an extremely unbalanced child could have been frightened into delusions by what I had told him. Besides, that story was what had persuaded Hugo to begin to put his confidence in me.

I heard footsteps hurrying after me. Glancing around, I saw Mr. Perowne, dishevelled, handsome, smiling charmingly.

"Miss Fabian, one moment—I must speak to you. In here, the Music Room."

He gave me no chance to demur but half-pushed me into a panelled room overlooking a green lawn, a shrubbery beyond, and to the right an orchard of grey twisted apple trees standing in drifts of egg-yellow daffodils. The room itself seemed haunted by the ghost of music, with a piano under holland covers, a shrouded harp, and cases of yellowed scores drying to dust behind cobwebbed glass.

I began to make some sort of protest, but Mr. Perowne interrupted me. "Poor Miss Fabian, I know you are upset though I am sure you have no need to feel guilty, and are probably longing to retire to your room—but I am sure you are the very person to help me. My nephew tells me you are half French. Can it be true?"

I looked at him warily, before inclining my head.

"Astonishing!" he cried. "I should not have thought my sister-in-law would have had you in the house, even though you are certainly a Royalist."

"My mother was French. I know nothing of her politics, I am afraid."

He brushed that aside. "You speak the language?"

"Yes, sir."

"Capital! I had hoped that as a governess you might know a little, but to find that it is your native tongue! You do read it also, I trust," he added rather sharply. "Could you translate a letter for me?"

"From the French? Certainly, sir, if it is legible."

"Ah, bless you." I felt warmly conscious of his approval and it came as something of a shock when his eyes narrowed, his mouth thinned and he added coldly, "But it is the most deadly secret and if you ever breathe a word of it you may be sure I shall creep into your room one night and slit your pretty throat from ear to ear."

He sounded remarkably serious and I stared at him apprehensively, wondering if I ought now to refuse to help him. But I did not want to arouse his displeasure, and said meekly that I would not mention it.

I could see the tension begin to leave him. He said more lightly, "I should not care for it to come to Adela's ears that I am corresponding with a French girl."

"You may be sure I shall not tell her ladyship of it, sir."

He smiled a little mockingly. He was undeniably attractive, I thought, and I wondered if I should have lost my heart to him if Mary had not given me her kindly warning. But then, I thought, such warnings probably counted for nothing if two persons had the impulse to fall in love. No, the fact was that Mr. Perowne was not quite my ideal and for that, no doubt, I should thank God fasting.

"You sound very cool, Miss Fabian," he said, as if he

could not bear to think of even one small fish escaping from his net. "You need not be jealous of Marie. There is nothing serious about my relationship with her—besides, she is not half so pretty as you."

I said even more coldly, "Have you the letter with you, sir?"

"Ah, the letter." He put a hand to his inner pocket and pulled out a crumpled piece of yellowed paper. It was ill-written, I saw, crossed with spidery writing and obscured with splutters of brownish ink. "I have muddled through most of it well enough but it seems a long time since I studied French and there is still a phrase or two that has baffled us—me, that is to say. What does '*Comme d'habitude*' mean?"

" 'As usual,' sir."

"Ah, I should say that sounds promising, should not you?" He frowned at the letter and then folded it carefully. "What is that phrase, there at the top?"

I leaned forward. "I can't see properly. Let me have the letter a moment, sir."

"No," he said peremptorily. "There are certain passages I prefer you not to see. Just tell me what it says there, where it begins, '*A côté de la falaise.*' "

I peered at it. " 'Beside the cliff, where the cleft comes down—' "

He snatched it out of my sight and studied it, frowning. "Yes, I see now. And this, about Yves?"

" '*Yves est toujours malade . . .*' 'Yves is still ill, it is always the same thing, no one to depend upon; however I will do what I can for you until Pierre the One-Eyed returns. I am sorry I have no English but—' " It was certainly a curious love letter, I reflected as again Mr. Perowne jerked his hand away, to stop me reading further.

"Enough of that." He folded the paper again. "And here?"

" 'Pierre'—Could you hold it a little closer, sir? Thank

you. 'Pierre will be at least another fortnight but he is not wasting his time—' "

"Very good, Miss Fabian. I am extremely obliged to you." Mr. Perowne raised the letter to his lips in a theatrical gesture. "*Ma chère Marie* . . . not that I am seriously interested in the wench; indeed, if she doesn't soon learn English I shall be forced to cut the connection. But it is tantalizing to receive a letter one cannot completely understand and I am glad you have made it clear to me. But not a word, remember. Adela would be furious, she would never understand my disloyalty, my brother having been killed by the French, you know. You will remember that you said you would not mention this—to anyone?"

His very intensity made me hesitate. "What if I did?" I murmured, aware of a novel sensation of power. "Lady Perowne cannot prevent your meeting whom you please —and you say you are not serious."

He smiled, but the smile was strained. "You need not concern yourself with my reasoning, Miss Fabian. Just rest assured that the consequences would be serious—for you."

"I am not likely to forget that you threatened to come to my room and—how do you know which is my room, by the way?"

"Why—why, it is the one the governess always has —the Stair Room."

A strange certainty came to me. "Mr. Perowne, it was you I heard last night, talking to some other man on the stairs outside my room."

His eyelids flickered. "I assure you, you are mistaken, ma'am."

"Just as Sir Hugo was mistaken today? There was another man with you last night, just as he thought there was with Old Nurse half an hour ago. Cumber's Familiar, I remember that you called him. A familiar spirit, I presume you mean. Did she conjure him indeed?"

"Enough, Miss Fabian," he said dangerously. "You

sound as overwrought as Hugo. Did my cousin Mortmain set you here to spy on me, by any chance?"

Now his face was not attractive at all. I licked my dry lips and hoped he could not hear my heart announcing to the world how frightened I was. "Does that sound like his lordship's way?" I asked, wishing I had not thought it might be instructive to play with fire.

"It might be," he said slowly. "You are damnably pretty and it is not usual to seek for prettiness in a governess." Suddenly he smiled and seizing my hand, pulled me towards him. "Unless you are his honey-bee?" he suggested softly.

"Lord Mortmain's?" I cried in a startled voice. "Are you mad? And kindly release me, sir."

"What the Devil goes on here?" demanded the Earl in icy tones behind me. Mr. Perowne stood back with a wink and a rueful look at me, as if we had been interrupted in the middle of a scene of dalliance, I thought angrily.

"Get out, Perowne," commanded Lord Mortmain. "I'll deal with you later."

He must have heard that last snatch of conversation, I thought unhappily, as Mr. Perowne gave an insolent shrug and left the room. I did not want the Earl to think I had any ambition to be his honey-bee, odious phrase; but at the same time I supposed I need not have protested quite so violently at the notion.

"I did not bring you to Holy Mote so that you might waste your time in flirting with Perowne," the Earl informed me coldly. "Innocent you may be, but have you no more sense than to allow yourself to be alone with such an accomplished philanderer?"

I shook my head. "You are mistaken, sir," I said in a low voice. "It was not what you suppose. Mr. Perowne had a trivial request to make of me. . . ."

The Earl regarded me rather grimly. "I wish to Heaven you were older," he declared.

I felt this was unfair of him, considering that Mrs.

Mason had done all she knew to try to dissuade him from employing me on that very score. "You knew my age when you appointed me, sir," I reminded him with what I felt to be truly adult dignity.

"True," he owned, his anger seeming to abate. "I dare say time will remedy the fault. I was coming after you to say that I shall be pleased to frank any letter you may wish to send. If you leave it on the salver in the hall, it shall be attended to with the rest."

As I had already intended to spend the afternoon in writing to Mr. Shorncliffe and as I was reluctant to burden him with the postal dues which the receipt of my letter would otherwise have entailed, I was obliged to accept his lordship's offer with good grace. He acknowledged my thanks with his twisted smile, looking as if he knew very well how much it cost me to profess myself in his debt; and I relieved my feelings by hurrying away in a manner which would leave him in no doubt of my eagerness to escape from him.

The next day Lady Perowne informed me that Sir Hugo was somewhat indisposed, and sent me to the garden to gather flowers for one of her arrangements. I was happy enough to postpone my first visit to the schoolroom, until, as I was walking back to the house across the lawn, I was suddenly aware of a face at one of the attic windows, a man, I rather thought. He stayed there, staring down at me, and I felt curiously uncomfortable under that steady gaze. Could Sir Hugo have been right after all about seeing a man in the attic? Could it be "Cumber's Familiar," whose voice I had heard myself? Should I report it to Lady Perowne? But she would not thank me for reviving the subject, I guessed; and besides, it might merely be one of the servants, wondering to see a governess so enjoyably employed at this hour of the day. Nevertheless, whoever it was had taken some of my pleasure from the task and I was glad to enter the house again and know myself invisible to him.

The following day Hugo was pronounced to be recovered, which was fortunate as it enabled Old Nurse to transfer her attentions to Lady Perowne, who was prostrated by one of her frequent headaches. The rain which had fallen the previous night had stopped, the sun was shining and small clouds were racing across the bluest of skies, so I was delighted when Lady Perowne sent me a list of one or two things she needed from the village shop, instead of instructing me to begin lessons with her son.

Sir Hugo was pale and heavy-eyed but soon revived under the influence of such a day enough to take up his curious skipping walk as we went into the wood, which was all slanting dancing light, rippling from the green leaves to the green grass. We paused by the stream to look at the flowers starring its banks, aconites, primroses, frail wood anemones, and scented violets all blooming together; and our conversation was all of such things, and why spring should make one feel like running and dancing. We laughed a good deal as we crossed the fields and climbed the stile. A few yards farther on we came to the lane and here Hugo suddenly stopped and clutched my arm.

"Born-To-Be-Hanged," he hissed in that chilling half-whisper of his.

There was nothing out of the ordinary to be seen except a waggon trundling towards us with wide wheels and a canvas hood, drawn by six horses and accompanied by a carter, a pleasant-looking fellow with a round red face and straw-pale hair.

"That—man," whispered Hugo, shrinking against me as the carter cracked his whip and then, seeing us, touched his hat and called out that he had just been delivering to Holy Mote.

"What is wrong?" I asked as the waggon passed us and turned the corner. "What did you call him?"

"Born-To-Be-Hanged—and Nurse says it is true and that the shadow of the gallows hangs over him."

I remembered now that he had been mentioned at breakfast on my first morning at Holy Mote. "But who is he?"

"The brother of Thomas Smithers, our second footman. He's a carter and horse-breeder," Sir Hugo explained more cheerfully now that the waggon was out of sight. "He brought Mama's harp from London a long time ago, when Papa was in Egypt. That was when he met Uncle Piers again, and now they are friends."

"Well, that is a little unusual, certainly—but what is there in that to make you so afraid of the carter?"

"He is one of them—one of the witches—" I shook my head, but Hugo rushed on, "He is! I saw him once with an athame—"

"And what is that, pray?"

"A witch's knife—the one they keep in the silk cloth. He looks so nice, too," Hugo added with a shudder.

"It is his nickname that troubles you . . . but, Hugo, reflect that people are not hanged for being witches."

"He will hang one day," Hugo said with absolute conviction. "It must be for something else. Perhaps he is a thief. Look, Miss Fabian," he cried, waving to a well-dressed man who drove past in a smart gig, "that is Mr. Curtis, the master of foxhounds."

"Indeed," I murmured, glad that he had been distracted.

"Lady Martha is his wife—she is nice and gives me comfits that she keeps in a little silver box, and Miss Curtis draws pictures for me on my slate—she is Mama's greatest friend, but she hates Uncle Piers. There is the village shop," he added, taking my hand. "Shall we go in? You can buy great peppermints there, and string, and oil, and candlesticks—oh, anything, indeed."

We passed the village pond and went up a sunken path between the daffodils. Hugo pushed open the blistered green-painted door. It was rather dark within, the light struggling inadequately through the thick bottle-glass window panes, and I stood for a moment waiting for my eyes to become accustomed.

"Another half-anker last night," a comfortable voice was saying. "Well, I believe I do deserve it, for I should hope I know how to hold my tongue after all this time. Methods may change, my dear, but the principle remains the same, as the Reverend Pole-Carter tells us."

The shop smelt of cheap tallow and tarred rope, I thought: like a ship chandlery.

"Oh, you are discreet, no question," agreed a thinner, sycophantic voice on the far side of a roll of fishing nets. "I was wondering about the gentleman at the Pines. Do you reckon he's in it?"

Yes, I could smell the oil now, that Hugo had mentioned; and another smell, like ink.

"Well now, my dear, that's hard to tell. Keeps himself to himself he does, and has done since he came to these parts; and old Mrs. Randall as cooks and cleans for him is as deaf as a post, as you know, for all her charitable work. One thing I can tell you, he wears a wig."

"A wig, Mrs. Hobden? Well, I must say I thought hair so dark and neat couldn't be natural, but then he's that close I didn't reckon he'd be willing to pay the guinea tax—"

"No more did I—but Mrs. Randall caught him putting it on the other day—powerful cross he was—"

"Bless me!" cried Sir Hugo. "That must be Mr. Holloway!" He chuckled at the notion of Mr. Holloway wearing a wig, while agitated gasps and rustlings came from beyond the fishing nets. "Hello, Mrs. Hobden. Good morning, Mrs. Butcher. Have you any of those big peppermints today? Miss Fabian is going to buy me one—or two," he added hopefully.

"Well, I declare," said the comfortable voice rather less comfortably, appearing in the person of a short, stout and breathless woman on the far side of the wooden counter, which was polished to a high gloss by innumerable arms having leaned upon it.

"Such a start as you gave us, Master Hugo—Sir Hugo, I should say. We never heard the door—Hobden oiled the hinges last week and a dratted nuisance that is." Her little eyes slid sideways to determine what impression her remarks had made on me, I supposed, and I endeavoured to look as though I had not heard her, even if Sir Hugo had done so.

It seemed that I had succeeded, for she continued in an easier tone, "Peppermints, sir? Now, where did I see that green jar? Just behind you, Mrs. Butcher—ay, that one, if you would be so good as to pass it this way. Here you are, sir—cooked yesterday. Was it one or two as you were wanting?"

We completed the transaction and I bought Lady Perowne her paper of pins and stick of sealing wax. A moment later, as we left the shop, the gossip broke out again behind us. "They do say, my dear, that his lordship—"

"Don't let us go home yet," said Hugo coaxingly, his cheek distended by the peppermint. "Old Nurse goes out this afternoon to visit her mother—"

"Her mother!" I repeated incredulously.

"Old Granny Cumber—she's ninety-two and they carry her about in a basket," said Hugo with relish. "We can visit her one day, if you like. She sleeps most of the time, but it is nice when she talks. She can remember everything—she was taken to see Jonathan Wild's execution in 1725, only fancy! Let's go up on the Downs and then by the time we get home, Nurse will have gone out."

"Don't you like her, Hugo?" I asked involuntarily.

"To be sure I do," he said dutifully. "But she frightens me sometimes . . . I have to do whatever she tells me even if I know Mama would not—anyway, it is more comfortable when she is out. Besides, I spend the afternoon with Mama, and she reads to me. This way, Miss Fabian, through the gateway."

He led me up a rutted track which soon dwindled to a sheep path as we began to climb the hill—Hollow Hill, I remembered Mr. Pole-Carter telling me, had been its original name, abode of Herne. . . . I narrowed my eyes looking up towards the sun. "Shall we explore that little wood right on the top of the hill?"

"That's Holly Wood," he mumbled, moving his peppermint on to the other side. "I don't like it there."

"Oh, do let us. I want to explore it and I need you to show me the way. Oh look, I had not realized the sea was so close to Lindenfold. How blue it is!"

"It reflects the sky," he said knowledgeably. "How many ships can you see out there? I expect those are merchantmen over on the horizon. Those are two Royal Navy frigates, of course; and the nearest is a Revenue cutter—they often come close inshore. Do you like sailing?"

"I have never even seen the sea before coming here—at least I came from France when I was about two so I have, of course, but I don't recall anything about it. Have you ever sailed?"

"Yes, of course—I used to go with Papa, and Uncle Piers takes me out sometimes."

We were climbing up the sheep tracks that criss-crossed over the short turf and the sun began to feel hot on our faces.

"There will be a wind at the top," said Hugo. "A south wind. I wish we had something to drink, don't you?"

"Perhaps there is a spring up there."

"No, but there might be a dew-pond. I could drink a barrelful—or a half-anker anyway, like Mrs. Hobden—five gallons," he added, suspecting my ignorance. "Look, you can see the quarry from here—and Lindenfold House. Cousin Mortmain has the builders in. Old Nurse says that means he is thinking of getting married."

I stared along the valley, inland from the village, and descried a large stone house in the Palladian style, set in a green estate ringed by a grey wall. A network of scaffolding veiled the outlines of one wing and figures moved like ants upon it.

"If there were a wedding," Hugo mused, "would I get a present?"

"Certainly not," I said sharply. "Come, let us go on. My legs ache and I want to get to the top."

It was not really very high, I realized, as our path straightened out along the shoulder of the Down. The square church tower did not seem very far below and I felt I could almost toss a stone into the chalk-quarry. I walked on, with Hugo skipping and shuffling behind me, into the shadow of the wood that clothed the hill crest.

I had been looking forward to the coolness and shade but the trees cut off the wind and it was rather stuffy. The track was dusty in spite of last night's rain, protected by the leaves above, and it seemed to have been tumbled by innumerable hooves. Twigs were broken all along both sides of the track, as if many horsemen had ridden that way. Perhaps it was they who had frightened away the birds, I thought, for the wood was strangely quiet.

I stood still. A little cloud of dancing midges began to form about me. I realized that I could hear nothing but the humming of their wings and my own breathing—not the crackle of a twig or so much as the rustle of a last year's leaf.

I spun round. "Hugo!" I cried, but he was nowhere to be seen. The silence persisted and I wished I had not disturbed it. Sir Hugo had said he did not like the wood, I reminded myself. He must have turned back as soon as I had entered it, and would be making his own way home, no doubt. I had only to retrace my steps and follow him down, but somehow I found myself unable to do so. Mr. Pole-Carter had been right when he promised me that I had only to go to Holly Wood to appreciate the full flavour of witchcraft. I felt as if I had been laid under a spell—a spell that urged me to go on and find the place where once human sacrifices had been made to pagan gods.

After a little while I could see sunlight through the trees ahead. The clearing must lie there, and I quickened my steps. As I walked out of the shadows I realized that I was no longer alone and stopped abruptly.

A man knelt in the glade with his back to me, before a great slab of stone sunk in the grass, the center one of three. He must have heard my gasp, for he leapt to his feet and whirled about to face me, and the sunlight ran like flame along the bright blade of the knife he held.

It was Mr. Piers Perowne.

8

"Miss Fabian!" exclaimed Mr. Perowne. Then, seeming to regain his self-possession, he added more calmly, "What in the world are you doing here?"

"I—I came with Sir Hugo, for a walk."

"Hugo? Then where is he? Besides, he knows he must not come here." Mr. Perowne glanced at the knife in his hand as if surprised to see it. He opened his coat and slid it into a scabbard fastened to a belt about his waist. I began to feel a little braver.

"Why not, Mr. Perowne?"

He stared coolly down at me, his light eyes shaded by the brim of his tall hat. "How does that concern you, Miss Fabian?"

I looked over my shoulder. The trees stood round us like a ring of silent enemies, watching, waiting. "Sir Hugo doesn't like this place. I don't myself. And Mr. Pole-Carter warned me . . . but if there is a real reason why one should not come here, I think I should know of it."

"There are always reasons for places being shunned by animals and children," he said evenly. "I could not say whether you would consider the reason valid. This is a very old and—holy wood. Even the Romans dared not burn it when they stripped the hills around to feed their fires. . . ."

While he was speaking he was leading me to the edge of the glade and under the trees again. "Here is your path,

Miss Fabian," he said with finality. "You will be able to find your own way down, I trust."

I thanked him briefly for his escort and hurried down the sheep-track. Clouds had covered the sun and the wind was cold. I felt no temptation to linger by the quarry when eventually I reached it and, hurrying past, I nearly missed Hugo, who was just climbing out of it.

"I saw Old Nurse," he announced with one of his sly looks. "I saw her, and she didn't see me until I laughed. She was like this." He pressed his left hand into his stomach. "She was saying *Adam betam alam betum*, or words like that, over and over. Then she touched the ground and spat, and started again. Poor Old Nurse, she has a stomach-ache."

"Is that what she does for it?" I told myself I should soon have lost the power of being surprised, if I stayed much longer at Holy Mote.

"Oh yes, she never sees a doctor. She can cure herself. She has the power."

He was frowning and his hands were twisting together as he skipped along beside me. I said sympathetically, "Was Mrs. Cumber cross with you for laughing?"

"No. She said that I must do . . ." his voice trailed away.

"Do what, Hugo?"

"I don't remember the word. But she meant, do something to make up."

I glanced at his pale strained face. "Are there any other boys and girls you are allowed to play with in the neighbourhood?" I asked abruptly.

He looked sulky immediately. "I don't like playing with other children—and Nurse doesn't like me to."

"I was wondering if there was not some child who might share your lessons. It is more interesting to learn in company. Well, we will leave it for the moment," I reassured him, noticing the extreme disapproval with which he regarded this idea. It was what he needed, however:

simple lessons to give him confidence, and the companionship of children his own age to offset the influence on him of an ancient twilit house, a mother in mourning who protected him too much, and an old nurse who believed she was a witch.

When we had nearly reached Holy Mote, Hugo muttered some excuse and ran ahead of me. I followed more slowly, thinking that it was pleasant to know Mrs. Cumber was out of the house even for so brief a time.

The hall seemed very quiet when I entered it. There were no servants in sight and the only sound was the ponderous ticking of the grandfather clock. I went up to the gallery, unfastening my cloak and the ribbons of my bonnet. Suddenly Hugo sprang out at me from behind a carved Jacobean chair. His face was white, his eyes glittering. I started back with a gasp of surprise. He lifted his hands and I felt a sudden tug at my hair.

"Hugo!" I cried. "What are you doing?"

He backed away from me. I caught his arm. He wriggled but I held him long enough to discover that in one hand he clutched a small penknife, in the other a thin ringlet of light brown hair that I had no difficulty in recognizing as mine.

"Oh, how could you do such a thing?" I exclaimed reproachfully. I was less concerned with losing the lock, which would grow again, than the fear that this meant Hugo had decided not to be friends with me. "You should not be allowed so sharp a knife if that is the use you make of it."

He twisted out of my grasp and ran to the end of the gallery. There he hesitated and looked back. "Don't be cross, Miss Fabian," he said in a pleading tone. "I won't do it again." He laughed on a high note and skipped away.

I went to my room wondering if there were any history of madness in the Perownes. Fortunately a diversion awaited me, for I discovered to my pleasure that what the carter had been delivering to Holy Mote was my ancient cowhide trunk. This contained not merely my adequate though hardly fashionable wardrobe but all the little absurdities which even a homeless orphan somehow contrives to collect in the course of years. There were slippers and sashes and ribbons, well-thumbed books, and scraps of immature poetry scrawled on crumpled pieces of paper, a couple of pottery ornaments, a miniature cup and saucer, a roll of half-finished embroidery, a bundle of old letters, and best of all, a note from Elizabeth Tremaine which served to bring the memory of her vividly before me.

After reading the note I laid it aside in my writing-desk to answer later. Then I put away my possessions on shelves behind the chintz curtain, and hung my few gowns in the convenient modern wardrobe which had been placed, on Randall's instructions, no doubt, beside the door on the other side of the room.

That done, I recalled an idea which had occurred to me as soon as Hugo had said Old Nurse would be out for the afternoon; to explore the attics in an attempt to discover proof of the existence of the man whom Mr. Perowne had called "Cumber's Familiar," whom, if it were the same person, Hugo had sworn he had seen several times, and whom I myself was increasingly certain I had observed watching me when I returned from picking flowers the previous day.

What I should do if I actually encountered this person, I hardly paused to think. I was much more concerned with the possibility of being discovered myself by some person who had every right to be there, for I had certainly no business to be in the attics and was reasonably sure that if I asked permission to explore them, I should be refused it. However, the most alarming of such persons was safely out of the house, and the absence of Mrs. Cumber seemed to give me an opportunity that should not be missed. Quickly, therefore, before my resolution could falter, I changed into my softest slippers, forced open the creaking door, and turned to the right, up the attic stairs.

"Oh, miss," cried Randall's voice behind me, alarmed, reproving, "you can't go up there, please, miss!"

My hand, which had clenched apprehensively upon the rail at the first word, relaxed. I looked around. "And why not, pray?" I asked with an assumption of calm.

"Oh, miss," she repeated in an agitated way. "Mrs. Cumber wouldn't like it, miss. She's terrible set on no one going up to the attics—this end of them, that is, the old part. Why—" she gave a nervous laugh—" 'tis even said she put a spell on her la'ship not to have more children lest she should have to share her sewing room with a nurserymaid. They do say," she added warningly, "that the place is full of rot and that you're likely to break a leg up there."

I began to walk slowly down. "But do you mean that no one goes up there save for Mrs. Cumber—and Mrs. Perowne?"

"Jackson does, sometimes. She has to, for she knows what is in the trunks, but she told me she's not partial to it, not partial at all."

"And did Mrs. Cumber set you there to prevent me from exploring them?"

"No indeed, miss—if I was not forgetting! Seeing you up there made everything else fly out of my mind. I came to tell you that her la'ship has visitors and wishes you to meet them. Come, do let me brush your hair, miss, for 'tis so wild as the wind."

"Very well. Thank you, Randall. Who are these visitors? Had I better change my gown?"

"Oh yes, miss. It is Lady Martha and Miss Curtis, and they are very fashionable."

She hurried me into my room and in a remarkably short space of time had me respectably gowned and coiffed; indeed, I should have been ready within ten minutes had I not been delayed by the mysterious disappearance of my favourite handkerchief, over the working of which I had spent so many weary hours at school. I was obliged to leave off looking for it at last, in response to Randall's pleas for haste and, consoling myself with the thought that I had somehow lost it among the contents of my trunk when putting them away, I hurried after her down the corridors to the parlour, where she announced me to the visitors.

Lady Perowne introduced me to them languidly, and as I rose from my curtsey I was immediately aware that they were startled by my appearance, which they no doubt thought unsuitable to a governess. As for theirs, I saw that Randall had not exaggerated: they were indeed a fashionable pair. I looked first at Miss Curtis, remembering that she was to be Lord Mortmain's bride. She was not beautiful, her neck was too short, her nose too large; but she was extremely pretty, with curling dark hair and bright, rather merry, brown eyes and was dressed in the height of the mode in a white muslin gown embroidered with little brown flowers, with two frills at the hem and very long sleeves, worn without a sash and fastened in front with a tiny muslin bow beneath the delicately tucked bodice. Her mother's gown was equally elegant, of plain white muslin with a great deal of intricate drawn thread work on bodice and sleeve. Even after Randall's efforts, I felt a positive dowd between them and moved away to sit in an inconspicuous corner.

"Oh do pray sit by me," begged Miss Curtis. "Sir Hugo has been telling me all about you; he thinks you are a good fellow, you know."

I thought of my cut ringlet, cunningly concealed by Randall's art, but Miss Curtis's manner was irresistible and I could not forbear to smile.

"I hear you were at school at Mrs. Mason's," she went on, when I had obeyed her. "I very nearly went there myself, and you must know my cousin Mary Bowers, who did—but Papa decided he could not bear to let me out of his sight so I shared a governess with the Carlingfords—Lady Perowne's two younger sisters. They are not at all like her, for they take after their father and she is just as her mother used to be, it seems. We rubbed along well enough, however, for we had our love of horses in common; but Adela is my dearest friend."

By this time, Lady Martha Curtis had resumed her absorbing conversation with Lady Perowne. Miss Curtis leaned forward. "I am so intrigued to meet you," she said confidentially. "I was like to die of astonishment when I heard that Mortmain had picked you out for himself from the girls at Mrs. Mason's. I vow I have teased him unmercifully for it!" I gave her a startled glance, wondering what special qualities of courage enabled her to do such a thing. "Had it been Mr. Perowne who came home with a lovely girl in his curricle," she went on, "no one would have wondered at it—but Mortmain! Women are forever chasing him, except those who declare themselves to be terrified of him, but he is singularly elusive where the fair sex is concerned, you must know. Have you no explanation of the mystery, Miss Fabian?"

In some confusion, I repeated what the Earl had said to Lady Perowne on the subject. Then, before she could enlarge on it, I changed the conversation to the subject of her horsemanship, which Mary Bowers had assured me was notable.

"Oh, I have ridden from the cradle, have not you?" She was shocked to hear I had never ridden, and suggested I should ask Mr. Perowne to teach me.

"Just what I intend to do," remarked Mr. Perowne, entering the room upon the words. "Your servant, Lady Martha, Miss Curtis. Yes, I should be very glad to teach Miss Fabian to ride. Don't you think she will look charming in the saddle, Thelma?"

"Don't call me that, sir," she snapped, her sunny expression fading as if it had never been.

"Why not, pray? If you are to marry my cousin—"

"Please be quiet, sir. I have heard no announcement to that effect and therefore you are hardly likely to have done so."

"Ha! But you are a most dutiful daughter, ma'am, and your father makes no secret of his intentions for you."

"Oh, how can you, sir?" she protested with a worried

glance in Lady Martha's direction, furious and embarrassed.

"Why, is not your mother a party to your plans? How very curious."

"I don't care for the tone of your conversation, sir," announced Miss Curtis, rising. "I am going to find Hugo."

Lady Perowne looked up quickly. "Oh, don't leave us yet, Thelma. Sit down and talk to Piers. He has bought some wonderful new hunter and was saying only yesterday that he would like your opinion of it."

"Really, ma'am, I don't believe I said any such thing," Mr. Perowne protested in some annoyance.

"Oh, Piers, I am sure you did."

"I may perhaps have wondered if Miss Curtis knew the horse," he allowed.

"I wonder if I do?" Miss Curtis said sweetly. "It is that showy stallion of Colonel Horton's, perhaps? The broken-winded one, with spavins?"

"Really, children, are you quarrelling again?" exclaimed Lady Perowne in some exasperation. "I don't know what has come to you both lately. You used to be such friends. Lady Martha, what do you think?"

Lady Martha smiled slightly. "I think that you are looking rather pale, Adela," she said calmly. "I came this afternoon with the intention of taking you for an airing in my carriage, and I think that now is the very moment for us to begin it."

"Oh, very well. It is true that a carriage is the only place where one may be almost sure of conversing without interruption. Miss Fabian, you will come with us, I hope. Lady Martha was saying that she would like to talk to you."

I thought that Lady Martha looked as if she now regretted having said so, but she could hardly contradict Lady Perowne and I could think of no excuse not to accompany them. As soon as we had put on our bonnets and cloaks, therefore, the three of us went out to the

Curtis landau, which soon moved off slowly down the drive.

"I must confess, I have been curious about you, Miss Fabian," Lady Martha owned. She turned to Lady Perowne. "I have heard the most curious tales of how Mortmain found her. Did you really send him off to pick up a governess for Hugo from a girls' school?"

Lady Perowne hesitated. "Oh, you know what Mortmain is," she said vaguely. "He is a law unto himself. I admit I do not understand him, but I know him well enough to do what he tells me."

"Is he such a tyrant, Adela?"

"My dear ma'am! You have heard of Quentin's Law, one assumes."

"But his mother was responsible for his upbringing—"

"He has still his father's blood in him."

Lady Martha glanced at me. "You will not object to it if we talk freely, Miss Fabian? I am sure you are discreet. Now, Adela, one of the reasons I abducted you just now was to ask you the following question: why do you persist in thinking Mortmain such a monster that you are flinging Thelma at Piers's head in order, one supposes, to save her from him? I know Ivo was a wild boy, a careless, reckless young man too, perhaps—who could blame him, with such a father? But I cannot believe there is any real harm in him."

"No?" An unusual look of determination hardened Lady Perowne's features. "Well, it won't do for me to be bearing tales of my own cousin, Martha, but in the circumstances I think I will just give you a hint. It was more than mere high spirits on Ivo's part that caused old Mortmain to die of shock."

"But I know all about that affair," exclaimed Lady Martha, while I stared out of the window and hoped they did not know how fast my heart was beating. "Old Mortmain could not bear the sudden realization that he had nearly lost his only son and heir in some foolish duel—that, surely, was the shock which killed him?"

"I am sorry, my dear ma'am, I can say no more. I have no proof, it is all hearsay—"

"Hearsay on the part of whom, may I ask?"

"His lordship's valet, Stafford—you remember him, surely?"

"Yes, indeed. A most respectable person. What did he tell you?"

"Why, that a letter was brought in to old Mortmain a few weeks after the duel, that upset him considerably. He stormed off upon the instant to his son's sickbed and dismissed his attendants in order to question him. He collapsed from the shock of whatever Ivo there admitted to him—was obliged to admit, one assumes, in the light of what was in the letter—and died later that night. Of the letter, for which Stafford subsequently searched, there was of course found no trace."

"And Stafford came to you with this tale?" Lady Martha cried incredulously.

"No—to Hugh, who told me later. He should not have done so, of course, but he was so worried I contrived to draw it out of him by degrees." Lady Perowne picked up her mourning-fan, tastefully decorated with funeral urns and weeping women, and examined it thoughtfully. "What it was all about we never discovered—only I think that it would have taken no ordinary shock to kill old Mortmain."

Lady Martha bit her lip. "Oh dear. If I thought there could be any substance—Stafford, of all people! A most reliable man. I knew you did not care for Ivo but I thought—forgive me, my dear—that you were jealous of his closeness to poor Hugh."

"Poor Hugh," echoed Lady Perowne. "He wanted to confront Ivo with what Stafford had said, but I dissuaded him. Ivo was too weak, and by the time he recovered, we were at war again and Hugh was so busy, raising the Militia, and then he was sent to Brighton." She sighed. "When peace was declared again, Hugh used to say, he would discover the truth of it. Alas!"

Lady Martha patted her hand. "This is all very disturbing. Mortmain is such a very eligible match."

"Oh quite, if that is all that signifies. You would have to go into the next county to find a better."

"Adela, you are cruel. Only wait—well, you have no daughter, sons are rather different. But the disposition of a daughter is a subject of immense anxiety, let me assure you. Piers Perowne," she added reflectively. "At the risk of offending you, Adela, I doubt if Mr. Curtis and I could agree to Piers as a husband for Thelma even if she could tolerate him for a moment. He is by far too volatile and I hesitate to say this to his sister-in-law, but his reputation—"

"Oh, I know. But someone of Thelma's character is just what Piers requires. I have given this a good deal of thought, you must know."

"I think young Perowne dislikes everything about poor Thelma."

"My dear Lady Martha, you exaggerate. If only they could be encouraged to see the best in each other, I have every hope—"

"Enough, Adela. Even if such an unlikely event were to occur, I could promise nothing unless Mortmain gave them his blessing—and that, you will agree, is even more unlikely. He, by the way, has given me permission to drive through his park occasionally, and I have instructed my coachman to do so today. I thought Miss Fabian would like to see it."

I murmured truthfully that I would indeed, and watched with interest as we turned between the stone gateposts which I had observed from the Downs that morning. A stout woman held the gate and curtsied as we went by; then the carriage began to traverse a well-kept park which, from the low browsing line of its venerable oaks, was obviously grazed by deer, though I looked for them in vain.

We pulled up for a moment before the house which, though not enormous, was yet sufficiently imposing, despite the scaffolding which temporarily marred the appearance of one wing. Lady Martha observed my interest and told me that the house, which was of grey stone with white Portland dressings, had been built towards the end of the seventeenth century.

The drive continued onward to another gate, but the coachman turned his horses to enable us to go back by the way we had come. I looked round for a last glimpse of the gardens, attractively laid out with many yew trees and stone terraces on which peacocks flirted their great coloured tails, and was rewarded a moment later by an unexpected glimpse of the Lindenfold deer rippling away like a russet tide under the green oaks.

We reached the gates and I leaned forward to smile at the stout occupant of the lodge. This time she was not alone, for a youngish man stood by her, soberly clad. On seeing us, he carefully removed his hat, revealing very dark neat hair, and bowed in a gesture so extravagantly polite as to appear sarcastic. As he straightened, our eyes met. Perhaps he interpreted the expression in mine correctly, for his narrowed with a flicker of antagonism, I felt. Then Lady Martha leaned forward to make some remark and the encounter, such as it was, was over; though it left me wondering if I had seen him somewhere before or if he merely reminded me of someone I knew. A moment later I recalled the conversation I overheard in the village shop and realized that my sense of recognition was probably due to this being the man they had mentioned, with the neat dark wig.

A few minutes later we were set down by the front door of Holy Mote, where Miss Curtis was impatiently awaiting the carriage. I thanked Lady Martha for having included me in the expedition, bade her and Miss Curtis farewell, and went inside. I was about to follow Lady Perowne up the stairs when Old Nurse rustled along the corridor from the kitchens. She appeared to wish to speak to me and I waited for her reluctantly.

"Good afternoon, Mrs. Cumber," I said nervously. "I trust you found your mother well?"

She ignored my politeness. Standing on tiptoe, she

pushed her white wrinkled face close to mine. "Those who walk out of place shall have knots tied out of place," she hissed cryptically, and turned to leave me.

"Wait, Nurse," I cried. "What do you mean? Where have I walked out of place? Oh, I suppose Randall told you—or—" my voice dropped as I remembered the morning's expedition, the holy wood, the quiet glade, and Mr. Perowne kneeling before the great stone, the knife in his hand—"or do you mean, up in Holly Wood?"

To my surprise, her features were immediately lit by an expression of Satanic glee.

"Oh, you went up to Holly Wood, did you, miss? No, I did not know that or I should not have spoken as I did. There is no need now for me to think of knots, for others more powerful than I will deal with you."

She went away chuckling. I shrugged, trying to dismiss the incident as it deserved, but I caught sight of my face in the blue-toned mirror as I went towards the stairs and I knew that my own fears would punish me that night for whatever there might have been of presumption in my trespass.

9

I had expected to dream that night, but I was not prepared for the dream which came.

It began comfortably enough, in the deceptive way of dreams. I was riding with Rowley on his great black horse, floating over the short grass in a garden of yew among fountains and peacocks. Suddenly a dark-haired man was bowing to me and Lady Martha was saying that he wore a wig and that was why his hair looked so neat. He bowed and bowed, and then he changed into a deer and as he leaned forward his horns threatened my eyes so that I screamed and struggled to get away.

This time, when I forced myself awake, the nightmare was still with me.

In the watery moonlight a shadowy figure stood at my bedside, the figure of a man with antlers on his head. He leaned towards me, a heap of some dark stuff between his hands, and I saw the gleam of his eyes behind his mask before I closed my own in terror and screamed again. A gust of wind sprang up as though I had summoned it. The casement rattled and a door slammed somewhere in the house.

When I dared to open my eyes again, I was alone. Only a faint scent of snuff, and the moonlight flooding in from the window proved that I had been visited, for I always slept with the curtains tightly drawn.

My first thought was to run to Lady Perowne, to

Hugo, to anyone, seeking comfort and reassurance, but a moment's reflection reminded me that I did not know where anyone slept in this rambling house and that the search might prove more terrifying than staying in my bed would be. Next it occurred to me that no one was likely to believe my story of an antlered man bending over me in the moonlight. I would not have believed it myself, if I had not known it to be true. It would be supposed that I had had a nightmare—and how absurd I should be thought, expecting to be comforted after a nightmare at my age. No, there seemed nothing for it but to wait for dawn, but I could take a few precautions against the return of my visitor even if their only value was to give me a false sense of security. First I lit the lamp and then, encouraged by its cozy glow, I slipped out of bed onto the cold polished floor and went to draw the bolt across the bedroom door.

I wished I knew what time it was. I looked out of the window but all was quiet and seemed somehow withdrawn and uncaring in the cool moonlight. I shivered and pulled the curtain across. My cold hands made such wretched work of striking the spark to light the fire, I was obliged to take a spill from the lamp. When I had the kindling crackling in the hearth I felt a little braver and sat on the rug, staring at the flames and telling myself that it was not only for the rest of this night that I must prepare, but for all the nights to come. How could I ever again bring myself to lie down calmly in the dark and wait for sleep—and perhaps the Devil's return?

But had my visitor really been a supernatural one? The Devil might wear horns but it was hard to believe he was addicted to Macouba-scented snuff—which I was almost certain I had smelt before at Holy Mote, though I could not remember when and where. And there was something else about that evil figure that seemed familiar: the eyes shining behind the mask. Or was it only their malevolent expression that reminded me of Mrs. Cumber?

Had Old Nurse—or Mr. Perowne, or even the Earl—really gone to such lengths to frighten me into leaving Holy Mote?

Frighten me, I thought, trembling again, aware that I must face what in my heart I knew: it was not merely to frighten me that the masked and antlered man had come. The stuff he held between his hands had been to suffocate or strangle me. It was my death that he had intended and if the wind had not risen when I screamed, causing a door to slam that he dared not ignore, my body might at this moment be cooling in that bed.

When I entered the Breakfast Room on the following morning my hollow-eyed looks occasioned no immediate remark, for it seemed that Lady Perowne had just announced, to her brother-in-law's displeasure, that she had invited her father to stay for a few days. "Partly for your sake," as she told Mr. Perowne rather haughtily.

"Well, you might have the goodness to consult me next time, ma'am. It so happens that I have a good deal to occupy me just now, and I'd find it deucedly awkward staying in to entertain Sir Jeremy. You say you have invited a party to dine tomorrow night, but I must inform you that I am already engaged. I might just find it possible to join you for the meal but I should have to leave immediately afterwards," he added, crumpling his napkin and throwing it down beside his plate, while I sat down in silence, totally absorbed in trying to determine whether the faint trace of tobacco in the air came from Mr. Perowne and whether it smelled of Macouba. I had an inkling that the answer to both questions was in the affirmative, but the rival odours of ale, coffee, and mutton chops were too strong to enable me to be certain.

I started when Lady Perowne addressed me.

"Are you quite well, Miss Fabian?"

"Why—why, yes, I think so, ma'am," I stammered, aware of Mr. Perowne's sharp bright glance.

"You look pale. I thought perhaps you might have been suffering from the nightmare as Mortmain warned us you might."

I stared at my plate, wondering if she had heard my scream and set it down to that; remembering with a sickening heart-thud that the Earl had warned her to take no notice if I should scream in the night.

The silence seemed to grow oppressive. I said uncertainly that I believed I had, in fact, had a bad dream last night. Then, seeing she was on the point of asking me what form it had taken, I hastily suggested that I should start lessons with Sir Hugo without further loss of time.

"Oh, but it is still holiday time, is it not?" she protested.

Mr. Perowne paused on his way to the door. "Of course it is. Besides, I was thinking of giving you a riding-lesson today, Miss Fabian."

"A riding-lesson?" I repeated rather faintly. "But . . . I have nothing suitable to wear."

"Oh, you may borrow my breeches and habit if you want to ride," offered Lady Perowne. "Jackson will find them for you. You had better ask her now, in case she has packed them away."

I could think of no further excuse and rose reluctantly. I found Lady Perowne's maid engaged in putting away her mistress's clean linen. She was a slender submissive little person, with a perpetual worried frown. When I told her what I required, she put her neat brown head on one side and clicked her teeth distractedly.

"I've laid the habit away in camphor, miss," she explained. "The breeches I didn't trouble so much over, being as they was leather. They only needed to be kept dry so I packed them in tissue and put them in a box in the powder closet, but the habit is upstairs, in a trunk."

"Upstairs? You mean—in the attic, Jackson?"

"Yes, in the trunk room, miss. It's not that I mind going up there—well, not in daylight, at all events—but that upsets Mrs. Cumber so. I wonder if she would fetch it down for me?"

"I will go," I said boldly, "while you look for the breeches."

Jackson did not seem eager to accept this offer, but I soon persuaded her and she meekly described the way, and told me which trunk to look for. My heart beating loudly at the challenge of this unexpected opportunity to explore the forbidden regions, I hurried through the tapestried archway she indicated, past Sir Hugo's room, which I identified by the martial sound of a drum proceeding from it, and climbed the narrow uncarpeted stair beyond.

I reached the top, to find a dimly lit corridor wandering away to the right. To the left, a wall blocked off the servants' quarters. I went quietly down the passage, opening every door. It was a disappointing experience, for at this end of the house, at least, the attics seemed to hold no secrets. Some rooms were large and swept clean, lit by dormer windows and containing nothing more sinister than rows of apples and onions; others had rotted floors and a peculiar smell and were obviously unusable. Then the passage dropped down three steps and became narrower. I had entered the older part of the house and turning a corner saw that I had almost reached the end. Another corridor branching off to the right led to the stairs that descended past my room. To the left, after one more curve, the passage ended in a little landing off which, as Jackson had informed me, three doors opened.

I listened carefully for any sound that might betray the presence of Old Nurse, but I could only hear the house-martins busy with their nests under the eaves. I seized and lifted the latch of the nearest door. Within was the trunk room, half-filled with cases and trunks, wig-boxes and portmanteaux. I wasted no time there but hurried on to the next door. In this room, perhaps, lay the answer to what I sought. It contained a truckle bed, heaped with old rugs, which it was certainly possible some intruder might have used. I felt them, but there was no discernible warmth. I looked about the room but there were only a few pieces of damaged furniture. The window, however, overlooked the garden and was certainly the one from which an unknown person had watched me walk across the lawn with my arms full of flowers.

I closed the door gently and descended the two steps to the last of the attic rooms. This was illuminated by a skylight, and panelled in worm-eaten oak. It contained a table, a chair, a workbasket, some pieces of sewing and an ancient loom on which a few inches of dusty cloth awaited some probably long-dead hand to send the shuttle flying to increase it.

This, then, was Old Nurse's sewing room, so jealously guarded, from which a man, if Hugo were to be believed, had mysteriously disappeared. The answer to that enigma, I decided with a thrill of terror and delight, might lie somewhere behind the dark, inscrutable panelling.

I crossed the room and, feeling rather foolish, began to tap the wall. It did not sound particularly hollow. Disappointed, yet relieved, I turned to the left-hand wall. I had thought this would be the end wall of the house, but a moment's reflection reminded me that the room with the bed in it had projected farther in that direction, so something must lie behind this wall, if it were only a chimney. I knocked on the wood. It seemed to me it did sound a little hollow. I was about to knock again when I heard footsteps in the passage outside, and the drag of a long skirt on the splintered boards. I sped across the room and had my hand on the door when it was pushed wide open to reveal Mrs. Cumber, her face flushed and her milky eyes narrowed in suspicion.

"Oh—good morning, Mrs. Cumber," I said rather faintly. "I am looking for the trunk room."

"Oh, you are, are you, miss? And might I be so bold as to enquire what you hope to find there?"

"You may indeed—a habit of her ladyship's that I am

fetching for Jackson. She seemed rather reluctant to come up here herself," I added for good measure.

"H'm. The trunk room is this way, miss. Curious you did not think to look in here as you passed by," she remarked, opening the trunk room door for me.

"Yes, is it not so? But then," I added on a sudden impulse, "there is so much that seems curious at Holy Mote." I hesitated in the doorway, smothering an artificial yawn. "Heigh-ho. Excuse me, Nurse, but I had a disturbed night."

She looked at me keenly. "Very likely, miss. I thought you would have nightmares."

"This was more than a nightmare, Mrs. Cumber. I had a visitor and a most unusual one, at that."

She looked alert and not so much puzzled, I decided, as concerned. Had I misjudged her? "A visitor, miss?"

"Yes—a gentleman, Mrs. Cumber, with horns upon his head."

No actress could have caused her face to blanch as Mrs. Cumber's did at that. She drew in a quivering breath, her lips quite blue, one hand to her heart as the other stabbed out two fingers towards me in an involuntary gesture. I turned away as if I had noticed nothing and bent over the nearest trunk to twist the key in its lock.

"Ah, here is the habit, right on top. Faugh, this camphor! Stand back, Nurse, while I shake it out."

She muttered something and stumbled away to her sewing room. I shut the trunk and went thoughtfully to my room by way of the nearest stairs.

Jackson was already waiting for me, with the breeches and a frilled shirt laid out on the bed. She dressed me expertly and turned me towards the looking glass. The sapphire-blue habit was a trifle too long, and the breeches an inch or so too ample in the seat, but her ladyship's boots fitted me to perfection and her feathered beaver hat was extremely becoming. I lifted my skirts carefully and swept off to keep my rendezvous with Mr. Perowne.

He was waiting in the hall, running his finger along the edge of a great pitted sword. Was he wondering if he could arrange for it to fall on me somehow? He turned at the sound of my feet on the stairs.

"Lord, ma'am, you look magnificent," he cried extravagantly.

"I hope I look as I do now at the end of my lesson. I should not like to take a tumble and spoil my lady's habit."

"Oh, there will be nothing of that sort. Brown Bess is quiet enough and I shall have hold of the rein."

I could hardly tell him that there was nothing to reassure me in such a statement. Besides, if he were my night's visitor he must know I suspected him—there were few enough, after all, who had threatened to steal into my room and cut my throat if I betrayed them.

"Come," he said smiling, "are you nervous? I should not have believed it. I thought you were eager to begin to learn."

He led me outside to the drive where a placid brown mare was waiting, held by a cheerful-looking groom. There was, I found, nothing alarming after all about the next half hour. It was full and satisfying and I was grateful that there was no time to think of anything but the position of my back, my head, my hands, my legs. We promenaded up and down the drive, the sun came out, the wind fluttered softly among the new green leaves, the mare tossed her head gaily, and such is the power of human recuperation, I was quite transported with delight.

"Are you tired, Miss Fabian?" asked Mr. Perowne at last. "Don't you think we should conclude the lesson?"

"Oh no," I cried. "I have never enjoyed myself so much. I wish I could ride on forever."

"Very well for you, perhaps," he said, half-smiling, "but it's cursed dull work for me, I must confess."

"I am sorry," I said stiffly. "I am afraid I was not considering your point of view. Let us go back to the stables, then."

"I have an appointment in the village," he was beginning, but I was annoyed with myself for having been about to plead with him and drove my heel into the mare's side in order to end the conversation. Even then all might have been well had not the Earl of Mortmain suddenly appeared, trotting along the drive on a handsome grey. Brown Bess pricked up her ears and pranced a little, jerking the rein out of Mr. Perowne's hand. I wrenched her head towards the stables, determined not to face the Earl just then, and when she resisted me I brought down my whip upon her flank. She gave a startled sideways leap. I felt myself falling, threw my arms about her neck, lost my balance and tumbled to the ground.

I opened my eyes. My head throbbed and I felt sick. Rough hands seized and lifted me. "Are you mad?" demanded Lord Mortmain.

I felt the tears rush into my eyes. "It's my first lesson," I muttered weakly. I became conscious that Lord Mortmain was holding me in his arms as we sat on the grass at the edge of the drive, while Mr. Perowne chased after Brown Bess into the trees. I gave a murmur of protest and pulled away from the Earl. He was staring rather grimly over my head, his face quite pale and a small muscle flicking above the scar in his cheek. As I moved, he looked down. His eyes narrowed. He said quietly, "Why do you stare at me like that, Miss Fabian? Does my old wound repel you so?"

"No. . . ." I said, and began to rub the tears from my eyes like a child. The Earl stood up awkwardly, brushing pieces of grass off his clothes.

"You looked at me then as though I were the Devil himself," he remarked.

I felt dizzy and confused. If I had been fully myself I should never have said, as I did then, "I am afraid of remembering where we met before."

He started, and I knew I had been right. I had seen him long ago in circumstances so unpleasant that I had expunged them from my mind—and he plainly remembered the occasion.

In a strained voice he said, "What is your recollection of me, then?"

I shook my head, and winced from the pain of it. I began to get up. The Earl did not help me. He limped towards his grazing horse and put the reins over the gelding's head.

"The deuce," I heard him mutter. "Here comes Perowne." He looped the rein over his arm, turned back to me and dragged me upright, releasing me the instant I stood upon my feet. I swayed and he caught me again, perforce, his fingers biting into my arms. Then he lifted me into his horse's saddle, slipped my foot into the stirrup and kept a firm hold on the reins. "I will lead you into the stables," he said tersely.

"Here, wait a moment," cried Mr. Perowne, leading the mare out of the wood. "What were you about, Miss Fabian, to behave so foolishly?"

"She was running away from me, I think," said the Earl evenly.

"Ay—and come to that, what are you doing here, Mortmain? I thought that knee of yours was too painful to allow you to ride."

"I have been trying it out in the hope of getting a few days' hunting before the season ends. Come, Miss Fabian, if you don't have a short ride now you may be put off the exercise for life, and that would be a pity."

"You had better take over her tuition, then," said Mr. Perowne rather sulkily. "It's a damned unrewarding game, to my mind. I'll leave her with you now, for I must be off."

Lord Mortmain turned to look up at me. "I shall be glad to teach you, Miss Fabian. We will begin tomorrow, if this leg allows me to do so."

"I think I don't want to ride again," I faltered.

"But you will, Miss Fabian."

As usual, I found myself helpless under his powerful air of command and I bowed my head. But did his decision owe itself to his lordship's desire to tyrannize, I asked myself, or had he seized on an excellent excuse to find himself alone with me in order to discover, if he could, precisely what my former memory of him was, and even, the thought oddly occurred to me, if it endangered him?

10

I was relieved the following day to hear that Lord Mortmain was unable to ride, and tomorrow would be Sunday, I reminded myself as I changed for the dinner party to which Lady Perowne had expressly invited me, so I would be safe from the ordeal of his tuition for a while. Even if I had not now been afraid of riding, I would have found it painful, since I was extremely stiff both from the unaccustomed exercise and the fall.

I stood up and thanked Randall for dressing my hair. My reflection showed me quite elegantly gowned in my blue sprigged muslin, which Lady Perowne had suggested would be correct for me to wear, but I was rather pale from a restless if uneventful night. I turned from the glass and began to make my way to the drawing room which, I had discovered from Randall, was just beyond the parlour. I found the room hung with green watered silk, and the light from the moat reflected upward dappled the beams and ceiling restlessly. The effect was rather dark and eerie, like being underwater, but it had the charm of originality and was certainly in keeping with the spirit of Holy Mote.

Despite Lady Perowne's assurances, I was not surprised to find myself quite underdressed in comparison with the other guests. Lady Martha wore white muslin, richly embroidered and cut extremely low, with a high belt distinguished by a large gold buckle; and Miss Curtis wore white gauze over a satin slip, banded in green ribbon, which would not have disgraced a ball, I thought.

Lord Mortmain entered behind them, lugubriously announced by Percival. He was limping badly and his face was pale and drawn, but his eyes burned as fiercely as ever, I found, when he had bowed to the ladies and approached me, unsmiling.

He greeted me curtly and announced that he would call for me at eleven on Monday, to take me riding.

"Please, my lord, I don't wish you to put yourself to so much trouble for me—"

"Say no more, Miss Fabian."

"But I no longer have any desire to ride."

"I am afraid I am not considering your desires, ma'am." He bowed slightly and moved away. I found my hands were trembling. I had feared him from the first, I thought, and not merely because I believed I had seen him before in unpleasant circumstances which I was beginning to feel myself on the brink of remembering, but because he was a man who decided on a course and followed it ruthlessly —a Quentin, in a word. I did not know what had been in his mind when he appointed me, nor did I know now, but of one thing I might be certain and that was that he would achieve his end at whatever cost to others, like myself, who might stand in his path.

Mr. Curtis entered next, with Sir Jeremy in great good humour at his side, loudly informing the master of how he could improve his hounds by breeding them with a hardy Yorkshire strain. He greeted Lady Martha and Miss Curtis boisterously, and flung himself down into a beautiful walnut Queen Anne chair with a tapestried seat, from which he was only just able to extricate himself in time as it collapsed about him.

Mr. Pole-Carter was the next to arrive, followed by the Bowerses, whom I would have known anywhere for Mary's parents. When I had made my curtsey to them I went to sit by Lady Perowne, who was admiring Lady Martha's diamonds.

"Yes, are they not brilliant?" Lady Martha was saying with unaffected pleasure. "My godmother left them to me, and when I received them from the lawyers they were dull and yellow with dirt—you would not have given five farthings for the lot of them."

I found myself nodding in agreement, for my guardian's collection had long made me aware of how dull diamonds could look. But Lady Martha's necklace of cleaned and faceted stones, sparkling with light and fugitive colour, made Mr. Medlicott's interest in diamonds understandable to me at last.

Dinner was now announced and we went down in order of the strictest precedence, the females led by Lady Martha and brought up by their hostess, followed by the gentlemen with Lord Mortmain at their head and Mr. Perowne at the tail. Once in the dining room we were seated alternately, but still in order of rank. Mr. Curtis proceeded to inform me, between mouthfuls of fried pike and stewed lobster, that Sir Arthur Wellesley had resigned his political office, given up his seat in Parliament, and had already embarked for Portugal; while on my left I could hear that Miss Curtis was engaged in a far more lively conversation with Lord Mortmain. At last, after a seemingly endless progression of roast mutton and cucumbers, veal boiled with bacon, a ragout of poultry and finally custards and jellies, I received the welcome signal from Lady Perowne to follow the ladies back to the drawing room. There, however, I found the conversation was no more enlivening. Lady Martha was rather cool to me and I somehow received the impression that Lady Perowne had convinced her that my mysterious arrival at Holy Mote under the auspices of the Earl made me a threat to her plans for her daughter, who also seemed less friendly to me than she had been. I was not made more comfortable by discovering from Lady Perowne that it was Lord Mortmain who had insisted that I should dine with the family henceforth. I felt rather unhappy and wondered whether it would be considered a slight to my hostess if I were to remove myself, but then the tea tray was brought in and I was able to busy myself handing cups.

The gentlemen eventually joined us, Sir Jeremy considerably the worse for drink. Mr. Perowne very soon announced that he must be leaving for his appointment. "Percival need not wait up for me," he informed his sister-in-law, "for I have the key to the side door—at least, it should be here." He pulled out his watch-chain and there it hung, a small brass key, intricately wrought.

Unnoticed by anyone, I pressed my hands against my sash behind which a sudden painful emptiness had seemed to yawn. I had seen a key like that before—a pair of keys, in fact: the keys to my guardian's safe. Because he kept his diamonds there, Mr. Medlicott protected them by having two locks to his safe and one of the keys had been kept on his watch-chain, the other on that of Steele, his clerk. Both had therefore to be present when he unlocked the safe—so how, it for the first time since then occurred to me to wonder, had his murderer been able to steal his papers and fortune? For Steele and I had been out that night, sent from the house on an errand my guardian had declared important but which Mrs. Featherstone's puzzled face, when she read his note, had proved unnecessary. There seemed to be a whirling in my brain as various thoughts occurred to me at once: that my guardian had contrived an excuse to get us out of the house that night—he too had mentioned some appointment—and—one moment, one moment more and I would have it! Yes! He had detained Steele as we were about to leave. . . .

"I have an appointment," he had declared rather self-consciously. I could see him now, stroking his chin in that way he had, pulling at his ear lobe with two fingers. "Nothing out of the way, but I may need to get a paper out of the safe. I'll be obliged if you'll leave me your key, Steele." And Steele, grumbling as was his wont, had complained that in that case he would have to leave Mr. Medlicott his watch because the key was fastened to the chain.

"Are you feeling quite well, Miss Fabian?" Miss Curtis asked in a low voice. I stared at her uncomprehendingly for a moment, observing her clear skin, her soft dark curls, the solicitous expression in her warm eyes, the smoothness of her glove against mine as she took my hand, her hostility forgotten.

"Oh . . . yes, I thank you. I am quite well. I am afraid my thoughts were far away." But why had they returned to that fatal night so clearly? It was the sight of that key, of course; but my guardian's keys had been different, a little larger, of slightly less intricate design. It must have been everything together, the key, the mention of the appointment, the earlier talk of diamonds, that had caused me to remember for the first time in such detail that little scene before my guardian's murder.

Suddenly Lady Perowne addressed me, making me start. "I think we had better excuse you now, Miss Fabian," she was saying. "Sir Hugo wishes you to bid him good night, and if you do not go to him soon, he will be so tired in the morning."

I was glad to obey her and lost no time in curtseying to the company, conscious of Lord Mortmain's dark eyes following me broodingly until I closed the door between us.

Sir Hugo was flatteringly pleased to see me. He reached under the bed and carefully brought out a model ship: a Revenue cutter, he informed me proudly. "Like the one we saw cruising past when we went up Holly Hill," he reminded me. "I made a drawing of it when I got home and started it that very afternoon. It would have been finished before this but that Old Nurse made me—do something else for her. In any event, I thought you might like to see it, for I've only the rigging left to do."

The shape of the hull was beautiful, as I told him, though I did not feel myself qualified to judge if it was correct for the class of vessel it was meant to represent.

"When I've finished it, we'll sail it on the moat," Hugo declared, his eyes shining.

"Are not the sides too steep to launch it?" I asked, going to the window, only to find that I could not see the moat from there.

"You can get down to the water at the other end of the house where it has been filled in," he explained. "I shall tie a thread to her, to pull her back when I have seen how she sails—or I shall ask one of the gardeners to find me a long stick. Then, if all goes well, we might try her in the quarry," he added with assumed carelessness.

I leaned my forehead against the cool glass and shaded my eyes with my hand, but a line of trees interrupted any view there might have been of the sea. I was glad I had looked out, however, for the curve of the Downs was beautiful, with the glow of the moon rising behind one smoothly rounded shoulder.

Then I became aware of movement on the Downs, a dark ripple thrown into silhouette against the moon. It was a line of horsemen riding at speed over the short turf, recklessly, for their path must have taken them very near the cliffs above the sea. I caught my breath and released it slowly. The glass misted and I blinked.

Why, oh why, I wondered, had I not taken the opportunity I had wasted this evening, of asking Mr. Pole-Carter what a Wild Hunt was?

I went back to Hugo, congratulated him again on the lines of his ship, tucked in his sheet, and daringly dropped a kiss on his limp brown hair.

I was distracted from the ordeal of walking down the shadowy corridors by the plaintive, trembling music of a harp and half-wished I might have stayed in the drawing room to hear Lady Perowne perform upon it. However, I had much to think about and moreover was still stiff from riding and unexpectedly tired. Once I was in bed, with the door securely bolted, I had hardly begun to order my thoughts when I felt myself falling . . . falling into a comfortable sleep.

A curious shuffling noise awoke me some time later. I stretched out my hand to turn up the lamp and stared about the empty room. After a moment it occurred to me that the sounds were coming from outside my door. I got out of bed and put a shawl round my shoulders. I moved to the door and stood listening. There were two kinds of footsteps ascending the stairs. They sounded like a man and a woman, walking rather unevenly—or a large man and a small, I thought, my heart beating quickly. The stairs led up from the side door. It would be Mr. Perowne and the other man with whom I had once heard him talking—Hugo's man, Cumber's Familiar. Now was my chance to discover his identity, and if I did not act swiftly it would soon be gone.

With shaking fingers I shot back the bolt and pulled open the creaking door.

A little way above me on the stair, two men halted and looked down. Even in the uncertain light of the guttering candle one of them held I could see that the other was indeed Mr. Perowne, booted and leaning on the arm of a thickset man in stockinged feet. This second man held the candle high and awkwardly since under the elbow of the same arm he was conveying a large and flattish leather bag. As he stared angrily down at me I recognized him as the man who was known as Born-To-Be-Hanged.

I gave a frightened gasp. Mr. Perowne, far from reassuring me, leaned back against the wall and closed his eyes as if he accepted no responsibility for anything that might follow.

The carter released him, took a firmer hold upon his bag and hurried on up the attic stairs.

Mr. Perowne opened his eyes. "Well, my little singing-bird," he remarked faintly. "Did you sing to some purpose? Did you betray me, fair Chantal?"

"N-no," I contrived to stammer. "Indeed, I have not breathed a word about—about what you showed me—what you said."

To my surprise, he seemed satisfied with my denial. "Be sure I shall discover if you lie . . . and if it is so, then I shall cut out your tongue before I slit your throat," he murmured. "If not—I must beg you to forgive us for disturbing your slumbers. We were looking for Old Nurse but I fancy I have not the strength to climb these infernal stairs . . . perhaps you will minister to me in her stead."

"What—what is it?" I asked, clutching my shawl nervously. "What is wrong?"

Mr. Perowne did not seem to hear me. "You won't give me away to Mortmain, will you, Miss Fabian? The fact of the matter is, I have sustained a wound to my shoulder. I need a little help—but if Mortmain should hear of it—magistrates are notoriously inimical to duellists even when they have themselves been guilty of duelling in their youth." He swayed slightly, but kept his bright eyes fixed on mine. "The fair Marie," he murmured. "She is to blame—but she is not here to succour me, alas."

"You had better come into my room," I said reluctantly, putting my arm beneath his and assisting him down the steps. I helped him into a chair and turned up the lamp. Then I caught my breath sharply as I saw the wound properly for the first time. It was a mess of scorched cloth and torn flesh, and I could see fresh blood welling into the cravat which Mr. Perowne was ineffectually holding against it. Somewhere in the back of my mind I was aware of footsteps above, but I had no time to wonder why Smithers should have gone up to the attic because it was all too apparent that Mr. Perowne required immediate attention.

In my last term at Horsefield House, when I had been the oldest pupil, I had often assisted our geography mistress and self-styled apothecary, Mrs. Simmonds, in dressing cuts and wounds, and though I had never seen anything as bad as this, I was not quite so dismayed by it as many another girl might have been.

Water was the first necessity, and I brought the jug and basin from the washstand.

Mr. Perowne almost snatched it from my hand and began to drink thirstily while I went to fetch my bedside table and the lamp to stand beside him. As soon as I could get the water away from him I began to soak the wound, trying not to notice that the blood began to trickle faster as I worked. When the padded cravat was sodden with it, I tore a strip off my petticoat, which was luckily rather old, and instructed Mr. Perowne to press it against the wound whenever I took my flannel from it. After a while all the cloth was soaked free of the flesh and I was able to cut away the sleeve of his coat. Mr. Perowne went very white as I pulled off the sleeve, but he bit his lip determinedly and did not lose consciousness.

I stared at the wound, now fully revealed. "You need a doctor, sir. This should be stitched."

Mr. Perowne stared at me. A light sweat had broken out on his pale face. "Impossible," he murmured. "Burton used to give me away to my father—why not to the magistrates? Do you want to see me in prison, Miss Fabian?"

"N-no," I told him, but for a moment I had been tempted by the thought. If he were out of the way then I would either be safe, or find it easier to discover who it was who threatened me at Holy Mote.

"Then you will have to make shift to attend to me yourself—unless you wish me to die of loss of blood here in your room. You might find that a trifle awkward to explain."

I sighed, and submitted. I went to the cupboard and found another petticoat to sacrifice, and some court-plaster which I began to cut in strips to pull the edges of the wound together. "When you are in bed you must keep the arm up on pillows, as high as you can—and you must stay there for a day or two, or longer if you develop a fever, as you well may."

"You are severe, Miss Fabian."

"You can say you have a heavy cold, or the influenza. Old Nurse will have to dress your wound after this. Make sure she keeps it clean and tightly bound."

"Don't you intend to visit me, then?"

I put a clean pad on the wound and now began to

bind it. "Certainly not, sir. It would be most improper." I stared at my handiwork. "Do you realize you are very lucky, sir? It has just occurred to me that the bullet scored the surface of your flesh and continued on its way. If I had had to probe for it, we might both have succumbed."

"I don't feel very lucky," he grumbled. "It is damnably painful. Have you finished torturing me yet?"

"I think you are extremely ungrateful. Are you sure it is not more comfortable now?" My dressing looked clean and neat and I had just decided I was very proud of it.

"Comfortable!" he exclaimed. "It hurts like the Devil. I suppose you don't keep brandy in your room—or wine?"

"Of course not. But—" I looked at him doubtfully—"I suppose I might fetch you some."

"Obliging of you, but old Percival keeps it under lock and key. Garland has a key, curse him, but he would not give it up to anyone." He stirred restlessly. "Perhaps I had better be thinking of getting back to my room. The trouble is, I must have lost more blood than I thought—I feel damnably weak. I think I should move while I still can . . . and must enlist your help again, I fear."

I was only too thankful that he felt able to make the effort. I got him to his feet and helped him up out of the door onto the landing.

"Is Smithers still upstairs?" I asked, thinking of my return, alone, and the night ahead.

"Upstairs? Oh, that was merely to return the—that is, he will be gone by now. Curse this pain. Have you any laudanum?"

"No, sir. No wine, no spirits, no drugs; I fear you will think me very ill-equipped."

He grunted, obviously concentrating his energies on reaching his room, and I said no more until at last we had arrived at it. He half-fell onto the bed while I lit his lamp, drew the curtains and offered to pull off his boots.

"No, no," he objected halfheartedly, "I cannot expect you to do that."

"Then who will, sir?" I asked reasonably, inspired partly by pity for him in his predicament and partly, I fear, by a cowardly hope that the more I did for him the less, surely, he would be my enemy, if it were he whom I had to fear. "You do not wish me to ring for your valet, I suppose?"

"No—but I am sure Old Nurse—"

"I rather suspect I am stronger than she. Lift up your foot."

It was a good deal harder than I had anticipated but somehow I succeeded in removing his boots, shook out his nightshirt and then left him to undress while I went in search of laudanum. I knocked on Lady Perowne's door. There was no response so I opened it and went in. The bed-curtains were drawn and Lady Perowne slept quietly. I went through the large apartment to the smaller room in which Jackson was heavily asleep, snoring faintly. I hardened my heart and touched her shoulder, intending to shake her gently, but she started up immediately, clutching her nightcap and gasping with alarm.

I apologized for waking her, half-covering my mouth with my hand. I mumbled that I had the toothache and could not sleep. She got up and after suggesting various remedies such as rubbing my feet with bran and placing slices of roasted turnip behind my ear, she produced a small bottle of laudanum. Then, of course, she wanted to administer it to me herself, but I muttered that I would try a small dose first and would keep the bottle by me, with her permission, in case I needed more. Thanking her somewhat incoherently, and clutching my prize, I fled back to Mr. Perowne and thence, as soon as I had given the bottle to him and wished him good night, to the housekeeping tasks that awaited me in my room, and blessed sleep at last.

11

The next day was a fine Sunday. Mr. Perowne made no appearance at the breakfast table and Lady Perowne informed me that his valet, Ashe, had reported him to be a trifle feverish.

"He will not be accompanying us to church, Miss Fabian. And how are you, by the way? Jackson mentioned your toothache. I was sorry to hear of it, but your face is not swollen, at least."

I assured her it was no longer troubling me; and presently we set off with Sir Hugo in the first chaise, escorted by Sir Jeremy on horseback and followed by Old Nurse, Jackson, and Ashe in the second carriage, just as the great bell, Clap Hammer, began to ring out from the Norman tower. The church was situated among fine yew trees in a green churchyard on the inland side of Lindenfold, beside a handsome grey stone rectory. We were led to a high-walled pew and Hugo, nudging me, drew my attention to the carvings and gargoyles of which Mr. Pole-Carter was so proud. It seemed to me as the service progressed that it was actually to one of these, a horned and goatlike figure, that Old Nurse was making her prayers.

During the sermon, which was very long, my attention wandered again. I found myself staring at a memorial tablet set into the wall above the pew in front, evidently that of the Quentins, for the tablet commemorated Ivo

Charles Fortescue Quentin, sixth Earl of Mortmain, Viscount Rolland, Baron Lindenfold, K.C.B., of pious memory; and his widow, Eleanor Sarah—the grandparents, I supposed, of the present Earl.

The service ended and we walked out into the sunlight, where half the congregation was already gathered conversing busily, and even, I thought, rather excitedly; but before I could hear enough to determine what their subject was, Lady Perowne had drawn me over to the Curtises.

While they talked, I looked about me, turning to avoid Lord Mortmain's eye to stare instead at an old woman with a hand cupped to her ear, whom I assumed to be Randall's deaf mother, talking to Wilson, Lord Mortmain's groom. Presently my assumption was confirmed when the man in the black wig, for whom she kept house, addressed some remark to her, very much the aloof employer.

The Curtises left, and Lady Perowne expressed her intention of visiting Granny Cumber, Old Nurse's mother. I was quite eager to meet the old woman, and Hugo revealed more of the delights to be expected from her acquaintance as we walked along to the village green.

With one eye upon his mother, he confided that Granny Cumber was accustomed to spit and swear—"And she falls asleep while you are talking to her—and she has a black cat who does everything she tells it, for of course she is a witch—"

"Here we are," said Lady Perowne, unself-consciously bringing out a handkerchief saturated with rose water and opening her fan. "We need not stay long, but she does appreciate the visit."

Old Nurse hurried up from behind us to open the door. We stepped down onto a worn brick floor, entering a stuffy ill-lit room in which two persons were dimly visible: the ancient crone in her chair, heaped with rugs and shawls and clutching the sleek black cat; and the pale freckled child of about fourteen who was employed to look after the old woman.

Granny Cumber made some cackling noises in response to our greetings, and whispered to the cat, which leapt from her arms onto the rag rug before the fire, where it proceeded to wash itself. Hugo lost no time in joining it; Old Nurse and Lady Perowne began to question the child in low tones and I was left to entertain Granny Cumber. After racking my brains for a suitable subject of conversation, I could think of nothing better than to inform her that I had heard she had attended the execution of the famous criminal, Jonathan Wild.

"Wild?" muttered the old woman, opening her rheumy eyes at me. "Wild, ay . . . too many Wild Hunts, is what I say."

I gasped and she seemed to wake completely. Her sparse eyebrows lifted and a gleam of mischief came into her face. Reaching out a trembling knotted hand, she touched my arm. "So that interests you, does it, miss? Zounds, you are one of those that likes to play with fire. And will you watch the Wild Hunt on May Eve, then, up by the Holy Stones in the wood? Ah, but I've said too much . . . your wits and your tongue do run away with you when you are old, 'tis Herne's truth. Time was, you could have burnt me at the stake before I would have mentioned such."

"I promise you I will not tell," I said nervously, and she cackled with faint laughter before spitting with practiced accuracy into the stained, cracked bowl at her feet.

"Come, kittikins," she said, and the black cat took a flying leap into her arms and began to knead her lap.

Lady Perowne now brought the visit to a close, and gave Old Nurse permission to stay with her mother until Sir Hugo's bedtime, for the rest of us, it seemed, were promised to Lady Martha Curtis for the afternoon.

"Is Granny Cumber not *prodigious?*" cried Sir Hugo

delightedly as he scrambled after us into the waiting chaise. "Only I wish you had seen them carry her in a basket—that is the best of all."

His mother fondly bade him hush and be a good boy. The chaise moved off, and she began to talk about the proposed visit to the Curtises, where her father would be awaiting us before going on to dine with Lord Mortmain. I suddenly realized that if I could contrive to stay at Holy Mote I should have the perfect opportunity to explore the attics more thoroughly than I had been able to do on the last occasion, for that there was some secret connected with Old Nurse's sewing room I was certain, or why should she protect it so carefully? Accordingly, I pleaded a return of the toothache and begged to be excused from visiting the Knoll. Immediately Lady Perowne became most concerned and said that if only it were later in the summer the gardeners could have brought me material from a wasp's nest with which to pack the afflicted tooth, while Hugo made me feel even more ashamed of myself by taking my hand in his and saying he wished he could make the horrid toothache go away.

The carriage set me down at Holy Mote and bore them off to the Knoll. I removed my outdoor clothes and, making sure that I was not observed, looked in on Mr. Perowne. He did not seem particularly pleased to see me, and assured me that Old Nurse had dressed his wound with cobwebs and that it was on the mend; as for the laudanum, he said crossly, when I asked him to return it, he had mislaid it and did not propose to look for it.

I closed the door, reflecting that he was one in whom adversity did not bring out the best; unless it was that he was warning me I need not expect him to treat me differently because I had cared for him once when he had needed help.

The clock striking the hour reminded me that there was more than Mr. Perowne to think about this afternoon. I hastened back to my room and lost no time in removing my gloves and changing into my softest slippers. Then, this time making sure that I was not observed, I began to climb the attic stair.

It was rather dark at the top, where the two passages met. Immediately on my right hand, above my bedroom, a blank lathe-and-plaster wall seemed to bulge slightly as such walls often did with age. I walked down the corridor to the small landing off which the three doors opened.

As a precaution, I glanced into the trunk room and the room with the bed. Both seemed exactly as they had been when I had visited them before. My heart beating fast with excitement, I turned towards the sewing room and went down the steps. This time I went at once to the left-hand wall. I tapped it in several places. There was no doubt that in one area, towards the corner farthest from the door, the panelling did sound hollow.

I pushed and prodded at it, suddenly certain that a secret room must lie behind it, but the wood resisted my efforts. I looked at it more carefully, seeking some irregularity which might conceal the clue to opening it, but there were no carved bosses or barley-sugar work, only plain panels of worm-holed oak. I was just putting my hand to it again when I thought I heard some muffled sound beyond. Had I unwittingly set some distance mechanism in motion?

I waited breathlessly. Then with a slight creak a section of the panelling began to slide across, revealing a dark rectangular opening.

I gave a little gasp and reached out to help increase the widening gap, but before my fingers could touch the wood I saw something moving in the darkness beyond—something pale that rapidly emerged and clung lightly to the panel's edge. It was a man's hand, with long white fingers.

I stood petrified, gazing at it. Wild ideas chased through my mind. Was this the apparition Hugo had seen, conjured by Old Nurse? Cumber's Familiar, guardian of the attic? Or was I about to meet the antlered man in daylight, face to face?

My dry lips opened to scream, but, as so often in my nightmares, I seemed incapable of either sound or movement. I could only watch in terror as the second hand materialized on the other side of the gap. I noticed lace at the wrist, a neat blue cuff—and then, above the hands, a pale face appeared.

Then I screamed—or rather, gave a frightened croak.

There came a smothered oath from the dark rectangle before me and the Earl of Mortmain stepped up into the room, from beyond the sliding panel.

"What the Devil are you doing here?" he demanded, his face blanched, his eyes glittering. "I thought everyone was out."

I discovered I, too, was angry. "Why did you not speak?" I cried accusingly. "Why did you make no sound? I—I thought—"

"You thought I was a ghost, I collect. But what do you think you are about, Miss Fabian?"

There was no time to reflect. "I? Oh—oh, I am just exploring the house," I said with an assumption of carelessness.

"Indeed," he murmured, undeceived.

"And—and what of you, sir, may I ask? Where does that panel lead?" Suddenly I realized the answer. "There are stairs behind it, are there not?" Stairs, I thought, that led down into my room.

"Yes." He was looking about the sewing room as if something puzzled him. "Hugo spoke of a mysterious apparition . . . and I thought I would take the opportunity, while everyone was out, of trying to resolve the matter. As soon as I realized that this was where Hugo believed he had seen a man, I remembered the secret stair. The disappearance could have been effected by the man's merely stepping back through the panel."

"You knew about the stair?" I said rather faintly. It was by that means the antlered man must have come and gone; and on its steps Mr. Perowne and Cumber's Familiar or another pretending to be such a person must have held the strange conversation I had overheard. The click would have been this panel closing behind them.

"Ay, but . . . this room is not as I remember it. There was a fireplace, surely, when Hugh let me into the secret. Have you tried the panelling here?" He tapped the wall opposite the door.

"Yes, sir. I thought it sounded solid. But do you mean you think there might be another room through there—a room above mine?"

"Yours?" He turned about and stared at me.

I was confused. "I mean, the one in which I sleep."

"You sleep down there?" he repeated in an astonished voice, on the verge of anger. "Do you tell me that they put you in the Stair Room?"

He was either a good actor, I thought, or else he had not known which was my room and I had no cause to suspect him. But how was one to tell who lied and who spoke truth?

"And Lady Perowne knows of it?" he asked, following my affirmative.

"It seems it is the custom . . . for the governess here."

"If that is so, it must be changed. The fact that these stairs lead down into that room—"

"There is a door, which can be locked," I pointed out, feeling a perverse affection for the Stair Room, in spite of this startling discovery.

He frowned. "Then be sure it is locked—as it was not just now."

"I can assure you that it will be," I said with a shiver.

"As for that other room," he added more easily, "I advise you to forget it. I may well have been mistaken . . . it was so long ago and at night, and we had only one candle between us. Now, I must ask you, Miss Fabian, whether you saw any sign of an intruder in the course of your—explorations?"

I did not want to give him any information but—it is

hard to explain quite how—his intent gaze seemed to draw it out of me.

"None, sir. There is a bed but it has not recently been slept in—I saw it on Friday in the same disorder."

"Very well," said the Earl briskly, coming to a decision. "Let us forget him, then. This attic is no place for you, Miss Fabian. Many of the floors are rotted through. You might break a leg and not be found for hours. I am sure my cousin would have forbidden you to come up here if it had for one moment occurred to her that you might think of doing so. Come, let me lead you down to your room."

He held out an imperious hand. I stepped forward, gave him a helpless look, and stopped before the gap in the panelling.

"Ah, now you are afraid, Miss Fabian," he observed.

"Well, yes, I must confess I am a little."

"There is no need. I will go first, if you prefer it."

He stepped down through the opening and turned towards me. "Take my hand, ma'am. There are three steps down."

I put my bare hand in his—a disturbing sensation and one that demonstrated to me the reason why gloves were *de rigueur*. I had removed mine in order to be able to feel the panelling more sensitively; now it was the warmth and strength of Lord Mortmain's hand that I was feeling and even as I tried to withdraw my own, his fingers tightened round it. I told myself severely not to be so missish and stepped down into the darkness.

Immediately I was aware as never before of the strange power that emanated from the Earl. The confined space seemed full of his presence and I was forcibly reminded that not mere modesty should have prevented me from finding myself in this situation, but self-preservation.

Suddenly I felt his other hand on my waist. He was drawing me towards him. I struggled but so feebly, overwhelmed by my peril, that it was as if he had put a spell on me. Was he the would-be murderer after all? Was it to be here that he would kill me? Was this where he kept the black cloth with which he had once failed to smother me? Would my body rot and my skeleton one day be found upon this secret stair? Why had I ever let myself trust him so foolishly?

I pushed my hand against his chest. At the same moment a slit of light appeared beyond him. He pulled me close and I felt his breath on my hair. Then the door was open and he released me.

"Excuse me, Miss Fabian," he said coolly. "There is very little room on this landing."

I let out my pent-up breath and made to pass him. He caught my hand again and I glanced at him quickly.

"Wait," he said. "Don't you want to see how the panel can be made to move? It is not simply a matter of sliding it, you know. That would not have afforded enough security to the recusant priests who fashioned it."

I hesitated, and then reflected that it might be useful to me to know the secret of the stair.

"Very well," I said, as calmly as I could. "I should be interested to see the—er, mechanism."

"Kneel down, then," he said, dropping stiffly to his own knees.

"It is in the riser of the first step beyond the door, just here," the Earl explained, leaning across me. "As you see, it is itself a panel which slides back . . ."

I was not attending to his words. I was staring at his stern profile, remembering how I had seen it etched against the light in Mrs. Mason's room. Why had it affected me so strongly, then and now?

I heard myself say in a queer, breathless voice, "I am afraid of remembering—but I must know! I have met you before somewhere, have I not, my lord?"

His head snapped round, his dark eyes were only a few inches from mine. For a few moments we knelt there in silence gazing at each other with the most extraordinary intensity which I, for one, felt powerless to interrupt.

Lord Mortmain blinked, and the spell ended. I felt his withdrawal as if to a distant place. In a light, almost sneering voice, he said, "Are you flirting with me, Miss Fabian?"

I gasped. I was shaken by a spasm of rage. I felt a sudden pain in my wrist and only then realized that I had actually moved to hit him and that he had prevented me by catching my hand in a cruel grip.

"Well, what a spitfire you are, Miss Fabian," he observed calmly, and then, dismissing the subject, "Now if you will behave yourself and look at this, you will see a wheel inside the stair behind this panel. When one turns it, so, a cable runs through a pulley—or several pulleys—up there in the wall and moves the panelling back. Turn it the other way, so, and it closes. It is possible to open and close it by hand, but hard enough to discourage any casual experimenter. I don't suppose you could do it, for example."

He stood up and I turned to stumble blindly down the rest of the winding stair into my room. I went over to the window and stared out unseeingly at the blue and green April afternoon. Behind me I heard the sound of the door closing across the stairs, and the click of the lock turning. Then the Earl clattered unevenly down, and the curtain rings rattled as he pulled the chintz firmly across the opening.

I felt as if I wanted to cry, without quite knowing why, and concentrated my thoughts on the task of rubbing the mist off one of the little diamond-shaped panes of glass with a dusty forefinger.

"Here is the key," said the Earl. "Make sure you keep that door securely locked—and don't try those stairs for yourself. You might become trapped in there."

He waited, but I said nothing, and after a moment he laid the key on the windowsill beside me.

"It seems as if the intruder, if there was one, has been frightened away," he said in a reassuring tone. "I believe you should be safe enough. Well, I must be on my way, for I am expecting Carlingford to dine."

The room felt very empty when he had gone. I dried my eyes and wandered restlessly about it. I looked for the handkerchief I had lost, without success. I picked up a book and laid it down. At last I found Elizabeth's letter and, somewhat soothed by re-reading it, sat down to answer it.

I wrote page after page, far more than I could ever send, or expect her to pay for the pleasure of receiving; but it seemed that once I had started I could not stop. There was so much to tell her, so much that it relieved my mind to be able to express. My hand grew cramped, the light changed as the dinner hour approached. I stopped writing at last, brushing my lips with the grey goose-feather as I began to read the letter through, vaguely aware of some sound I could not quite identify. And then a voice seemed to speak quite close to me, though I was alone. It was like the first night when I had woken to hear voices in the room, but this one was more distorted, echoing, but deliberately softened, I thought. After the first few words I realized it appeared to be loudest by the chimney, and as I knelt on the hearth rug I recognized the voice that filtered down to me as one of those I had heard on that first night, the one who preferred Artemis to Selene, or Hecate, goddess of the dark moon.

"—can never rest until it's done!" he declared emphatically. Some murmuring suggested another speaker; but who could that be, since Mr. Perowne was confined to bed and Lord Mortmain had already left? To this person, the first presently replied, "To be sure I have reflected—ay, and planned, for years. Do you think I am a fool?" Gone was the suggestion of servility that had been present on the earlier occasion. "Nothing can go amiss. My Master—" Some murmuring followed, either an interruption or the same person lowering his voice. Then he continued, "And if it fails, who could swear to Herne?" Or so I thought he said. The speaker must then have moved, for only by straining my ears could I make out the words, "—came here to wait . . . knew that my hour would come at last . . ." before the voices faded altogether into an indistinguishable muttering.

I went back to the table and stood staring down at my letter. It now seemed foolish beyond belief to have committed my thoughts to paper, and I began methodically to tear the sheets and then to burn them in the hearth while telling myself that the plans of the man who had spoken in the room above mine, where there was no room, might not have concerned me in the smallest degree. Nevertheless, in that moment I seemed to come closer to believing in the danger that threatened me than ever before, so close, indeed, that I wondered if the time had not come when I should be leaving Holy Mote, whether or not I could find another post; or when I should at the least be writing to inform my trustees that this situation was far from the sort of thing they had surely had in mind for me.

But what, I thought more practically, if they decided their only course was to send me back to school? Were not the half-suspected thrills and probably imaginary dangers of my life at Holy Mote preferable to the stultifying routine of school, which I had been so thankful to leave behind me? And was I to impose the burden of my fees upon my benefactor, whether Mr. Shorncliffe or another, when he had just been relieved of it after so long?

The answer, of course, was that such considerations were paltry so long as I was seriously threatened; but was not my position safer now than it had been, forewarned as I was by suspicion, and forearmed with a bolt to one door and a key to the other? The chimney, I satisfied myself, was too small to allow anyone but a little sweep's boy to pass through it, and no one could climb up the few creepers that reached out between my window and the moat.

At that moment, Randall knocked on my door and entered with a tray of soup, chicken and ham, and a glass of red wine, when I should have preferred a white, that I soon decided had not come from one of the more prized bins in Percival's cellar.

After Randall had returned and removed the tray, I addressed myself to writing a new letter to Elizabeth, for I was conscious that she would be waiting anxiously to hear from me. I told her, circumspectly, that I had arrived safely after an uneventful journey; that I had not instructed Sir Hugo in a single lesson but felt we were already friends; and told her that Mr. Perowne was a handsome young man, "just the kind we used to dream of."

I blinked, aware that I was growing sleepy. The fire was not lit but the room felt warm. It was nearly May, after all. As soon as I had finished the letter I would lock the room and go to bed.

I dipped my pen again in the ink and wrote, "but alas, Elizabeth, as Mary warned us, he is too flirtatious, too thoughtless, too mischievous, I think, for perfection—"

Yawning, I laid down the pen. It was no use, I could not keep my eyes open. I should have to finish it in the morning. I stood up, swayed dizzily and sat down again, breathless. What was wrong? Was I ill? Why was I falling like this across the table? Why . . . ?

Time had passed. It was dark, but a lamp burned softly. A shadow loomed between it and me, the shadow of the antlered man. I forced my heavy-lidded eyes to focus on him as he reached out towards me. He wore a hood to which the antlers were attached and I knew it was the same man as before because his eyes gleamed through the slits in the cloth in the same vindictive way, and I could smell the scent of Macouba before my mind floated away again into the whirling darkness.

His hands gripped me and I moaned in protest, plucked back towards the light and consciousness I no longer had the strength to seek. There was pain and confusion and fear—yes, that was certainly fear beating great wings against my heart—and then I was falling in a dream more terrible than any I had known before, falling in coldness and terror down into the shock of an icy element, down and yet farther down into it, and it pressed on me and forced its way into my nose and mouth and consciousness so that I suddenly knew I had been pushed out of my window into the moat, and realized that as I could not swim, I must surely drown.

12

Somewhere down in the dark swirling water I came to my senses, so that when I came up to the surface I knew enough to let out my breath and fill my lungs again with the sweet night air before I was sucked down again into the depths. This time my horror of drowning in the stagnant moat inspired me to struggle wildly. My hand struck out against the slimy stone foundations of the house and my feet sought vainly for a ledge. I began to rise again towards the surface of the water, more slowly this time, impeded by my saturated clothes, and knew I had little time left to me.

My head came up into the air; I choked and sobbed and coughed. I was close against the wall. My desperate hands slipped over it, one of my nails caught in a crack, gaining me a second's respite. As I slipped back into the water I remembered the creepers that I had thought so evil-looking when I first saw them reaching out their sparse twisted tendrils as if they sought to strangle the house in their clutching grasp. I flung up my hand in search of them, but my fingers had touched only the stone when the waters closed again over my head.

This time I almost despaired. I was limp in the water, bursting with the sense of suffocation, blinded by what seemed to be dazzling lights behind my eyelids, deafened by noise, sinking, sinking—and then something touched my foot, some stick or broken root perhaps, and I kicked out violently without conscious decision and that was enough to send me to the surface a third time.

Again I reached for the creepers that I could not see but knew must be somewhere above me. I touched and caught at something thin and brittle, clawed with my other hand and found another branch as the first broke off in my fingers. The second one began to come away from the wall, the little rootlike tendrils that held it to the stone parting under my weight. With a last despairing effort I seized another creeper and this was thicker. It sagged and drooped as I clung to it, a drowning man to a straw, but as long as I stayed still it seemed as if it might support me with my head barely above the lapping ripples that fanned out around me.

The water stank, I suddenly realized, and I began to retch. Gradually I got back my breath and coughed out some of the water I had swallowed.

I had been shocked into wakefulness but my eyelids were still heavy and this was a new danger, that I felt as if there were a suffocating cloak of sleep waiting to smother me in its folds the moment I allowed my attention to wander.

After a while I blinked the water out of my eyes and stared up at the house. I was able to see a little in the dark by this time, enough to make out a splash of white fluttering below my casement, which swung wide, soft yellow light streaming from it, high and inaccessible above me. No head, antlered or otherwise, leaned out anxiously to discover whether I floated or drowned; nor did there seem to be any activity elsewhere. It appeared as if no one had heard the great splash I must have made. What should I do now?

I opened my mouth to call for help; and then, again remembering the antlered man, was silent. He had tried to kill me, I reminded myself. He had disguised himself in case, after all, I lived to tell the tale, for "who could swear to Herne?"; and he had, no doubt, set something of mine up there among the creepers to suggest that in reaching for it I had overbalanced and fallen of my own accord. If he heard me scream he would know that his attempt had failed. Might he not then finish what he had set out to do, if no one came to rescue me? He would, I thought; for it was surely he who had said in the secret room over mine that he would never rest until it was done—and who else was there who would hear a cry for help? The kitchens, the servants' quarters, were all at the far end of the house; everyone else slept on the other side.

It seemed that, if I were to escape from the moat, I should have to accomplish it alone.

Meanwhile the water was pulling at my clothes, trying to drag me down, and the creeper had begun to stretch again under the increasing strain. My arms ached agonizingly and I realized that I was very cold. My hands were becoming numb with it. It seemed a long time ago that I had thought it a warm night and reminded myself that summer had nearly come. If I did not get out of the moat soon, I should probably die of exposure.

It seemed madness to let go of my sole support but I had to free one of my hands in order to feel for another, stronger creeper. Gradually I persuaded one of my cramped hands to release its grip. Immediately the shift of my weight caused the branch to slip still farther, until it was only by holding my head back that I was able to keep my nose above water. It was with the desperation of a gambler staking all upon a final throw that I flung myself sideways and found and clung to another stronger tendril of the tangled plant. I dared allow myself only a moment to rest before reaching out again for another branch, because I knew my strength was failing and though I was almost at the corner of the house, there were still several such moves ahead of me before I reached the place where the grassy earth tumbled into the water at an angle gentle enough, Hugo had said, to enable him to launch his boat.

I wondered afterwards how long it took me to accomplish that frightful journey. All I knew at the time was pain and fear, the weight of my clothing in the water, and that I had no strength left for gratitude when the moon rose and made my task comparatively easier. Eventually my feet found the muddy bottom, and then I was conscious of an almost overwhelming relief, but it was soon forgotten in the agony of the blood rushing back into my too-long-extended arms.

Soon afterwards I began to crawl up the grassy slope on hands and knees, shivering violently and wondering if my enemy were waiting to dispatch me at the top.

There was no sign of him, however, in the garden which was now quite brightly lit, and no sign of anyone else. Holy Mote slept in the moonlight.

I rested on the top of the embankment a little while, hugging my arms about me and wondering what I should do next. I could see the side door to which Mr. Perowne had taken the key the previous evening. If it were unlocked I felt strongly tempted to hurry straight to my room, to clean water and dry clothes—first making sure to lock and bolt the doors. But would it not be more sensible to show myself to Lady Perowne just as I was, and tell her the whole story while I still bore the evidence of it on my person? Yes, I had better do that, I determined wearily, for if I waited until the morning she would remember my history of nightmares and refuse to believe me.

I stood up. The wind seemed to whistle through me as if it were winter. My long wet skirts clung to my legs, impeding me, and my knees trembled so much I nearly sat down again. Though the night was quiet and still, I found myself as nervous as an unbroken filly, starting at the slightest sound, shying at shadows, my teeth and fists clenched in the expectation of a sudden attack that I would be unable to prevent if it came. Nothing happened, however, and I staggered unmolested to the door.

It resisted my first attempt to open it and for a moment I thought it was locked. Then it gave and I half-fell inside. I stood there, breathing heavily, and then began the weary ascent. I paused on the half-landing before my door, and was reassured to hear no sound within. Then I started down the silent corridors, quiet save for the steady beat of the great clock below and the little sound of the lamp flames flickering in the darkness.

I came to Lady Perowne's door and knocked upon it. Hearing no response, I opened the door and looked in. A lamp burned low upon the dressing table, illuminating a tidy room, an empty bed. Beyond, the door stood open to Jackson's room and here too all was silent. Lady Perowne and Sir Hugo were still out, it seemed, and Jackson was no doubt passing the hours of freedom with her peers below stairs.

On a sudden impulse, instead of returning to my room, I turned and forced myself towards the chamber that was Mr. Perowne's. Here, too, all was silent, neat and, I thought for a moment, empty. Then, as if the sound of the door opening had disturbed his slumbers, I was startled to hear Mr. Perowne speak in an agitated voice from the bed.

"Look out to the right there," he muttered. "Break for it, lads. Each man for himself. Oh, God!"

He was talking in his sleep. I tiptoed to the bed and peered through the curtain, wondering if his wound had taken an infection and made him feverish. He was lying quite still, however, and looked cool enough. I was not the only person in the world to suffer from nightmares, I thought, as I withdrew.

I went back into the corridor. A weakness began to overtake me. Great shudders convulsed me. Suddenly I began to realize what it was I had escaped, and by how narrow a margin. I had only one thought then, to find the sanctuary of my room before I broke down entirely. I clutched my sodden skirts in my shaking hands and stumbled along the passages, half-falling down the steps into the Stair Room at last.

My lamp still burned. The room was empty. The casement gaped open on the night but there was no other sign of the violence that had taken place. If my chair had been knocked over in the struggle, someone had righted it. If my pen had fallen to the floor, it had been picked up.

I began to blunder about the room like a dying moth, going first to the window where I looked out to see my shawl caught on the creepers just out of reach. I slammed the window shut and dragged the curtain across it. Then I made sure that the door was locked across the stairs. I bolted the other one and secured both with chairs wedged across them. I was too tired to light the fire. I stood as near the warmth of the lamp as possible while I stripped off my sodden weed-dank clothing with clumsy hands. I set my chattering teeth and washed myself all over with the water from the jug. I even washed my hair, icy though the water felt, and dried it roughly in a towel. I found my thickest nightgown, tumbling all the others onto the floor in the search, and crawled, crying and gasping, between the sheets, expecting to drift back immediately into that unnatural swamp of sleep that had been dogging my footsteps for hours, it seemed.

But the bed was cold and sleep eluded me. I sat up, wondering if I should light the fire after all, and my eyes fell on a small bottle of familiar shape upon my bedside table. It was Jackson's laudanum, but what in the world was it doing there? I had been drugged, of course, I thought dully. I had been drugged with the wine that accompanied my supper. The presence of this bottle here in my room would lead people to assume it had been by my own hand. They would think that I had taken laudanum for my toothache and that I had already been confused by it when I had leaned from the window to try to recover my shawl that had been skillfully placed out there, just beyond my reach. It would seem the most natural thing in the world that in those circumstances I should have overbalanced and fallen into the moat.

I lifted the bottle. Yes, the level had fallen considerably, even allowing for Mr. Perowne's having taken some.

An idea occurred to me and I uncorked the bottle and tilted it to my lips, for if ever I needed the healing oblivion it contained, it was now.

Afterwards I lay clutching my pillow, shaking and praying, until at last a soothing warmth crept over me, and I slept.

My awakening was late and painful. I came almost as slowly, reluctantly, to the surface of consciousness as I would have done if I had sunk one more time in the waters of the moat. My heart beat frighteningly loud, my mouth was dry, my eyelids heavy. When I moved I found I ached as if I had been beaten and my arms were particularly painful. But I had suffered no serious ill effect, I decided thankfully. I had no broken bones, no sore throat or congested breathing that might herald pneumonia. I wished fervently that I could spend the day in bed but I was obliged in any event to get up to unfasten the door for Randall, so I felt I might as well make a further effort and dress myself.

As it happened, Randall came in almost immediately with the hot water, and I enlisted her aid. She referred to the heap of wet clothing in the slop-pail and I told her, with a sigh for my own untruthfulness, that I had been lifting up the washbasin to empty it into the moat when it had slipped out of my hands and deluged me. This explanation she immediately accepted, merely remarking that it was to be hoped I would soon recover from the cold I had apparently caught, for thus I explained my helplessness.

I asked her, too, to help me recover the shawl, which I said had fallen during the same incident, and she took down the curtain rod and fished for it successfully.

"Don't mention it, miss," she said to my thanks. "Oh, and I did just notice your gown was torn—ay, that wet one. The sewing-woman won't be calling until the beginning of next month but I could set a stitch in it myself if so be you'd like it mended sooner."

"That is very kind of you, Randall, but I can do it

myself." I began to draw on my gloves. "Where is the sewing room? For I suppose you don't use the one upstairs?"

"No indeed, miss, that is only for the nursery sewing. There is another room for the maids to mend the household linen, near the dining room, and that is where the sewing-woman does her work. There, miss, just hold still a moment while I tie your hair-ribbon, so."

I looked in the glass, and sighed. I cast a longing look at the bed, visible in the mirror, and went as slowly from the room as an old woman, and so downstairs to the Breakfast Parlour.

Would anyone seem shocked or disappointed to see me there? I wondered as I fumbled with the handle, listening to the voices within.

"Disgraceful!" I heard Lady Perowne exclaim. "Everyone was discussing it after church. I am surprised you did not hear them, Piers—oh no, of course, you were not there."

"I call it exciting, rather," drawled Mr. Perowne. "Naturally, your position obliges you to take a more serious view of the affair, Mortmain."

Mortmain! I flung the door open upon the word and stood trembling upon the threshold, while the Earl sat there at his ease, his hand upon a pewter mug of ale, his once-handsome face rigidly controlled as he looked expressionlessly at me.

I let out a quivering breath. Impossible to speak to Lady Perowne before him, and the longer I put off the story of the previous night, the more incredible it must seem. Indeed, I thought, stepping forward uncertainly, I should soon cease to believe in it myself, no doubt, if my arms stopped aching and there were no further attacks.

"Good morning, Miss Fabian," said Lady Perowne, echoing the others, and a moment later, "Is anything the matter?"

I shook my head, all too well aware of what I looked like.

"I wondered," she said reprovingly, "because you are late for breakfast, and here is his lordship come especially to take you riding."

I caught in a breath that hurt my ribs. "It is very good of your lordship," I declared falsely, "but I am sorry to say I shall be unable to ride today. I have the headache, and I seem to have strained my arm."

"You are in need of exercise, no doubt," the Earl suggested, holding out a chair for me. As I stepped in front of him I smelt tobacco. I breathed in, closing my eyes for an instant in concentration. When I opened them I discovered it was his turn to stare, one dark eyebrow raised interrogatively, his mouth twisted in what could have been either scorn or amusement. I took my courage in both hands.

"Do you scent your snuff with Macouba, my lord?" I asked quietly, looking round at him as I seated myself.

Both eyebrows went up then. "Why, are you an authority on snuff, Miss Fabian?"

"I know a little about it, sir. What sort do you use?"

"Mortmain's Sort, and since I have always refused to divulge the secret of my blending, I fear I cannot gratify your curiosity, Miss Fabian. Macouba, however," he added with infuriating deliberation, returning to his place, "does not add its fragrance to my snuff—nor does Violet Strasbourg, despite the example of the Queen, which I understand most females prefer. I believe I do possess some Macouba, still, if you want some. I used it at one time, in my impressionable youth. It was the only kind a friend and I were able to obtain, when we felt we had reached an age for such adult delights. . . ."

All the time he was speaking, he kept his eyes on me as if he were really interested to know why I had asked that question.

"Perowne there uses it, do you not, Piers?" he added carelessly. "Having neither nose nor palate, he has remained faithful to his first snuff and never felt the desire to experiment."

"Oh Lord, it is all nonsense the way you fellows talk of snuff," said Mr. Perowne easily. "It must not be too dry or too damp, or too hot or too cold; you must not add this, you must not drench it in that. I have better things to do with my time, I hope. I don't take it often—only when we have company, I suppose."

"And it is scented with Macouba, sir?" I persisted, stirring my coffee.

"Oh, ay, I believe it is, now that you come to mention it."

"I hope you are not using delaying tactics, Miss Fabian," remarked Lord Mortmain. "Nor attempting to draw a red herring across the track? For I have more patience than perhaps you are aware and shall not so easily be distracted from the purpose of my visit."

"If it is to take me riding, sir, I have already explained why I shall not be able to join you."

"You have enumerated your reasons for preferring not to do so," he agreed, "but I have failed to accept them, ma'am."

Mr. Perowne set down his cup. "What a cursed brute you are, Mortmain. Can't you see how pale Miss Fabian is?"

"Certainly I can. Her pallor is induced by fear—and it is my intention to banish that fear forever."

Lady Perowne glanced at me with sympathy. "Do you think it right, Cousin, to force people to overcome their fears?" she protested.

"Indeed I do," he replied implacably. "Miss Fabian is trembling like a jelly now but by dinnertime you shall see how firmly she is set."

"I am not afraid of riding," I assured him, and indeed that fear seemed nothing now. "It is the pain in my arm which disables me for the exercise."

Lord Mortmain looked skeptical, and I remembered that he had to overcome pain when he was riding—and yet he still managed to look a part of the horse. For some reason then I suddenly remembered the horseman I had seen on the Heath, that day at school. I had barely glanced at him, being too busy reproving Emily for boldly returning his stare, but now I wondered if he could possibly have been the Earl. That man had been a consummate horseman, so much I had noticed; but otherwise I had thought him nondescript, which the Earl certainly was not. But then the rider had worn a high cravat and the brim of his tall black hat had curled low over his brows—an effective disguise. No, I thought, it was impossible to be sure.

"Have you a black horse in your stable, sir?" I asked abruptly.

He was startled indeed. After a moment he answered warily, "I have horses in my stables of every colour, Miss Fabian. Horseflesh is my greatest extravagance."

That was inconclusive, then; but I was thinking that if I saw the horse I might recognize it, and wondering whether or not to say any more on the subject when Mr. Perowne remarked in a grumbling tone that the only black horse in these parts was Holloway's Midnight. "And a rare animal he is, too; and it will be long before I forgive Smithers for not allowing me a first refusal on him—"

"Holloway is a warm man and has more to offer," the Earl remarked. "Your friend Smithers is more influenced by money than friendship when it comes to horse-coping, I believe. Now, Miss Fabian, that was a noble try at a change of subject, but it will not do, for I mean to know how you hurt this arm of yours."

I stared at my plate. For the life of me I could think of no good reason to account for the kind of strain I had suffered in hanging from a creeper which was above my head.

"Exactly so," he murmured. Before I could divine his intention, the Earl had actually lifted one of my arms and was beginning to put it through a series of movements.

"Please, sir," I cried. "You are hurting me."

"But I don't believe you, Miss Fabian."

"Lord, ma'am," said Mr. Perowne, "if your great eyes won't move him, you can't expect your words to do so. Come to think of it," he added casually, "I've a stiff shoulder myself today. Must have been a change in the wind last night."

As if the wind had blown open the door on his words, Hugo and Old Nurse entered. She looked wretchedly ill, I thought, while Hugo by contrast was flushed and seemed almost like an ordinary healthy boy for once.

"Uncle Piers," he shouted, "you have bought my new pony!"

"Oh, has he arrived? What do you think of him?"

It was plain that Hugo was delighted with him. "He is so friendly—not like Snapdragon—"

"Have you hurt your arm, miss?" asked Old Nurse in a curious tone. I became aware that Lord Mortmain was still holding it loosely, and jerked it free.

"Yes, Mrs. Cumber. I have strained it—reaching up."

"Have you indeed, miss?" she said with ill-concealed satisfaction. "Would you like me to attend to it for you?"

Lord Mortmain spoke in the cool way which even Old Nurse did not dare to question. "No need for that, Mrs. Cumber. Miss Fabian seems merely in need of rest. Mr. Perowne could better avail himself of your services, perhaps."

"Ay, you do looked peaked, Mr. Piers," she said severely, and just as if she had no knowledge of his wound. "Come up to my room directly and take a dose."

"Very well, Nurse," he said meekly.

Lord Mortmain turned his unfathomable eyes on me. "Since it seems that you are incapable of eating anything this morning, I am sure Lady Perowne will excuse you," he suggested.

"Oh, yes, indeed, Miss Fabian," she said quickly. "First your toothache and now this—I am sure you are quite worn out. I shan't need you today, and as for Hugo, nothing would distract him from his new pony, I fear.

You had better go to your room and have a quiet day."

I sighed, aware that much as I wished to reject the Earl's prescription, it would suit me perfectly. With mixed feelings, therefore, I curtsied and retired, to spend the day in dozing and asking myself fruitless questions, falling asleep at last through sheer exhaustion.

13

The next day I woke refreshed, to find my fears had dwindled proportionately. I went downstairs determined to begin to teach Sir Hugo; but again I was to be frustrated in my intentions, as I discovered when I asked Lady Perowne where the schoolroom was.

"The schoolroom?" she repeated abstractedly. "Oh, it is opposite Old Nurse's room. But Sir Hugo is not fit for lessons today, poor lamb. He is ill—so heavy-eyed we could hardly wake him. Old Nurse fetched me to him quite early this morning. She says he has been overtaxing his strength, with all this riding and walking."

I stared into my coffee cup. "You told Mrs. Cumber that I intended teaching Sir Hugo today, I suppose?"

"Certainly. I am in the habit of discussing what he is to do the following day each evening with Old Nurse—unless I have company, of course."

"And is he often ill like this, ma'am?"

"Pretty often, poor child—but a day abed generally puts him to rights."

An ugly suspicion was growing in my mind. I recalled some of my own symptoms of the previous morning. "Is Sir Hugo's pulse fast and uneven?" I asked in a low voice.

"What was that, Miss Fabian? Oh. I really could not say. I did not take it."

"May I visit him, ma'am, when we have breakfasted?"

"I don't believe Mrs. Cumber would allow it, Miss Fabian."

I decided to concoct a story, for Hugo's sake, to alert his mother. It was one, moreover, that might well have a grain of truth in it, I reflected. I said slowly that I had borrowed a bottle of laudanum from Jackson when I had had the toothache. "Unfortunately I mislaid it and when eventually I found it a good deal more had gone from it than I had used. I thought little of it, assuming it had spilt, perhaps, but now it occurs to me that Sir Hugo may have found it and taken some for a game of his own. He might have poured it into another bottle, and drunk it in the night."

"Good Heavens! Do you think it possible?"

"From what you say, it sounds possible at least that he took some drug," I said carefully, remembering that according to Mr. Perowne, Old Nurse kept laudanum "by the gallon." "If he were my child, ma'am, I should call the doctor."

Lady Perowne twisted her napkin, looked at it in astonishment, and fell to twisting it again. "But Old Nurse detests Dr. Burton," she pointed out. "She prefers to treat Sir Hugo herself when he is ill. She is so good with him, she understands his constitution."

"She has not cured him, however—and even the best doctors sometimes seek a second opinion. In this case, ma'am, when Sir Hugo may have been—may have poisoned himself, I think you should take no chances. I entreat you, if not for his sake, for mine, who may unwittingly have been the cause—"

"Yes, you are very right," Lady Perowne declared, ringing the bell. "Dr. Burton is a good man, though Nurse dislikes him. Percival!"

"Yes, my lady?"

"Send a groom for Dr. Burton at once. Tell him that Sir Hugo's life may be in danger."

I jumped up as the butler hurried off. "Ma'am, I hope I have not alarmed you unnecessarily. It is very possible that Old Nurse does know what to do for the best, having had your son in her care so long. But even if he took no laudanum, it can do no harm for the doctor to see Sir Hugo in one of his attacks—"

"Don't disturb yourself on that score, Miss Fabian," Lady Perowne told me as she crossed the hall. "It is better to be on the safe side and I shall not blame you if this turns out to be a false alarm." She added more breathlessly as we hurried up the stairs, "Nurse will not welcome us. She says he needs to sleep undisturbed, and that is all he needs."

We entered the Great Room. It was dark, all the curtains closely drawn and no air stirring there on this fine early summer morning. A fire burned on the hearth, increasing the closeness of the room despite its large proportions. Old Nurse was setting a nightlight on the bedside table, her shadow thrown upward onto the intricately moulded ceiling where it hovered malevolently, curiously dislocated by an insistent shaft of sunlight.

She turned slowly as we entered, and sketched a curtsey. Then, putting one finger to her lips, she began to approach us. "Pray don't wake the young master, ma'am. He is sleeping, as I told you he would."

"But I must just look at him, Nurse," Lady Perowne explained apologetically but with a certain authority. "Be so good as to draw back the bed-curtain."

Mrs. Cumber did not dare protest, but her stays creaked as if in sympathy with her evident reluctance as she pulled back the velvet curtain. Within his father's great bed, Sir Hugo looked very small and pale as he lay among heavily embroidered pillows, his mouth slightly open as he snored, oblivious to all.

Lady Perowne caught a sobbing breath. "I don't like the look of him, Nurse. I must tell you that I have sent for the doctor. If he was at home when the message reached him, he should be here in ten minutes or so."

Mrs. Cumber looked angrily from Lady Perowne to me. "The doctor, ma'am?" she repeated incredulously. "I should like to know why, when I have nursed Sir Hugo all these years without complaint, so far as I am aware."

Lady Perowne looked rather taken aback by this outburst. "You forget yourself, Nurse. I suppose I may call a doctor to my son if I so wish?"

"If you do wish it, ay, with the best will in the world, my lady—but I suspect it was miss here that asked for him and not yourself, ma'am, that has always been perfectly satisfied with me before this."

"You are not to be offended, Mrs. Cumber. The fact of the matter is that Miss Fabian was aware, as you were not, that a bottle of laudanum was missing in the house and it is possible that Hugo drank some, which would certainly account for this unnatural sleep."

"There is nothing unnatural about it, so please you, ma'am. If miss had not walked the poor young gentleman off his feet and his lordship had not forced him into the saddle for hours at a stretch—but there, I could see how it would end but I knew no notice would be taken of me. His body won't stand any more, that is all that ails Sir Hugo, and he'll be well enough tomorrow or the next day, if so be I have the minding of him, and not some pert miss just out of school."

"Oh, Nurse, look," cried Lady Perowne. "He's so restless now. A moment ago he was sleeping like—like one dead, and now he is tossing and turning as though he can't stay still. There, there, my darling. Hush now, my poor Hugo. Oh Nurse, I am so worried about him."

I felt as if I were intruding on her distress and went over to the window. I pulled the curtain aside and looked out. It was a fine day. The dark circle of Holly Wood was starkly outlined against the innocent-looking azure sky. Beneath it the Downs swept in shadow to the village where the roofs again caught the sun, some gleaming slate, others a softer gold where red tiles had been overgrown with lichen. I could see the square tower of the church among the black yew trees and the bright contrast beyond of an apple orchard, pink with blossom. Nearer home, a flock of sheep grazed near the quarry; they scattered and then bunched together as a man rode up the meadow path, cantering easily on a handsome horse. It was Lord Mortmain, not the doctor as I had thought at first. A bird flew up under his horse's feet and the big bay shied violently, but his rider controlled him easily despite the fact that he had one foot out of the stirrup. He rode on into the wood and I turned my attention instead to watching the birds wheeling in the unblemished sky.

Some minutes later a sudden commotion broke out in the room behind me. I turned, letting the curtain fall from my hand, as Lord Mortmain strode into the room demanding to know what was the matter with his young cousin.

"Good God," he exclaimed in disgust, "no wonder the child is ill in such an atmosphere as this. Can you not open the window on such a fair day?"

"I'll not be responsible for Sir Hugo if you do, my lord," Old Nurse informed him, bristling.

"I vow I can't understand how he can have ailed so suddenly," Lord Mortmain said to Lady Perowne. "I believe I never saw him in better health and spirits than he was yesterday."

Lady Perowne put a hand to her mouth and glanced at me as if she knew not what to say. Lord Mortmain looked in at Hugo while Old Nurse hurried protectively forward.

" 'Tis often so, sir," Mrs. Cumber took it upon herself to reply, "as your lordship would know, if you had but half my experience."

"Well, well," he said in a softer voice, "I don't much like the look of him, I must own. Do you know, Cousin, I would send for Dr. Burton if I were you."

"I have already done so," Lady Perowne admitted, with a side glance of gratitude for me.

"I am glad to hear it, and shall be interested to learn what he has to say."

At that appropriate moment the doctor arrived. He too strode into the room, exclaiming at the heat and gloom. He took a quick look at his patient and then pulled back the curtains from the windows, remarking that he could not be expected to examine a pig in a poke.

As he turned back to the bed, his eyes met those of Old Nurse in a glance of the frankest animosity. " 'Tis not the first turn of this kind the young fellow has had," he said brusquely, "but it is the first time I've been called in to attend him for one. I am glad you had the good sense to do so, for I'm more accustomed to having my advice disregarded in this house." He bent over Sir Hugo, who was again stirring restlessly, and raised one of his eyelids. "What's this?" he cried, and looked into the other eye. "Who has been giving the child an opiate? Is this your doing, Nurse?"

"Oh, Miss Fabian," cried Lady Perowne impulsively, clasping my arm. "I am so grateful to you!"

"You are too generous, ma'am," I assured her, remembering that I had pretended to be guilty of the negligence that accounted for Sir Hugo's condition.

"No, no—I do not blame you—oh, the naughty boy, it was his own fault—" She sobbed.

"Why should you blame Miss Fabian?" Lord Mortmain quickly asked. "Or praise her, for that matter?"

"Miss Fabian insisted on my sending for the doctor—she lost some laudanum—"

He frowned. "And it seems that Hugo found it? Damnably careless of you, Miss Fabian." He stared at me. "Are you commonly so forgetful, ma'am?"

"No, sir," I murmured, wishing that Mr. Perowne were here to take some of the blame which was being apportioned to me, unjustly but by my own fault, I had to admit.

"When did you lose this bottle?" the Earl went on, drawing me to one side.

"Oh . . . a few days ago. Saturday night or Sunday morning," I added more particularly under the influence of his questioning stare.

"The night I came to dine?"

I nodded, telling myself that now I knew what it was to be a rabbit entranced by a weasel.

The doctor had begun pinching and slapping Sir Hugo,

explaining forcibly as Lady Perowne cried out in protest that the boy must on no account be left to sleep but must be made to walk about and drink quantities of coffee, strong and black, as soon as it could be made.

Lord Mortmain drew me from the room. "Well, Miss Fabian, in plain words, how did you come to lose that bottle of laudanum?"

"If you must know," I said, goaded beyond the point of discretion, "I lent it to Mr. Perowne. You know he was complaining of a stiff shoulder yesterday. . . ."

The Earl released my arm. "A stiff shoulder," he repeated in a thoughtful tone. "And little more than twenty-four hours earlier—" He broke off, frowning. "There could be a connection—but would he be such a fool?" He stared at me as if I had dared to contradict him. Then with an air of extreme irritation he answered himself. "Yes, damn him, he could. What do you know of this, Miss Fabian?"

But on the subject of Marie my lips were sealed. I gave the slightest of shrugs and bade the Earl a cold farewell. He detained me a moment longer. "Very well—for the time being, Miss Fabian. As for Hugo, I beg you not to torment yourself. If he swallowed the stuff last night and is still able to be roused this morning, then he could not have taken enough to kill him. The danger now, I imagine, is of a congestion of the lungs, but Burton will see to it that it does not come to that. I advise you to put the whole affair out of your mind, for there is no more you can do at present."

I showed him by my look that I thought him heartless. In fact, as soon as I returned to the room I discovered that Lord Mortmain had been right, for Sir Hugo was now being walked about the floor and the doctor was preparing to take his leave.

By the next day, Hugo was well enough to ride. I left him cantering about a field in circles, looking quite cheerful despite the merciless lash of his instructor's tongue, and walked across the meadows to the church, where I said grateful prayers for his recovery and begged divine protection for myself. As I rose from my knees and sat back on the oaken pew I realized precisely what it was that had been worrying some corner of my mind with regard to my own nearly fatal adventure. I had somehow been convinced that Old Nurse must have been concerned in the attempt, that while attending to Mr. Perowne she had found Jackson's bottle of laudanum and drugged me with it, but though I knew she disliked me extremely I could not reconcile her action with her equal dislike of Lord Mortmain. Even to rid herself of me, whose presence at Holy Mote undoubtedly aroused her jealousy to a dangerous degree, would she ally herself with one she hated and who, she knew, longed to restrict her power? And by the same token, would Lord Mortmain risk such a dangerous alliance?

The answer, I thought, must certainly be that neither of them would consider helping the other, even for a common cause; but that conclusion either cleared Lord Mortmain as the villain, or cleared Old Nurse of being my enemy's accomplice. In the latter case, I did not see how Lord Mortmain could have found the opportunity to administer the drug to me, yet if Lord Mortmain was entirely innocent it must be Mr. Perowne who was guilty.

Mr. Perowne. . . . He had some motive, in that he might suspect me of betraying him in the matter of the letter from Marie. His alibi of seeming to be heavily asleep just after the attempt on Sunday night, I had already reflected, was not absolutely convincing. And there were other factors that seemed to point to him: his involvement with the affair of the intruder in the attics, whether real or imaginary, and the fact that Old Nurse no doubt would lay down her life for him. And always, running like a black thread through the whole pattern, was the question of witchcraft.

I looked at Mr. Pole-Carter, a neat figure in his black coat and starched bands who, as I stepped down from the pew, was standing at the lectern apparently glancing through next Sunday's lessons. He knew a good deal about witchcraft, I reflected. If my knowledge equalled his, would it lead me to the truth?

I sighed. Mr. Pole-Carter looked up and hailed me affably. He asked after Lady Perowne and when I had dealt with that question in sufficient detail to satisfy him he enquired for Sir Hugo, who, Mr. Pole-Carter had heard, had eaten something that had disagreed with him.

"You could say so," I returned discreetly, "but he is now quite recovered and having a lesson in riding this morning from Lord Mortmain."

"Ah! And Mrs. Cumber? You do not know what I would give to have your opportunity of living beneath the ancient roof of Holy Mote with the chance to discover the secrets of an authentic witch—though she would tie knots in me, doubtless, if I tried to learn from her."

"Is tying knots a—a witch's pastime, sir?"

"Oh, ay, certainly. Knots have great significance in the black arts."

I ran my gloved finger round the dusty crevices of a *fleur-de-lis* carved in the end of the Quentin pew. "I am so ignorant, sir. I suppose being a witch must certainly affect one's moral sense?"

"Oh, yes, indeed," he replied cheerfully. "Their concern is with power, which should properly not interest Christians at all. Witches may use their power for good as well as evil, but in either case I am certain they become corrupted by the one that tempts them to it—I mean, of course, their Master."

"So the one who rides Herne is also likely to be—corrupted?"

"Why, certainly. Herne is the greatest position of power in the coven. Only a properly initiated witch can lead the Wild Hunt."

He closed the great Bible and stepped down from the lectern. "May I escort you home, Miss Fabian? I have been asked to dine at Holy Mote in any case."

"Oh . . . yes, if you please, sir," I said abstractedly.

"Perhaps you would be good enough to tell me, what is a Wild Hunt?"

He closed the church door and we walked into the graveyard, all sparkling with dew in the spring sunlight. "The Wild Hunt," he repeated musingly, and for a moment I thought he was not going to satisfy my curiosity. However, he went on as he opened the lych-gate. "It is precisely that, Miss Fabian—a wild ride, a gathering on nights of the full moon or those occasions such as Candlemas, known as the Witches' Sabbats, when persons of that persuasion saddle up and ride the countryside after Herne, their leader. Some believe Herne to represent the Devil, but others, myself among them, know him as the horned god of the Old Religion and quite distinct from Lucifer. I have heard that they dress up in strange clothes and masks of animals to avoid discovery, and that they cry out to frighten away any unwary traveller who might otherwise chance to see them, for, as perhaps you know, if any person who is not a witch or of their following happens to set eyes on such a hunt, his death, tradition has it, is certain within three days. Unfortunately," he sighed, "though one has heard that Wild Hunts are common enough in these parts, I have never met anyone who dared to confess to having witnessed one, so I am unable to tell you what truth, if any, there is in such a superstition."

I parted my lips, about to tell him that I had witnessed a Wild Hunt on Saturday night, and survived it; but then I decided with a slight involuntary shiver that perhaps so distant a view as I had had when I glimpsed that ride from the window of Sir Hugo's room could hardly be said to count.

I remarked instead, "It must be almost time for the full moon. Does it not occur to you, sir, to lie in wait for the Wild Hunt yourself?"

"The moon is at the full tonight, which is occasion for an Esbat, Miss Fabian. I don't deny the thought of spying on it has occurred to me, but it is a temptation I feel bound to resist. There are some things, I believe, into which one may enquire but to take positive steps might be construed as either interference or condonation, either of which, it seems to me, would be a grave mistake."

I turned the conversation there, but though we passed the remainder of the walk in discussing horseflesh and snuff and painting, my thoughts were involved with the forming of a great and alarming resolution. It seemed to me that if I hid myself on the outskirts of Holly Wood near the track that led to the inner glade I would be in a position to see the Wild Hunt and perhaps identify its leader. Even if I failed to do that, for he would be masked, I might well be able to discover if he was the same antlered person who had twice invaded my room with the intention of killing me. It seemed a chance, and the only one that occurred to me, to take a positive step towards what I might well call the unmasking of my enemy. If I lacked the courage to take this chance, I thought with hardening determination, then perhaps I would almost deserve to die, for was not some measure of blame due to the rabbit for failing even to attempt to escape the weasel's wiles?

All too soon, dinner was over, darkness had fallen and we retired to the green drawing room to find firelight and candlelight reflected elusively in the silken walls and highly polished furniture. At Lady Perowne's request, I sat down to play the piano, while Mr. Perowne flung himself into a chair and declared his intention of taking up sailing again this summer.

"I suppose Mortmain will not object to it if I have the keel laid down for a small cutter? I was thinking of commissioning White to build her, over at Cowes."

"But, Mr. Perowne, do you think a pleasure yacht in time of war . . ." Mr. Pole-Carter's voice died away uncertainly.

"I think it is a splendid idea!" Lady Perowne cried, clapping her hands. "Only think of how Hugo will enjoy it, later in the summer, when the weather is warmer."

I played on softly while they discussed the project, wondering if Mr. Perowne would ask for the key to the side door, but since he did not I decided he must have thought better of joining the Wild Hunt tonight; and of course I had no proof that he had ever done so, beyond a fragment of overheard conversation. His plans, however, did not affect mine. Mr. Pole-Carter took his leave, Mr. Perowne retired, Lady Perowne kissed me good night with unusual affability, and I closed the door of my room behind me, suppressing a cowardly wish that I too were going to my rest.

I dressed in a plain grey gown, with my dark cloak over it, and sat down to wait. When it seemed late enough for everyone else to have gone to bed, I turned down my lamp and slipped quietly down the stairs to the damp, ill-lit stone passage which led to the side door. I was fortunate here, for the key was in the lock. I took it, locked the door behind me and put the key in my reticule. The night felt cool after the warmth of the day and I wrapped my cloak closely about me as I crept through the grey garden, past the head of the moat and onto the mown verge of the drive. I went a little distance away from the house before crossing the noisy gravel. Once I had done so there were trees to shelter me from discovery if anyone should come past.

I found no enthusiasm in myself for the mission on which I was engaged. If someone else could have been found to watch the Wild Hunt and report on it faithfully to me, I should have been enormously relieved; as matters were, however, I tried to put the grim climax to my adventure out of my mind and concentrate instead on what I did enjoy, quite unexpectedly: the excitement of being outside at night, alone and feeling at one with nature. I was nervous, to be sure, starting when an owl swooped silently past; I flinched when shadows moved across my path and leapt at the cracking of tiny twigs, but I was not nearly so afraid as I had sometimes been beneath what the vicar had enviously called the ancient roof of Holy

Mote. I found myself delighting in the feel of the cool dew soaking my light boots and the hem of my skirt. I revelled in the scent of the wood which seemed so much stronger at night, a smell compounded of young bracken, green twigs, old leaves, trodden grass, a sudden breath of primroses. It seemed to me that to feel this kind of harmless untrammelled ecstasy was justification enough for leaving the security of school.

When I came out onto the meadow path I felt more vulnerable. I knew that here I could be seen from the house, for the moon, though not yet risen, was casting its radiance before it, filling the valley with opalescent mist which was, if not by any means as bright as day, almost as clear as twilight.

I hurried down the path and when I reached the broken railing by the quarry I paused, breathless, and clung to the scant cover of the fence as enthusiastically as if I had found sanctuary there in that unholy place. When I had regained my breath and my courage I continued more sedately down the path.

There was a green ride that skirted the village and I took it thankfully, entering it through a broken gate hard by the church. It occurred to me that it would not be long before moonrise and if I did not hurry I would be exposed by its light on the shoulder of the Down before I could reach the ominous shelter of Holly Wood. I had just begun to mend my pace, therefore, when I became aware of a throbbing in the ground which soon resolved itself into the steady drumming of hooves.

Overtaken by a sense of panic, I flung myself sideways into a thicket of brambles and almost missed seeing the first horse that passed me. But I did look up just in time to feel, first, a sense of shock that the rider was wearing a horned mask, and secondly a sick surprise as I recognized his horse.

It was the chestnut mare with three white socks, recently acquired by Mr. Piers Perowne.

Then I could see him no longer for the ride was

thronged with horsemen, all masked, a phantasmagoria of cats and dogs, sheep and goats, urging on their horses with human legs in the stirrups, and some with wild-haired women riding pillion, clinging close. As for the horses, they seemed as excited as their riders, eyes rolling, nostrils flaring, their sharp hooves flicking up tufts of grass and clods of mud into the air as they thundered past smelling of leather and of sweat; saddles creaking, bits jingling, breath panting, some bounding forward, others slow and already straining eagerly to keep up. There were twenty or thirty or forty of them, perhaps—and then suddenly they were all gone as if they had never been, except for the fact that I was crouched among the thorns with a labouring heart, and the green ride before me had been turned to plough by over a hundred hooves.

I burst into tears, I know not why. Some small part of it may have been the sort of envy felt by a tinker's child who looks through the rich merchant's window on Christmas Day—they were together and knew where they were going, while I lay outside, cold and alone. Then, I had expected to have a long walk before me, an arduous climb and a nerve-wracking wait in that hateful wood—and it was all over so soon. But perhaps most of all, I had prepared myself to find that Herne was the Earl of Mortmain, though I hardly knew if the shock of discovering myself to be mistaken was one of relief or disappointment.

And yet, I reminded myself as I dabbed at my eyes with a corner of my cloak and stumbled out of the bushes, it was Piers Perowne whose voice I had first heard in the secret part of the attic, who had sworn by Herne and spoken of the moon goddesses and of the Wild Hunt. Then, too, he had been sent down from Oxford for heresy; he had been brought up by Old Nurse at Holy Mote. What more likely, then, than that he followed the Old Religion that was the central core of witchcraft, and that he of all the coven was the one judged fittest to enact the part of Herne?

I thought of other things then: the soft leather bag

that Born-To-Be-Hanged had carried the night that Mr. Perowne had been wounded, and the fact that he had gone up into the attics instead of leaving when Mr. Perowne went into my room. That bag was large enough to have contained an antlered mask, it had the very spreading shape of it, and it seemed reasonable to suppose that the paraphernalia of witchcraft was kept in the attics of Holy Mote. I had no great love for Mr. Perowne, it was true, and yet it seemed terrible to think of his being involved in such a thing. There was no doubt also, I reflected, that the fact that he could bring himself to ride Herne made it easier to believe he might be the antlered man who was my enemy. As for his motive—suddenly I realized I had told him myself that I had overheard that first incriminating conversation.

I gave another gasp and the tears ran down my face and blinded me as I stumbled down the ride. My scratches all began to sting together and my ankle ached as if I must have twisted it when I threw myself into the brambles. I began limping along as quickly as I could until I realized all at once that I was still moving away from Holy Mote, and the tears sprang out afresh at the thought of having wasted so much effort. I stood there with my head in my hands for a moment, sobbing like a child.

Then a hand grasped my arm and I leapt back with a strangled cry.

14

"Who are you?" I whispered, staring at the bent old woman who had accosted me. Even in this dim light, her wrinkled but strangely blank face did seem slightly familiar, as were the straggling grey locks escaping from her cap, the black shawl bunched about her stooping shoulders.

"Who are you?" she echoed, but tonelessly, as if she had not much expectation of being answered. "What do you here on such a night? Best get you to cover, ay. My gentleman? No, he'd not like it for all he's out. The Vicarage? But best not, the Reverend being a bachelor. The Knoll, ay, that's best." She took my arm in a surprisingly strong grasp and began to hurry me along, deaf to my protests.

Deaf, it suddenly occurred to me, was precisely what she was. She was Mrs. Randall, of course, the mother of Randall, my chambermaid, who worked for a gentleman at the Pines, the one who wore a wig and had bowed to me by the gate of Lindenfold.

I struggled to free myself from her grip, but Mrs. Randall had apparently formed some distinct idea in her stubborn mind as to who I was and what she must do with me, and she had no intention of parting with me just yet. We must have made a curious sight as the moon rose to illuminate us, myself bedraggled, limping, dragged onward against my will by the small determined, bony woman at my side.

We came out onto the road again just beyond the village. A few paces farther on we passed between neat white gates and down a well-kept drive. I began to wonder what in the world I should say to the Curtises when I found myself abruptly thrust into their family circle, as I promised shortly to be.

Passing through a luxuriant shrubbery, we approached a white-pillared front door. Mrs. Randall changed her grip and held me no less a prisoner while she drew out and released the large brass bell-pull. A tremendous clanging within the house seemed to communicate itself even to her, for she nodded sagely several times and stood back a little in anticipation of the door's being opened.

This happened rather suddenly; certainly long before I was in any way prepared for it. A pink-faced butler appeared with something of the bearing of a bishop and stood staring at us in mild astonishment only expressed by the slight elevation of his neatly brushed grey eyebrows.

Mrs. Randall broke an embarrassing silence. "This poor young thing," she announced in her monotonous voice. "Found wandering, crying her heart out—is my lady within?"

"Her ladyship is entertaining the Quality," the butler replied austerely and in a loud tone, to which Mrs. Randall listened anxiously, a knotted hand cupped to her ear.

"What's that you say about jollity?" she demanded aggressively. "Speak up, young man. This ain't no time for jollity, let me tell you. Repent ye, for the Kingdom of Heaven is at hand. Fetch my lady here, for she has helped me before with my young persons. Powerful good to 'em, she is—specially them as did ought to know better."

The butler, whose complexion had by this time assumed the colour of the port under his care, as no doubt had mine also, shouted at the top of his considerable voice, "My lady is not at home, I say! Be off with you, Randall, and call tomorrow."

Mrs. Randall shook her head. "Not at home, you say? But there she is, yonder!"

The butler turned with ponderous annoyance, and I

shrank back as Lady Martha crossed the hall towards us. She looked extremely regal, decked in silk and diamonds, and her handsome forehead was creased with an uncharacteristic frown.

"What is all this disturbance, Manley?" she asked, mildly enough. "Oh, is that you, Mrs. Randall? Good gracious, can that be Miss Fabian? What has happened? Has there been some accident? But don't stand there on the step, my dear. Come in, come in."

Mrs. Randall, with an incoherent mutter of satisfaction, disappeared into the night, leaving me to explain matters for myself. The butler shut the door disapprovingly behind me, and Lady Martha drew me towards a room from which streamed the warm light of many candles. I hung back, aware of the murmur of conversation within, pressing my hands to my tear-stained face, but she drew me on as inexorably as Mrs. Randall had done.

The next moment I understood Lady Martha's strange insistence on exposing me, dishevelled as I was, to the gaze of her dinner guests, for one of these, I saw immediately, was Lord Mortmain, himself immaculately turned-out, as usual, and apparently quite as horror-struck by my appearance as his hostess could have wished him to be.

He jumped to his feet, demanding to know what had happened to me, while Mr. Curtis rose more slowly and with every appearance of irritation at the interruption. There were other guests in the room, a military gentleman, and another who seemed to be rather drunk, with a bewildered-looking wife.

Miss Curtis hurried forward. She flung her arms about me and asked if I would not rather go up to her room. I murmured a heartfelt affirmative, and she turned protectively. "Pray don't stare at Miss Fabian so, Papa, and Lord Mortmain," she exhorted them. "Have pity on her. Can you not see she does not like it?"

Lord Mortmain disregarded this appeal. "I must and will know what has caused Miss Fabian's appearance here, in this deplorable condition."

I shook my head, feeling myself quite incapable of satisfying his curiosity, and Miss Curtis cried that I was in no state to be questioned. "May Miss Fabian not stay with me tonight, Mama?" she entreated.

It was her father who answered sternly, "Miss Fabian's duty is to remain at the side of Lady Perowne."

"You are very right, sir," said Lord Mortmain swiftly. "If you will excuse me, and you, Lady Prendergast, Sir Abraham, Colonel Horton—I shall drive Miss Fabian immediately to Holy Mote."

Lady Martha assured him that it was not necessary for him to go, one of their grooms could perform that office; but Lord Mortmain insisted that he was in some sort responsible for me. Accordingly, his carriage was sent for and a hot drink was brought for me to sip while waiting for it to be driven round to the front door. After a few moments I recovered sufficiently to speak to Lady Martha.

I owed her some sort of explanation, I told her, after apologizing for breaking into her party. "Nothing dreadful has happened to me," I insisted, looking down into my glass of mulled wine. "I was foolish enough to think it sport to take an airing in the woods—such a lovely night—I tripped and fell into some brambles. Mrs. Randall found me—she would not heed my protests—could not hear them, indeed, and insisted on bringing me here—she does not know me, of course, and I could not make her understand—"

"No, well, it is hard enough sometimes. She works hard for good and has been of help to many poor creatures." Lady Martha patted me bracingly on the shoulder. "I am sure when I saw you there upon the step I took you for another of them! But I am glad to hear it was no more than a foolish escapade, as you say."

"Come, Miss Fabian," commanded the Earl, receiving his cane, hat and gloves from the disapproving butler. "The carriage waits."

He bade his hosts farewell, I murmured another apology, Mr. Curtis came out politely to see us into the carriage, the door closed behind us, the coachman cracked his whip and we began to move forward.

Lord Mortmain at my side folded his arms and stretched out his legs. "Now, Miss Fabian," he said grimly, "that explanation was good enough for Lady Martha, perhaps, but I am far from satisfied with it. What in the name of thunder were you doing out-of-doors alone at night?"

I knew I was going to cry in a moment. As I could not answer him, I told myself there was no necessity for me to do so and turned my head to look out of the window. As the silence grew heavy, I shrugged one shoulder slightly.

I heard his indrawn breath. Then the Earl's hand fell on my arm and he pulled me round to face him.

"Miss Fabian," he said again between his teeth, "I swear I am in no gentle humour, ma'am. Why did you leave the house tonight?"

I bit my lip and shook my head.

Lord Mortmain groaned. His arms slid round me. He said in an odd voice, "Do you want me to kiss you, Miss Fabian?"

"No!" I gasped, and began to struggle. But somewhere at the back of my mind was the thought that perhaps I should like it after all.

He held me more tightly. "I thought you would say that," he murmured with apparent regret. "Is it because you are a well-brought-up young lady, I wonder, or because you are revolted by my scar?"

With a slight feeling of surprise I realized that I had forgotten he was scarred. "It does not revolt me, sir," I informed him. "You are extremely handsome." As soon as I had spoken I wished I had phrased my reassurance less encouragingly. Hastily I added, "But of course I should have refused to travel with you if I had known you intended to insult me thus."

"No insult was intended, Miss Fabian. But if you do not desire me to kiss you, then you had better tell me all."

"You would blackmail me, sir?"

"I would, Miss Fabian."

I said in a stifled tone, "Then—if you release me—"

"With reluctance, ma'am. However, if you choose to look on what feels so right and natural as an insult—"

He took his arms away and, absurdly, I felt cold and shivered. "I am waiting for your explanation," he prompted me.

"Yes. . . . It was a fine moonlit night—is still, as you can see. I—could not resist the temptation to go walking, foolish though I knew it to be. After so many years of school, of continual discipline, there is a temptation to be foolish—"

"Do I not know it!" he muttered. "Go on, ma'am."

"I thought—it seemed rather a thrilling notion to look at the church by moonlight. I began to go down that ride in order to see it from the other side. Suddenly a—a troop of horses galloped past. I would have been trampled underfoot if I had not cast myself into the bushes. I was horribly scratched and very frightened—"

"Frightened?" he repeated swiftly.

"The riders were all masked like animals. . . ."

Lord Mortmain caught his breath. "Did you see their leader?"

I closed my eyes briefly. "Yes," I said in a low voice.

"Did you recognize him?"

"No, sir. That was not possible, with his mask."

He asked the question I had been dreading. "What colour was his horse?"

"What colour, sir? Colour at night . . . is difficult to determine."

"Miss Fabian," said the Earl in a silky tone, "you are in danger."

I started nervously. "It had white socks."

"Ah. A chestnut, perhaps?"

Recalling the horror of the moment at which I had recognized the mare, I began to shiver again.

"Answer me, Miss Fabian," said my tormentor coldly.

I shook my head, and gulped. He muttered something that sounded impatient. He pulled me roughly into his arms again and began kissing me all over my face. I gasped and cried. His lips descended inexorably on mine and my tears ran down into our mouths.

"Interesting," he murmured as he released me at last. "I never experienced such a kiss before. . ."

I wondered if I ought to slap his face, but the truth was I felt so bewildered and exhausted I had no wish to do so. Indeed, I believe I was incapable of movement just then.

I suddenly awoke to the fact that the Earl was addressing me sternly. "I must insist that you never walk alone again, either by night or day. Lady Perowne's chaperone is supposed to be above reproach, but now old Randall has dragged you to the Knoll under the impression that you are eligible to join the many unfortunate girls Lady Martha has saved from destitution, I fear your reputation has been thrown to the four winds. I am really very angry with you, Miss Fabian. I think I would have beaten you if we had been alone when you were shown into the drawing room just now."

I said in a shaking voice, "You knew from Mrs. Mason that I was impulsive and thoughtless when you chose me to be Sir Hugo's governess—and I see no signs of him using my services as a governess, by the way. It is all like some tangled plot. I wish you would tell me the truth as to why you employed me. Then, indeed, I might—"

"Might what, Miss Fabian?"

I shook my head slightly. "I was going to say, might tell you things."

"You can tell me anything, you foolish child."

Almost, he convinced me. He was cleared of riding Herne, all the evidence pointed to Mr. Perowne's being the antlered man who wanted me dead, and furthermore, Mr. Perowne had what must appear, to himself at least, to be a convincing motive. The moment the Earl had kissed me I had realized that a part of the power which I had felt surging from him was an attraction so great that it threatened my defenses. In the relief of discovering him at the Knoll and not riding in the Wild Hunt I had been ready to pour out my heart to him. In the moment of his kissing me I had thought that I could deny him nothing. But still something held me back. With a distinct shock I remembered that it was because I had associated him with evil from the moment of our first meeting. He had been right when he said I did not trust him.

I looked out of the window and, seeing we were already on the drive to Holy Mote, seized at an excuse to distract him. "Oh, please, I don't wish to go to the front door," I cried. "Percival—the fuss—I have a key to the side door."

"Have you indeed? Very well, ma'am, though I dare say the news of your escapade is already halfway around the county by now." He called out to the coachman and the carriage halted.

I said faintly, "What shall I tell Lady Perowne?"

"What you told Lady Martha, I suggest, if she should ask. I dare say Miss Curtis is bound to mention it. I think it would be best, on second thoughts, to speak to Lady Perowne before she hears of it elsewhere." He stepped down and held out his hand. I hesitated on the step. "Well, Miss Fabian? Are you so loath to part from me?"

"It is—I was wondering," I stammered, "that is, am I to be dismissed?"

"Certainly not," he said calmly, "though you deserve it. Your punishment must be to remain at Holy Mote."

I sighed. "I do not understand you, sir," I murmured truthfully.

"I do not intend you to understand me, ma'am. Come now, for your crime grows more heinous by the minute."

I knew my ankle had swollen and was throbbing but

I was not prepared for the stab of pain I experienced when I put it to the ground.

"What ails you now?" demanded the Earl as I gasped and swayed, clutching his sleeve. Despite my faint protests he lifted the hem of my skirt and examined the ankle in the yellow light of the carriage lamp. Without another word he swung me up into his arms and began to carry me towards the house.

"But, sir, your knee—"

"Thank you, but it is improving daily, despite all the unaccustomed strains it has experienced since your arrival here. Now be quiet, unless you wish me to drop you in the moat."

I was quiet enough after that. His words brought back all the horror of that experience and the contrast of being carried so steadily, so tenderly it seemed, was one to savour.

The Earl halted before the side door and I lay warmly where three nights earlier I had shivered on the threshold. He looked down at me and I waited expectantly, but he only asked where the key was.

"In my reticule," I told him and added proudly, "but I can manage now."

"You certainly can not. At the risk of compromising you, I have no intention of leaving you anywhere but in your own room."

I did not argue with him, but asked curiously what he would say if we were caught.

He found the key and inserted it in the lock. "I believe there are three possible solutions—or four, to that situation. One"—he turned the key—"would be to explain the truth and leave you to face the consequences. Somehow I do not think I could be so coldhearted as to do that. Which way is it now? Down this passage and to the left? Very well. The second solution? Ah, yes. Well, if you were any other governess-turned-companion I suppose it would be simplest to remove you from Holy Mote immediately and offer you carte blanche as my mistress—you need not struggle so, Miss Fabian. Crippled though I may be, I believe I am still a good deal stronger than you. Besides, I do not mean to do it, for I am aware that you would probably not accept my offer and that your pride would oblige you to starve rather, so *bien élevée* as you are. Up these stairs, I assume? What a rabbit warren the place is, to be sure—and confoundedly damp."

He adjusted his grip and began to climb the stairs. "The third way out of the tangle, then, would be for me to preserve your reputation by marrying you off to some blameless and obliging person. My secretary Charles springs to mind. You do not know him, but he is personable enough, good at his work, devoted to my interests, and not so high in the instep that he would object to marriage with a person who superficially might appear somewhat ineligible. How difficult you are making my task, Miss Fabian. I wish you would lie still. Ah, here is your door. Eyebrows would be raised at the fact that I recognize it, I fear. As I was about to say, the third solution would be well enough but for one insuperable objection."

He opened the door with some difficulty and carried me inside and down the steps. "Yes," he said thoughtfully, "how could I ever bear to see you married to another? Especially since, if you were my secretary's wife, I should be put to the pain of encountering you almost every day. No, Miss Fabian, if we are discovered here together there will be nothing for it but the fourth way out."

He paused as if expecting me to ask what that might be, but I was quite incapable of speech.

"I should be obliged to marry you, Miss Fabian," he said softly, lowering me onto the bed.

I stared up at him as he half-turned away to pick up the tinderbox. It was very dark. I could not make out his features, still less his expression. I had no notion if he had been trying to distract me from the real embarrassment of my predicament, if this were all some callous joke, or if he had been talking to cover some other feeling—even, I thought wildly, to prevent me from noticing some murdering accomplice standing in the shadows.

One thought occurred to me—to try to preserve the pride of which he had spoken, the only luxury, since my birth was not acknowledged and I had no fortune, that was left to me.

I said as calmly as I could, "And is it your opinion, sir, that marriage with you would be preferable to me than continuing through life with a damaged reputation?"

The flame leapt in the lamp, illuminating his Satanic smile. "Yes, Miss Fabian, it is. However, I understand the notion is as yet strange to you. No doubt the advantages of the arrangement will appear more obvious to you the longer you dwell on them."

"I will not need to dwell on them if you remove yourself before we are discovered," I said rather sharply.

"But—against my better judgment I am finding it rather hard to leave you, my love. If it were not for that wretched coachman out there—"

"Do not—call me that!" I gasped, deeply shaken.

He was silent a moment, adjusting the lamp. "Perhaps it is a little premature," he said gravely, "not for me to feel you are my love, but for me to express it aloud."

I bit my lip, wondering if I were dreaming. "Are you not forgetting Miss Curtis, sir?" I murmured.

He glanced at me in surprise. "Has Lady Martha succeeded in convincing you that Miss Curtis is the lady of my choice?" he asked.

"It appears to be common knowledge."

"Ah! And does that thought arouse your jealousy?"

"It arouses my pity for Miss Curtis, to think how mistaken she probably is in the character of her betrothed," I said stiffly.

"But Miss Curtis is not my betrothed," he informed me, smiling again. "I do not say that the thought of marriage with her has never crossed my mind but fortunately, as it happens, I have not been so rash as to express it. I have written her no love letter, have not danced exclusively with her at balls, have stolen no kiss—and certainly have never carried her to her bed-chamber. In a word, Miss

Fabian, to set your mind at rest, Miss Curtis has no reason to expect an offer of marriage from me."

"You have not set my mind at rest," I muttered, all too truthfully.

"Don't disturb yourself unduly, ma'am," he said slowly, seeming to retreat a little. "Recall that we were discussing a solution to discovery—and we have not been discovered."

"I should not hold you to it if we were," I assured him with a trace of bitterness.

"Do you mean you would refuse my offer?" he demanded haughtily. "May I ask why?"

"There would be a thousand reasons for my doing so," I cried extravagantly. "I would never hold a man against his will to such an impulsive decision—"

"Against his will?" he repeated as if surprised. And then, frowning blackly, "An impulsive decision? The intention—the possibility, that is—was not born tonight, Miss Fabian. I have been thinking of marrying you for longer than you know."

I turned my face into the pillow. "You will be saying next that you loved me from the first," I suggested, near to tears again.

He hesitated, longer than was gallant. "I cannot say that, precisely," he admitted. He walked to the door and stood by it, fidgeting with the latch. At last he looked up. "I can assure you, however, that I had it in mind to marry you the day we stood in Mrs. Mason's parlour."

The words, the way he stood, the angle that the light fell on his face, perhaps most of all his pallor and the tenseness of his expression, rolled back the curtains from my clouded memory. Before I had even had time to examine it I heard myself saying firmly, coldly, "Your lordship has done me great honour, but I must assure you that I should rather die than be wedded to your lordship."

He moved sharply, so that his face was in shadow. He said nothing for so long that I almost regretted the uncompromising manner in which I had refused an offer that, I had to remind myself, he had not even made.

What was he thinking now? Was it the first time his wishes had been flouted? He was a Quentin, after all, and against their decisions tradition had it that there could be no right of appeal.

He moved again, and I caught my breath. But he only bowed, and left the room in silence.

15

My ankle throbbed, my scratches felt like hot wires laid on my skin. Perversely I longed to feel the Earl's arms about me again, his lips on mine. If only I could have trusted him! But I had remembered now where I had seen him before and in the light of that significant fact, all else must fade.

I turned over on my back and stared up unseeing at the heavy black oak beams, the yellowed sagging plaster, the concentric rings of light above the lamp. For the first time in six years I found myself able to relive almost every moment of the fatal night, the night of my guardian's murder, the night that had ended for me when I looked into his room to bid him sweet sleep and found his tumbled body on the floor, an open window, a few unimportant papers blowing in a chilly wind, a painting hanging askew before an empty safe, and blood—blood lost in the rust and rose of the treasured Persian carpet, blood on the grey fallen wig, blood everywhere about the shattered head of Medlicott, my guardian.

I had stared into that room of murder for only an instant before Steele, my guardian's clerk, had pushed me aside; and yet that scene was etched forever on my memory, so deep, indeed, that it had obliterated until now all the details of what had preceded that dreadful moment of discovery. Now they returned to me, neatly and in order, as fresh as if I were recalling the experiences of yesterday.

Mr. Medlicott had been abstracted that afternoon. I had noticed him staring at me, stroking his chin, pulling at his ear lobe, muttering to himself two or three times that he wondered if it was for the best, without giving any indication of what he meant by this cryptic remark. Once I heard him say, "Bad blood on one side, good on the other—who can tell which will prevail?" and again, "Well, well, no harm can come of it while I live and for later—I can make conditions. Ay, that will be best. Tie the thing up—safeguard it." And off he had gone to consult his brother solicitors, Underwood, Jones and Underwood.

We had dined when he returned, on boiled mutton with onions, a pair of fowls, and a gooseberry tart of which I had been proud, for, as I had pointed out to him, I had bottled the fruit for it myself the previous summer, when I had been only twelve. After the meal, instead of banishing me to the sitting room while he enjoyed his port alone, my guardian had seemed restless, asking if I would not like to go early to bed and when I refused in some surprise, suddenly insisting that I take a message to Mrs. Featherstone on the far side of the town.

"Now, sir?" I had asked. "But it will be dark before I return."

"You will take Steele, of course," he had replied, and to himself, "Yes, that will be best."

"But why should not Steele take the message, then?" I had innocently wondered.

He had flown into a sudden rage. "Come, girl, who do you think you are, to argue with your elders and betters? Mrs. Featherstone was asking for you only t'other day, complaining that she has not seen you for an age. 'Tis you she wants to see, not Steele—Steele is to watch over you, and carry the lantern—and it happens that I wish

Mrs. Featherstone to receive this message tonight. Hasten now, and find your cloak."

I was glad to remember that I had gone to him for his kiss before I left, and that there had been no trace of his anger then.

"You're a dear, good girl, Chantal," he had murmured, pushing back my hood. "Pretty as a picture, too. Your parents would have been proud of you—only wish I had known them, God rest their souls."

"Amen," I had whispered, dragging back my threadbare memory of Maman: a breath of eau de cologne, the feel of soft furs, swift comfortable hugs, a smooth velvety cheek, the tickle of long lashes—and the smell of snuff, rasp of a scratchy chin, strong hands, and alarming piggyback rides which were all I could remember of Papa.

"Well, smile, girl," Mr. Medlicott had cried. "You've still got your old guardian, eh, for what he's worth. Run along now, and don't eat too many sweetmeats, mind, for you must not disgrace me at Mrs. Featherstone's."

So I had left him, though he had called back Steele for a moment, self-consciously explaining that he had an appointment and might require Steele's key to unlock the safe. I had opened the door and the clerk, grumbling at having to leave the fire on a windy evening, had hurried me along the greying streets to No. 3 The Crescent, where lived the fat, effusive Mrs. Featherstone. She had been surprised to see us, expressing horror that I should have walked so far with such a trivial message, and then she had made much of me, plying me with cocoa and bread and butter until I had been as reluctant to leave her cheerful fireside as Steele had been to leave ours.

We had walked back through the crisp darkness and I had thoroughly enjoyed what was, for me, a new experience. On the first part of the journey there were the neat modern buildings of the rich, tall in grey stone with elegantly pillared doorways and shining sash windows, few people to be seen walking like ourselves but several carriages on the road, an occasional chair, and a few horsemen. Then we had crossed the windy park and entered the older part of the town, where life spilled over from the tumbledown houses glowing with rushlights at their curtainless casements, across the streaming gutters, onto the cobbled road. I remembered ragged children playing at knucklebones outside a crowded inn, a woman pushing a gin-soaked rag into a baby's mouth while nursing another, boys chasing a lean dog, their wild cries echoing down the street, two old women pausing to rest their baskets, on their way home from the bleaching ground. There had been a girl in tawdry finery only a year or two older than myself, rolling her painted eyes at Steele, who flushed angrily at her insolence, and then turning haughtily from my interested gaze as if she feared I might corrupt her.

Steele, still muttering, had hurried ahead as we turned into our cul-de-sac. From the livery stable opposite our corner a groom had been leading out a fine roan horse towards a gentleman who waited impatiently, tapping his whip against his shining boot. I had seen his profile, illuminated by one of the new gas lamps of which Horsefield was so proud, and thought him good-looking. But he seemed agitated, this gentleman, I had thought. His stern face was taut and pale, his fists unconsciously clenched, eager to be gone, his thoughts already seeming to be far away. I glanced back at him as I approached our house and saw him snatch the proferred rein and seem to look impatiently down our road. He is waiting for his servant, I had thought—and at that moment a man, youngish, slim, dressed in some sober livery, had pushed past me, his collar turned up against the wind, his round hat pulled well down, one hand thrust within his waistcoat, the other carrying his master's bag.

I had heard the scrape and slip of shod hooves on the cobblestones behind me as I had hurried to join Steele, crossly beckoning from our doorway. He had stood back with his usual grudging politeness to allow me to precede him. It had been warm in the house after the windy street. I had thrown my cloak carelessly across the banisters and danced up the stairs—to find the door ajar and, within, my guardian foully murdered. All else had been forgotten thereafter . . . except that I had flinched when I first saw the profile of the Earl of Mortmain in Mrs. Mason's parlour. Then, when he had turned towards me and I had seen his livid deforming scar I had felt that I could not possibly have seen him before without remembering the occasion. Now it seemed obvious that he had been unscarred that windy night, six years ago in early spring; unscarred, but tense with fear or some similar emotion rigidly controlled, and he had been hastening to leave a cul-de-sac in one of whose houses a man had been murdered but minutes before. The other houses in our road were all lodgings, I reflected grimly. Which dwelling in that half-derelict place, once fashionable and now sliding fast into dissolution, was it likely that one of the mighty Quentins would have been visiting? There was only one respectable house in the place and that was the one belonging to the stubborn old solicitor, Medlicott, who would not change his address because people had known for nearly fifty years that they could find him there.

Oh yes, I thought, turning stiffly on the bed, reminding myself in parenthesis that I must get up and lock and bolt the doors before I slept—yes, it seemed obvious now. " 'Bad blood on one side . . .' " That referred to the Earl's father, presumably. And now that I came to think of it, had not Lady Perowne mentioned some evil act on the part of her cousin the discovery of which had caused his father's death from shock? And what would have shocked the wicked old Earl so grievously? Very little, one must assume; but surely violent murder for gain must have done so? It seemed incredible that Lord Mortmain could have murdered my guardian, when I came to put it into plain words; but it fitted, yes, it fitted so well. The old Earl had not seen much of his only son, I remembered. No doubt he kept him on a short allowance. What more likely than that such a wild, high-spirited young man had got himself into debt so deep that only a fortune would tow him out of the river Tick, as such blades put it? And my guardian had possessed a fortune and one that could be easily carried and realized, moreover. It was known in the town that Medlicott kept his diamonds in the safe; what could have been more simple than for the Earl to make excuse for a private appointment in which some legal papers would be concerned so that the safe would be sure to be unlocked?

The wind howled in the chimney suddenly and I realized that despite the burden of my thoughts I had almost fallen asleep. I rose stiffly and hobbled to the door. When it was secure, I soaked a handkerchief in cold water and bound it about my ankle. Slowly I prepared for bed.

As soon as I was settled between the sheets my thoughts became active again. How extraordinary it was, I reflected, that I, of all persons in the world, had come to be involved again with the Earl of Mortmain. But then, with a sure conviction, I began to think it could not have been such a coincidence that Lord Mortmain had received my name and description from the lawyers, that he had driven to Horsefield House, abducted me, and forced me on his cousin in a trumped-up post. What a fool I was! Surely it was more likely that he had seen me just as I had seen him that windy night; that he had lived in fear of my bearing witness to that effect, and that he had determined, once I had left school, to keep me under his eye and—destroy me.

I was shaking now as though I had an ague, despite the warmth and comfort of the bed. All was now explained, I thought, except his talk of marriage. But upon reflection, even that seemed to become clear to me. Marriage was the Earl's alternative to murder. A married woman, I remembered, could not give evidence against her husband.

The last pieces fell into place. The Earl had sought me out, intending to silence me. Marriage or murder, either would serve so long as he was not discovered. Marriage had been in his mind, he had admitted, when we met in Mrs. Mason's parlour. He had installed me at Holy Mote in order to have me under his eye while he courted me—or killed me, if the opportunity arose. No wonder he had not cared that I was so unsuitable for the post! No wonder he had had so few questions to ask Mrs. Mason about me! He had snatched me away from school before I could change my mind, and had forced me into his cousin's household.

He had not loved me from the first—no, indeed! He had regarded me as a dangerous enemy, whether or not he found me as attractive as, to my sorrow, I was bound to own I found him.

I had only one crumb of comfort left to console me and that was that I had not allowed him to suspect how I felt about him; and I could even be glad I had refused him in the words I had unwittingly chosen, the words that might very well turn out to have been prophetic.

For it seemed to me now that the choice had lain between marriage and murder: and I had clearly informed the Earl of Mortmain that rather than marry him, I would prefer to die.

I do not know if I dreamed the voices I thought I heard in the dark morning before dawn. Later, when I reviewed my conclusions of the previous night, it seemed as if I must have done so. But at the time I thought I woke and heard a man half-groan the words that later seemed significant.

"Alone, I cannot do it," he seemed to say. "Herne must aid me . . . a sacrifice to Artemis . . . or to Ashtaroth . . ."

And then another voice, faint and old, murmured that the season was past for sacrifice.

"Ah, if it were but Candlemas!" they seemed to sigh together, or was it the wind lingering in the chimney? "If it were but Candlemas. . . ." And then I thought I heard the stronger voice say firmly, "Even so . . ." before it faded quite.

I slept ill after that, not surprisingly, perhaps; tossing and turning, dozing for a while and then starting awake to wonder when I should leave Holy Mote, and where I should go, how I should protect myself and whether it would not all be in vain, since the Earl of Mortmain was powerful enough to track me down wherever I should try to hide.

I woke again at my usual time and found myself so exhausted that I felt drained of emotion, remote and uncaring like a dried leaf tossed in the gale. I looked from my window at the loveliness of late spring and saw only that the wind seemed to be trying to destroy the new green leaves.

My face was not scratched and it was cool enough for a shawl to occasion no remark, so it was only by the fact that I hobbled a little on my swollen ankle that one would know I was not quite as I had been the previous day.

I limped down to breakfast. In the Great Hall the wind lifted the corners of the tapestries and rippled behind them, set the swords swinging on the hooks that held them, and snatched the door of the Breakfast Room out of my hand as soon as I opened it so that it crashed into the back of Thomas the footman and made him drop the coffeepot.

The resultant disturbance covered my pallor, the awareness of horror that must, I thought, show in my eyes; and I sat down relatively unobserved to breakfast.

Upon my plate was a letter addressed in my godfather's scholarly hand. I snatched it up and passionately kissed it, thereby drawing on myself the attention I had hoped to avoid.

Mr. Perowne chuckled. He seemed none the worse for his adventures of the previous night and I stared at him, wondering how he could look such a healthy ordinary young man, if rather more handsome than most, when only a few hours earlier he had been leading the Wild Hunt, with horns on his head, to who knew what wicked excesses in the haunted wood or the deserted churchyard of All Saints.

I said, coldly, defensively, "I see my action amuses you, sir, but I suppose I may kiss a letter from my dear godfather if I wish?"

"Oh, is that all? How disappointing, ma'am. I thought it must be from the Unknown with whom gossip reports you to have had an assignation last night."

"Really, Piers," cried Lady Perowne, "I wish you would explain yourself. What can you mean?"

I glared at him, wondering that he should take the risk of teasing me when my knowledge of his adventures was surely more significant than his of mine. I said rashly, "I suppose you saw me?"

He looked surprised. "Not I, ma'am. I did not leave the house last night. It was Percival who told me Mrs. Randall had dragged you from your evil ways and delivered you to Lady Martha for reformation."

I gritted my teeth, reflecting that it was curious such paltry things could still affect me. I had thought myself dead to feeling when I rose, but the sight of Mr. Shorncliffe's letter seemed to have freed me from that spell. "I fear I must confess to taking a walk—alone—last night," I said, staring at my plate, while Lady Perowne exclaimed and gasped. "I suffered a slight, a very slight, accident when I stumbled into some bushes in the dark, which you will think a very proper punishment for my foolishness. I sustained some scratches and a twisted ankle. I was wondering how I should get home without help when old Mrs. Randall accosted me and took me to the Knoll by main force. I could not stop her, she was unable to hear my pleas. Judge how shaming I found it to be impelled by Lady Martha into the Curtises' drawing room, all torn about and dishevelled as I was—"

"Mortmain was there," Lady Perowne cried. Her apprehension was quite quick in some matters, I reflected. "That is why Lady Martha exposed you. She wanted him to have a disgust of you—she fears he is attracted to you, you see. What happened then? How did you get home?"

"Lord Mortmain took me in his carriage."

"Ah! She would not care for that! Was he angry with you?"

"Yes, ma'am." My conduct had been unbecoming in his future wife, of course.

"Poor child," she cried sympathetically.

"Thank you, ma'am. But I was in the wrong and am sorry for it. I am well aware that my thoughtless conduct might have laid me open to worse consequences. I have learned my lesson, I assure you." Indeed, I thought, scratching at the seal on my letter, it would take a good deal to lure me out-of-doors at night after such an experience as that had proved to be.

Sir Hugo was brought in and while he was greeting his mother I took the opportunity of reading my letter. Mr. Shorncliffe was agitated, it appeared. He did not like the speed with which my post had been arranged. He did not like what he had heard of Lindenfold. He did not like the tone of my letter to him, in which I had apparently made it clear that though materially comfortable at Holy Mote, I was not happy there. He ended by assuring me that he would come to see me as soon as he had rearranged his parish duties.

I folded the letter, finished my breakfast and, rising, instructed Sir Hugo to lead me to the schoolroom. He looked mutinous, and Lady Perowne seemed surprised. Mr. Perowne asked if Hugo was not meant to be riding with his cousin, who had taken over his tuition, and his nephew reluctantly admitted he had already concluded his riding-lesson. Perhaps if Old Nurse had been there I should not have been able to keep up my determined air, but as she had not come down this morning I excused myself firmly and swept my pupil out of the room.

"I was going to show you something this morning," Hugo said sulkily as we went upstairs. He yawned suddenly, and I asked him as a joke if he had been very late to bed. To my surprise his face lit up. "You will never guess what happened," he said eagerly. "Uncle Piers came into my room when your dinner party was finished and we—we had a midnight feast! He brought meat and trifles and custards and we ate so much—and we were so frightened Old Nurse would find us out, and she nearly did—she came into the room and Uncle Piers had to hide under the bed. And then we talked and I showed him my boat—he didn't like it, isn't that horrid? But he told me stories about pirates and—and it was the best fun I've had since Christmas."

I had paused to listen. I said carefully, "I suppose you do not know what time it was, when your uncle came in to you?"

"No, but Mama was just retiring. We could hear her talking to Jackson for a while."

"Oh. And how long do you suppose Mr. Perowne stayed with you?"

"Hours and hours—he is a great gun, isn't he? It was past midnight at least, because he said he wanted to make it a proper midnight feast, so we listened to the grandfather clock strike the witching hour, he called it."

"Oh," I said again, faintly. We went on up the stairs while I wondered what to make of it. Had Piers Perowne not been riding Herne last night, after all? But perhaps he had somehow established an alibi either before or afterwards. It would have been quite easy for him to have put the clock ahead and changed it back when Hugo was asleep.

Reluctantly I abandoned my speculations for the time being, as we reached the schoolroom door. I opened it, but Hugo hung back. "Well, come in, sir," I commanded.

He grinned apologetically. "Not on your life, Miss Fabian. You will have to catch me."

He darted nimbly into the passage as I reached out for him, causing a throb of pain in my ankle. "Oh, please, Hugo," I exclaimed crossly. "Do come back."

"No, no," he cried, skipping nimbly up the attic stairs. "Only if you catch me. Fiddle-de-dee, you can't catch me!"

If my ankle had been well, I felt that I would have caught him easily. As it was, however, I was tempted to let him go. Then I decided if I did not catch him, Hugo would never have any kind of respect for me. I limped quickly up the stairs and just touched his coat before he whisked round the first corner. Angrily ignoring the discomfort in my ankle, I hurried after him, determined to drag him back by the ear if necessary and lock him in with me until he had completed that morning's lessons.

"Fiddle-de-dee!" he called again, looking back round the corner at the top of my stairs to see if I were still following him. I gritted my teeth and hobbled towards him. I gained on him down the next corridor and then he ran sideways into the sewing room and I thought I had him trapped. But to my surprise the sliding panel was standing open and as I hesitated, staring at it, Hugo slipped through the aperture and down the secret stair. That I knew was blocked by the door to my room, which I had locked the night before, so it was with something of a feeling of triumph that I followed him through the gap and down into the dark.

"I have caught you now," I exclaimed, feeling with my hands before me as I heard him rattle the door latch. "Where are you, sir? You need not think—"

For an instant it seemed as if he must have got behind me. But the hands that seized my arms were stronger than his, and it was a grown man who, silhouetted for a moment against the light from the sewing room, turned me about and thrust me roughly forward, through another hole in the wall, into a black room in which a blacker fireplace yawned. I had just time to discern the outline of a table crowded with strange objects, a curving knife, goblets, a gleaming skull, a tangled spread of antlers, before I heard Hugo scream and a panel slid shut behind me, leaving me in darkness so complete I might have been enfolded in black velvet.

16

My surprise was so complete I had not even found time to cry out. Now that it was too late, I found I was terrified, even more frightened perhaps than I had been when struggling for my life in the stagnant waters of the moat. That had been a simple fight for survival with the choice of either keeping my head above water or drowning. Here I did not know from what quarter attack might come, nor even what it was that threatened me. The glimpse I had had of the horns, the skull and the knife suggested that this might have nothing to do with Lord Mortmain's intentions. It might be some joke of Hugo's—some punishment devised by Old Nurse for my trespass, or for having witnessed a Wild Hunt.

But then I remembered that Hugo, too, had screamed. He had led me into this trap but he had not realized what it implied. And Lord Mortmain, I recalled, knew not only of the existence of the sliding panel but also of this other room, the room with the fireplace that must lie directly above mine.

Suddenly through the pressing darkness came a voice. Distorted, indistinct, I realized it was coming from the chimney, the chimney that had conducted the other snatches of conversation I had overheard. But this was different, more like a man muttering and cursing to himself.

I forced myself to move through the unknown towards the sound. I lost it, then all at once it came again. "The spark, the spark," it seemed to be saying, as far as I could make it out. "In the name of Baal, what ails the flint?"

There were scratching sounds, and muffled movements; then all was quiet again.

I stood very still, afraid of what my hands might touch. Some primitive instinct told me that I was, at least, alone—but I had not liked the little I had seen of the inanimate objects in the room. What was my fate to be among them? Was I to be left here to starve?

Baal, I suddenly thought, and a passage from Jeremiah came into my mind: "They have built also the high places of Baal, to burn their sons with fire for burnt offerings unto Baal. . . ." and then I smelt the faint acrid odour of smoke and knew that it was to become a burnt sacrifice that I had been lured into the secret room.

Then I moved swiftly enough. There was air coming from the chimney but it was already tainted with smoke and it took no courage at all to get myself across the room and cast my weight against the panelled wall, beating upon it, pushing, pressing, crying aloud, praying that I would discover the trick of the panel before it was too late.

It was like a miracle when I heard Hugo's voice, high with excitement. "There—in there somewhere," he cried, and when I called out, a man's voice shouted to me to stand well back. Two violent concussions followed and with the second, light burst into the room.

It was Lord Mortmain who climbed through the splintered panel, while Hugo hopped up and down excitedly behind him, a smoking lantern held high in his hand.

"Look, look," he cried, "she is here—here in the secret room! This must be where I saw the knife that time, and Born-To-Be-Hanged—and the horns! Look, Cousin Mortmain, look at the horns! I told you there was a room here somewhere, didn't I, Cousin?"

"I knew there was one but I had forgotten how to find the entrance, for it is hidden by the stair door, whenever that is open," said the Earl expressionlessly. He was still breathing hard. "Come, Miss Fabian," he added curtly.

"I found the curtains on fire in your room, below. I want to make sure we have succeeded in putting them out."

I expected him to make some show of helping me to climb through the panel but he stood back aloofly, only saying when I hesitated, "You have taken no hurt, I trust?"

"No, none," I took pleasure in disappointing him. "It is only that my ankle is not perfectly recovered."

I held my head high and made to pass him. I became aware of an unpleasant smell of singeing. "Are you burnt, my lord?" I demanded incredulously.

"A little, but it is no matter. It was important to put out the fires without delay, for they had been started in several places in your room and if they had once caught hold in an old house like this, all but the stone work would have gone in no time."

I raised my brows, unseen by him, and stepped out of the room. Hugo lifted his lantern and led me down through the open door to my own chamber.

"Are you very cross with me, Miss Fabian?" he asked, peering anxiously over his shoulder at me. "It was only a game, you know. I had to lead you up to the attics and down these stairs—"

Lord Mortmain spoke behind me. "And whose idea was that, sir?" he asked. He sounded very grim.

Hugo was an old hand at keeping secrets, however. He merely looked stubborn and shook his head. "It seemed better fun than lessons," he murmured sullenly, pushing aside the scorched curtain.

"Good Heavens!" I cried in horror, looking about my room. The window- and bed-curtains were also charred and smoking, the fire roared in the hearth with half a chair thrust into it, smouldering, and there were scorch marks on the rugs. The smell was disgusting: appropriately evil, I thought.

"I threw everything that was burning on the fire, except the curtains," Lord Mortmain explained. "That chair was pulled out into the room and was obviously intended to set the rug alight." His voice was curt and remote. I looked at his veiled dark eyes, his closed and secret face, and wondered where Hugo had encountered him just now, and what he had felt at being obliged to undo all his work.

Rather peremptorily, I suggested that he should show me his burns but, as I had expected, he refused. "Hugo, pull the bell," he said, turning away.

Randall came anxiously in response to it, smelling the smoke and crying out to the footmen to bring pails of water even before she had seen the damage. Lord Mortmain left her at work and marched me along to Lady Perowne's bedchamber, where we found her sitting gazing at her reflection while Jackson buffed her nails.

"Madam," he said brusquely, "today you nearly lost your companion." In a few more words, while she gasped and exclaimed, he told her what had occurred.

"Lord, ma'am," said Jackson shrilly, "I never did like that attic, and I told miss as much, not long ago."

"But what were you and Hugo doing up there?" asked Lady Perowne perplexedly. "You know Mrs. Cumber does not like it."

I looked round for Hugo, but he had vanished. "Sir Hugo did not feel like lessons today," I explained. "He led me a dance and—someone pushed me into that secret room."

"Someone?" she repeated incredulously. "But who? You are not suggesting it was Mrs. Cumber, I hope."

"No, ma'am. I think it was a man." Carefully I refrained from glancing at the Earl. "But I believe someone had suggested to Sir Hugo that he should behave as he did. To do him justice, as soon as he realized what had happened he went for help."

"The house might have burned down!" she cried. "It was a narrow escape. But who could have tried to start a fire in your room? Mortmain, what shall we do?"

"The attics are being searched now for an intruder," he replied. "I intend to give orders for that odious room to be swept out and everything in it burnt; and then it should be exorcised by Pole-Carter before the secret panels are nailed up and the stairway properly blocked off. Naturally Miss Fabian cannot sleep in that room again. You must choose another for her, as I have asked you to do before."

Lady Perowne flushed delicately. "I have considered the matter, but Old Nurse pointed out that there is really no other room so eligible—"

"What, in a house the size of this? I dislike to contradict you, ma'am, but I dare say there are ten other rooms unoccupied at the present time."

"Well, perhaps—but as I was about to say, they are guest rooms or principal bedchambers and hardly suitable for—"

"Hardly suitable for a Fabian?" queried the Earl imperiously. "You cannot realize what you are saying. Make one of them available for her at once, if you please. It should have been done before this. Ah, Percival," he added, turning to the butler who entered just then. "Did you find anyone?"

"Not a living soul, my lord, beyond them that has a right to be here—and we've searched the whole house, secret ways and all."

"As far as you know, that is. Well, it is no more than I expected. Will you send a man down to Butcher immediately—"

"Butcher?" queried Lady Perowne.

"He is the constable this year," the Earl explained impatiently. "Inform him, if you please, Percival, that this outrage has occurred and ask him to report to me if any stranger has been seen in the village recently. Except for Mr. Shorncliffe, that is," he added with a side glance in my direction, "for whose probity I can vouch. Your godfather, Miss Fabian, is awaiting you in the library. I fear he will be growing impatient."

I cried out with pleasure and surprise but did not stay to hear more. With a breathless apology I hurried from the room, hardly aware now of the ache in my ankle. What was my dear godfather doing at Holy Mote, immediately upon the heels of his letter? I had no time to do more than wonder and to thank God that he was come so opportunely, before I saw him standing uncertainly in the doorway of the bookroom, looking shabby, lost, and bewildered.

Then he saw me, and his lined face seemed illuminated. With a muffled sob I almost flung myself into his arms.

He recoiled in mild surprise. "There, there, my dear. Has our parting been so long? Or is this a measure of your unhappiness here? Whichever may be the cause of this emotion, I wish you will calm yourself or we shall have the servants coming to find out what is amiss."

"I am—excuse me—so sorry, Godfather—but I am so pleased to see you."

He smiled and said whimsically, "Is this how you express pleasure, dear child? I shall hope never to be welcomed by you when you are displeased."

For a few minutes he allowed me to weep within the safe circle of his arms. Then he gave me a gentle push and I took a deep breath, wiped my eyes and blew my nose. "Please be seated, sir," I said, regaining control over myself. "Oh, I have so much to tell you! How shall I begin? But first, how come you here?"

"A young man brought me in his curricle," he replied literally, seating himself on a hard-looking chair of carved and polished oak. "He is one of the Quentins—yes, the Earl of Mortmain, I believe. He chanced to overhear me ask his groom the way to Holy Mote. Before that I travelled in the gig of a kindly farmer and earlier—yesterday, that was—by stagecoach—not an unpleasant journey, thanks to the clement weather, though I think I should not care to try it either in the depths of winter or in high summer, when I imagine the chalk mud turns to clouds of dust . . . but I digress. What was it you were asking me?" He took a pinch of snuff, savouring it appreciatively. "Ah yes, to be sure. Why did I come? It was your letter which brought me, Chantal. When I first read it I was disturbed by it, as I believe I must have made plain when I wrote to you—you did receive my epistle, I trust? Ah, I am so glad.

Really, the mail is very reliable nowadays. Then, when I came to read your letter over again, it seemed to me that there was even more in it to alarm me than I had at first supposed."

"There was cause for alarm, Godfather," I assured him, my hands trembling so that I was obliged to clasp them together. "Right from the first—when Lord Mortmain came to Mrs. Mason's—it was so singular—"

"Calm yourself, my child. Begin at the beginning and tell me what has frightened you."

Staring down at my hands, I told him of that meeting in Mrs. Mason's parlour, of how I had come to Holy Mote and what had happened to me in this house, keeping back only the effect Lord Mortmain had upon my treacherous feelings and my conviction that he was not only the man who had murdered my guardian but also the one who had attempted to kill me. It should, I thought, be obvious enough, but I felt it wisest to allow my godfather to arrive at his own conclusions.

He heard me out with few interruptions. Once, taking another pinch of snuff, he murmured reflectively, "Macouba, you say?" and on another occasion he asked me to repeat what description, if any, Sir Hugo had given of the man in the attic. The only moment in which he allowed himself to appear intensely surprised was when I told him where it was I had remembered seeing Lord Mortmain first. Even my tale of the antlered man did not surprise him more, though he was as sympathetic as I could have wished over my hurried description of my experience in the moat and again this morning in the secret room.

"It is bad, very bad," he pronounced at last, shaking his head. "I would take you away from here but that one wonders if that would satisfy the person who is—persecuting you. Besides, it seems so s ange that there has been so much violence centered on one young and innocent life that the implication cannot be avoided that there is some connection between this and poor Medlicott's untimely end. Now you are on your guard, however, I wonder if there might not be a chance here of apprehending one who certainly should be brought to justice." He mused a while. "You must be safeguarded meanwhile, of course. I wonder who is the nearest magistrate?"

"A gentleman to whom you certainly cannot apply," I cried. "It is Lord Mortmain."

Mr. Shorncliffe regarded me thoughtfully. "You are quite positive that it was his lordship whom you saw at the livery stable that night?"

"Positive, sir. I wish I were not!" I added involuntarily. "But did no one think to question the people who were about at that time? What investigations were made, what steps taken to apprehend the murderer? I was sent immediately to Mrs. Featherstone's, you will remember, and then to Horsefield House, so I have no knowledge of anything that happened afterwards. I suppose the Bow Street runners were called in?"

"They were, my child, they were," he said heavily, staring into the past. "They ascertained that Medlicott was known to keep a fortune in diamonds in his safe, together with his legal papers, and that on the fatal night he had mentioned some appointment and had sent out his clerk and yourself, my dear, on some flimsy pretext, retaining both the keys to the safe. They could not find anyone who remembered seeing either a horse or a carriage stopping at the house, nor any person entering it, though some while later a boy reported having seen a man with a leather bag and wearing a round hat hurrying away down the street. Steele, the clerk, could not at first remember having noticed such a person, but later testified that he believed he might have done so. Several men were shown to Steele and the boy but they were unable to make a positive identification and it was assumed the person, if he existed, must have left the area."

"And what about the diamonds? I suppose no more was heard of them—and yet one would imagine, if they were stolen, that the thief would have tried to convert them into money."

"Ah, yes, you do well to remind me of it. Some uncut diamonds did come on to the market in London which, as Bow Street was eventually able to discover, were among those owned at one time by Arthur Medlicott."

"And the—appointment?"

He shook his head. "No clue was ever found as to the identity of the man with whom the appointment was to be, whether it was the person in the round hat or another."

"Did you attend the inquest, sir?"

"I was obliged to do so. I testified that poor Medlicott had no enemies. It seemed to be established that the motive was for gain, though the jury was reminded that legal papers had also been stolen from the safe. A verdict was returned of murder, by person or persons unknown."

I reflected on all this. Why had the man in the round hat been noticed, and not the Earl, who was surely more memorable? I supposed it must have been because the other man had carried a bag. I still could not imagine why Lord Mortmain would have taken a servant with him on such an affair, if it had been premeditated—and if not, what could his business with my guardian have been, and what could have inspired him on an impulse to kill his lawyer? Perhaps the safe had been opened, as Mr. Medlicott had anticipated it would be, and the Earl had been overwhelmingly tempted by the sight of the diamonds within, or the knowledge that they were there. It was even possible that he had owned to an interest in such jewels and that Medlicott had displayed them for his admiration. After the murder he had swept them into the bag—why had he brought the bag if the murder were unpremeditated? And then he would have carried it down the stairs and handed it to his servant on the way out, as would have been natural, if he had had nothing to hide. This would have had the added advantage to him that the servant carrying the bag would have attracted the most notice in the street and, if apprehended, would have been found in possession of the incriminating gems. Why had the servant never betrayed his master, once he had learned of the murder? No doubt because he had been heavily bribed not to do so—an income for the remainder of the Earl's life would have been the best way of managing that. He might also have been threatened with the fact of his being an accessory to keep silent. In any case, silent he had remained. Perhaps the Earl had killed him! Perhaps it was in this connection that some discovery had been made which had caused the death of the old Earl from shock. . . .

"Why were the papers also stolen?" murmured Mr. Shorncliffe. "If one of them contained some clue to the identity of the murderer, why did he not take only that one?"

"Perhaps to hide the fact that only one paper was missing," I suggested. "Steele would probably know what was in the safe."

"Ay, no doubt."

"Well, sir, what do you think of all that I have told you?" I asked, impatient to hear if he had arrived at the obvious conclusion.

"I think that I shall be interested to meet everybody here," he replied. "Old Nurse, for example. I hope I shall soon make her acquaintance."

"Why, sir, do you believe, as Mr. Pole-Carter seems to do, that she is a witch?"

"I believe she thinks she is one," he said dryly. "She sounds a thoroughly dangerous woman to me, and most unsuitable to have a child in her charge, let alone a succession of them. From what you say, I suspect her of being an opium-eater, and possibly a reluctant accessory to attempted murder. But witchcraft—if that has gained an ascendancy over her mind, that is worse than all."

I said in a low voice, "And what do you make of the Wild Hunt, sir, that I have witnessed?"

He shook his head. "I like it not, Chantal. It is an indication of the hold that witchcraft has on this part of the country, I fear—or, at best, it could be a pack of young men with nothing better to do than to over-ride their horses and frighten honest folk. But I should have thought your pupil's uncle, Perowne, too intelligent for either, by the sound of him. However, no doubt he finds country living vastly dull, and money, thanks to Mortmain's hand on the reins, too short to allow him to go and make a fool of himself in London—the ambition of most young men, it seems—so he might have devised this somewhat curious means of entertaining himself at home. Mortmain should give him more to do on the estate, perhaps—but again, one should not judge, particularly since I have never met Perowne. Ah, here is my kind Jehu," he added, rising.

I looked round, to see the Earl standing in the doorway.

"I hope I do not interrupt you," he said gravely. "I was wondering if I could be of service, sir. I am shortly leaving for the village."

"Why, you can certainly advise me, sir. I wish to pay my respects to a magistrate without delay."

I got up, looking reproachfully at my godfather.

"I have the honour to sit on the Lindenfold Bench," owned Lord Mortmain. "May I be of assistance?"

Mr. Shorncliffe fixed the Earl with his piercing eye. "I should prefer to consult with you in private, sir. But first, do you think I should remove this young lady from Holy Mote?"

Lord Mortmain hesitated, as well he might, looking from Mr. Shorncliffe to myself. His hands had been bandaged, I noticed, and I should probably never now discover whether he had really burned them, or only pretended to have done so.

"Remove her?" he repeated slowly. "No, I think not, though I confess I am strongly tempted to send her to the other side of England, or back to school."

"There is the question of the fees," Mr. Shorncliffe said diffidently.

Lord Mortmain appeared to regard him with close attention. "Yes, I understand that is a consideration. No, I think it better for your goddaughter to continue earning her own living, for the present—and we will engage to keep a close watch over her."

Mr. Shorncliffe bowed. "You relieve my mind extremely."

"I have just ascertained, by the way," Lord Mortmain remarked, "that no strangers apart from yourself, sir, have been seen in the village for a fortnight past."

"You do not surprise me, sir. This affair seems too consistent, too contrived, to be a stranger's work."

"I am in agreement with you, sir."

"I wonder if you can further advise me, my lord. I have made arrangements to be absent from my parish for a while and would like to be at hand for a time, but I am a poor man. Is the village inn expensive, do you know?"

"You must not think of staying there, sir. Indeed, I fear it would incommode the landlord who is an unconscionable rogue and makes his living in more rewarding ways than that of dispensing hospitality for gain. Lady Perowne cannot, of course, invite you here, as you are unconnected with her family. I propose to introduce you to our vicar, Pole-Carter, who will be glad to have your company for a time, no doubt."

"That is exceedingly kind of you, sir. Such an arrangement would answer the circumstances perfectly, if Mr. Pole-Carter permits."

"My curricle is outside. Perhaps you will come out to it when you have said good-bye to Miss Fabian." He bowed to me and left the room.

Mr. Shorncliffe seemed about to follow him, but I caught him by the hand. "Wait, sir," I whispered urgently. "After all I have told you—how can you trust that man?"

He sighed. "Believe me, Chantal, I shall trust no one foolishly. As for your safety, I think it would be a foolhardy villain who would make a further attempt so soon after we have all been alerted, but I shall suggest the outside locks are changed this afternoon, if no one has thought of it. I am confident no harm will come to you now in the house."

"But—what if my enemy is one who has the right of entry, sir?"

He pressed my hand. "You are no longer alone, Chantal," he assured me. "I shall not be far away and forewarned is forearmed, you know."

I gazed after his shambling, departing figure with tears in my eyes, wishing I could feel strengthened by his words, but grateful at least to know that there was now someone in Lindenfold whom I could trust with my whole heart, helpless though he might be to protect me.

17

That night, when I had moved my things into the Blue Room, I reflected that though I did not like this large gloomy chamber one half so much as the other, it was undoubtedly a relief to know that I was within calling distance of Lady Perowne. The next day it was a further relief when the clearing of the attic was concluded, though Old Nurse took to her bed from the shock or shame of having all her nasty secrets dragged out into the open after so many years. Percival looked quite rejuvenated by the task of supervising the great bonfire on which dusty bones, withered bats, dried frogs, herbs, and nameless but horrifying objects in green glass jars were burnt before an audience of sullen villagers, who watched the destruction of the gruesome and perhaps once-powerful relics without comment or demonstration.

Only the horned mask seemed to have escaped Percival's net and I wondered sometimes who had it now.

As the days passed, Sir Hugo became more even-tempered and a better companion. I had even begun to teach him some writing, reading, French, general information by means of Magnall's Questions, and history from Goldsmith's *History of England*, which I doubted if I would have been able to accomplish if Mrs. Cumber had been free to interfere in my arrangements. As for my own education, I had begun to ride regularly under Lord Mortmain's instruction and had come to no harm as yet.

Three days after the bonfire the doctor announced

that Old Nurse had suffered a stroke, and implied that her wickedness had caught up with her at last.

"Old Nurse has performed some remarkable cures," Lady Perowne objected quickly. "My headaches . . . Hugo's bouts of lassitude and vomiting . . ."

"I have no doubt they were all induced by Mrs. Cumber in the first place," the doctor returned bluntly. "Well, you'll get no more work out of her now, if I'm any judge—no work, though mischief enough, I'll wager, if you don't keep a sharp eye on her."

I bore this in mind when I went to pay my daily visit to the old woman, to see if there was anything she wanted.

There was no reply to my knock. I hesitated before the narrow door, wondering if she slept, or if she had recognized my step and preferred not to see me. I softly opened the door and peered into the room, which was very dark. The thick window curtains were drawn closely, but not those of the high tent bed in which she reclined, sitting high against the piled pillows, motionless and breathing audibly.

I thought she was asleep and was about to close the door again when I realized that her light clouded eyes were open and gazing directly at me with their customary malevolence.

"Oh," I exclaimed, rather embarrassed to have been staring at her. "I hope I don't disturb you, Mrs. Cumber. Is there anything you need?"

Her breathing became a little heavier and I saw that she was making an effort to speak. At last she did so, indistinctly, seeming to move only one side of her face. "Give—" she said slowly, thickly. Her eyelids flickered. One fell, as if she winked. "There," she added, making a faint gesture with one hand. "Cloth."

I looked towards the chest she had indicated. There was a small white bundle on the top and Old Nurse tilted her head to one side as if to show that this was what she wanted. I crossed the room and picked it up. It appeared to be something hard, wrapped carefully in white silk from which emanated a scent of herbs. I wondered what it could be—a bottle, perhaps, or no, it was more like a decanter.

I laid it on her hands and her fingers moved protectively about the bundle, like claws, disturbing the silk and revealing part of an under-layer of lawn, heavily embroidered. I remembered Hugo talking of a knife wrapped in white silk and herbs and realized that whatever this was, it had to do with witchcraft.

I caught my breath. For a moment I hesitated, then turning on my heel I left the room without further farewell.

I had recognized the embroidery as belonging to the fine handkerchief I had lost.

As I closed the door I heard a whisper of a chuckle from the bed, a faint and gleeful cackle that seemed to follow me as I hurried away down the tapestried corridors, not caring where I went so long as it was far from Mrs. Cumber.

At the head of the stairs I encountered Miss Curtis, looking very flushed and pretty.

"Oh, Miss Fabian," she cried impulsively. "How fortunate we met! I was looking for Lady Perowne but it is much better that I should vent my spleen on you. If I do not complain to someone, I vow I shall run mad!"

"Miss Curtis, what in the world is wrong?"

She drew me into the Music Room. "It is that odious wretch, Mr. Perowne."

"Why, what has he done to displease you, ma'am?"

"What has he done? He has found every opportunity he can to—enrage and humiliate me!" She cast her muff into a shrouded chair.

"Do you wish to be more particular?" I asked cautiously.

"I had better be—I mean to tell you all! Well, then, I went to watch Lord Mortmain teaching Hugo on his new pony—how that child has improved! But that is by the way. Mr. Perowne chanced to be there for the same purpose. I was rash enough to offer a few words of advice—and I am not entirely ignorant of the art of horsemanship, I think. I could ride before I could walk, which is more than Mr. Perowne can boast, and was hunting by the age of six. I should like to see Mr. Perowne take some of the fences I have cleared! I say nothing of Mortmain, for he is something of a *nonpareil*, but not every great exponent has the art of teaching—"

"Well, ma'am, you are a splendid horsewoman, I am sure, but what has this to do with Mr. Perowne?"

"So far from thanking me for my advice," continued Miss Curtis stiffly, her cheeks flushed at the memory, "Mr. Perowne was pleased to be extremely rude to me—so rude, indeed, that if I had chanced to have been carrying my riding whip, I do not know but that I might have slashed him in the face."

"Miss Curtis, I did not know you could be so violent. One must be glad that you were empty-handed."

"Do not say so, for now my violence has been frustrated I shall doubtless treat him worse at the next opportunity which offers."

"But what can he have said, to have put you in such a rage?"

"Oh, what does it matter? He gave me to understand that I was interfering, domineering—unfeminine, I believe he said—"

"You, unfeminine? Mr. Perowne must be mad—or drunk."

Miss Curtis laughed angrily. "Neither, I assure you. He was precisely as usual. Oh, how I detest him! But the worst of it was that Mortmain heard it all—and spoke not one word in my defense!"

She forced a smile and went to pick up her muff. "And now I have to ask you to excuse this storm in a teacup, Miss Fabian," she said with a pretty air of apology. "Only think how I should have offended Adela if I had said it all to her! I feel much better having vented my ill-humour, and hope you will do me the kindness to forget it. How is your godfather? I understand he is staying at the rectory."

I explained that Mr. Shorncliffe was not really my godfather, though I had always called him so; but he had stood a proxy for whoever my godparents had been. The conversation went on pleasantly enough until Miss Curtis remembered that her mother would soon be calling for her and had particularly requested her not to keep the carriage waiting as they were expected at the Prendergasts, some miles distant.

I went downstairs, thinking about Mr. Shorncliffe and wondering if there were any significance in the fact that he had made no effort to speak with me alone since our meeting in the library, or whether he had merely forgotten all about my problems.

The carriage was not yet in sight, but Mr. Perowne was walking towards the house from the direction of the stables. Miss Curtis took a sharp breath, turned her back on him, and began to tap her foot.

Mr. Perowne halted. "Good morning, dear Miss Fabian," he said in a caressing voice. "And hail to you, Miss Vinegar," he added in a different tone.

"You are not very polite to Miss Curtis, sir," I protested rather faintly.

"You think not, Miss Fabian? Well, you may be right, but have you never heard that familiarity breeds—a certain emotion? I have known Miss Curtis long enough to treat her as I choose and she has given me small reason to admire her. Should you like me to enumerate her faults, so you can understand why I am unable to respect her?"

"No, indeed I should not!"

"But I have a notion to do so, for it may be good for her to hear them. Item, as she has been spoilt from the moment she was born, she thinks of nothing but to please her family, would sell her soul to win a nod from her mother or an approving smile from her father—"

"It is not true!" cried Miss Curtis, wheeling about

with scarlet cheeks and shining eyes. "Oh, how I hate you!"

"No doubt," he said with an appearance of calm, but I could tell from his clenched hands that he was just as angry as she. "It is always unpleasant to hear the truth about ourselves—when it is unpalatable. No doubt you tell yourself that you are a good and docile daughter when the truth is rather that you are criminally weak—"

"Weak? And not an hour ago you were describing me as overbearing!"

"That is another fault in your character, Miss Vinegar: that you are inconsistent. Charming one moment, odious the next. Too submissive to some, too arrogant with others—honey and vinegar. You are complaisant when it comes to marrying for power and position, stubborn and obstinate when it comes to looking into your own heart. You deceive yourself, Miss Sweet-and-Sour, a hundred times a day. No doubt you think it fascinating to blow hot and cold, to be as fickle as a weathercock—"

"Fickle!" she cried, now very pale. "When have I been fickle? It is obvious that you do not know the meaning of the word, but the remedy is at hand. Look in the mirror, sir. There you will see one who is fickle in truth, one whose eyes follow every slender ankle, every pretty face, who prides himself on raising expectations he never intends to fulfill, who deceives with every breath he draws, who—"

"What a little cat it is, spitting and scratching at one moment, sleek and soft the next—to the hand that holds the purse-strings of a fortune."

"Oh! You—"

Goaded beyond endurance, Miss Curtis flung off my restraining hand and leapt forward as if to strike her tormentor. With an air of mock terror he jumped back—and crashed against the doorpost with considerable force.

I seized Miss Curtis by the arm. To my relief she immediately halted. "You are no gentleman, sir," she informed him in tones as icy as her former ones had been hot.

He spoke between his teeth, one hand pressed against his shoulder. "And you, ma'am, are certainly no lady." He closed his eyes. He was alarmingly pale. Miss Curtis could not fail to remark it.

"What is it?" she asked uncertainly. "Mr. Perowne—are you all right?"

"I am—just as you would wish me to be, no doubt," he murmured.

"No, but—if you have hurt yourself—"

"Go away, Miss Curtis. Miss Fabian will help me." He opened his eyes and stared at me rather blankly. "Miss Fabian, if you please—"

"Piers!" shrieked Miss Curtis suddenly. "What have you done? Your shoulder—your arm—they are all over blood!"

"I think," said Mr. Perowne faintly, "that I have opened . . . an old wound."

"Oh, how dreadful," she cried. "And it is my fault! To think that I should be the cause—but never mind that! Miss Fabian, fetch water, bandages, plaster, and a styptic. Piers—no, don't argue! Lean on me and come into the hall."

"Interfering," he murmured as she put her arms about him, "arrogant, overbearing . . ."

I fled.

By the time I returned with all she had asked for, Miss Curtis had somehow removed his coat, had torn the sleeve out of his shirt and was holding it hard against the wound, while he gazed thoughtfully at her expression, gentle and absorbed.

She looked up as I laid down the tray. "It is very bad," she said in a voice which trembled slightly. "There is so much blood. . . ." Her voice trailed away and Mr. Perowne caught her as she subsided to the floor in a faint.

"Mr. Perowne!" I cried sharply. "Remember your shoulder—"

"Damn my shoulder!" he cried roundly. "Fetch

feathers—sal volatile—anything! Oh, thank God, she is coming round. Look, her lashes are fluttering—are they not the longest you ever saw? And has she not the sweetest little face—when it is not contorted with rage—"

"Piers," breathed Miss Curtis, "are you still provoking me?"

"It has succeeded in rousing you, has it not? Now what will further restore you, I wonder? What the Devil can a man do for a girl when he is encumbered by having to hold her in his arms?"

"What indeed?" she murmured, with a faint but teasing smile.

"Little cat!" said Mr. Perowne, and began to kiss her lingeringly while I did my best to ignore them and concentrated on attempting to stanch the wound.

There was a commotion in the doorway. "Oh, look, Cousin Mortmain," cried Sir Hugo in a voice shrill with surprise, "whatever is Uncle Piers doing to Miss Curtis?"

Miss Curtis sat up so suddenly she was obliged to cling to Mr. Perowne for support. "Oh, my lord!" she exclaimed in great confusion. "Sir Hugo! I—how you startled me!"

"Hush, little fool," Mr. Perowne rebuked her sternly. "Never apologize for being found in a compromising situation, as I see you are about to do."

"You are the expert, to be sure," she accused him tartly.

"Be quiet, or I shall kiss you again. Now, Mortmain, be a good fellow and clear off, will you—and take that brat of a nephew of mine with you. Can't you see when a fellow needs a little privacy, eh?"

"Certainly I can," Lord Mortmain conceded calmly. "I was about to retire."

Miss Curtis cried out, "Oh, but I think Lord Mortmain deserves a little more in the way of an explanation than that!"

"Silence, girl," said Mr. Perowne sternly. "I make the

decisions for the two of us from this time on, do you understand? Mortmain is well aware that I have first claim on you, however hard your parents may have been throwing you at his head—"

"First claim? What in the world do you mean, sir?"

"Don't call me 'sir' when you are lying in my arms, nonsensical female," he rebuked her. "My claim is the greater because I am the one you love, not he."

"And you love nobody but yourself, I suppose," she cried, struggling. "What a—an unchivalrous thing to say!"

"It was, indeed. What a splendid thing it is that we have no illusions about each other." He turned his head to look up at me. "Have you finished your ministrations, Miss Fabian?"

"As well as I can in the difficult circumstances," I told him primly, wondering what his French Marie would make of this. "The bleeding has not stopped, but it is easing."

"Good. Then I think Miss Curtis and I will excuse you, also."

Hugo, who had been staring with round eyes, said, "Are you going to be married? But I thought you did not like each other!"

Mr. Perowne appeared to consider this remark. "What I most disliked in Miss Curtis, I see now, was that she did not allow herself to consider me a worthy suitor to her hand," he announced judiciously.

"And what I most detested in you, Mr. Perowne," she began warmly, "was that you were always criticizing me, while behaving abominably yourself—"

"It is certainly time for us to go," Lord Mortmain declared. "But not before I have congratulated you on your forthcoming nuptials, however."

"Very handsome of you, Mortmain, in the circumstances," Mr. Perowne allowed. I saw a calculating gleam narrow his eye. "Would you like to celebrate the occasion by giving your blessing to another project I have in mind?"

"I will certainly consider your request."

"Then say that you will allow me to purchase a yacht of my own, or have one built? I am sure Miss Curtis will make a very model of a sailor, with tuition."

Lord Mortmain was silent a moment. "And do you believe you will be able to procure a license from the Admiralty, to say nothing of being able to afford the vessel, and the not inconsiderable stamp duty on the bond?"

"I would spend less upon my horses if I had a boat. As for the license, I am certain a person of your influence, if he exerted himself a trifle, would have no difficulty in persuading their lordships to license me to navigate, so long as it complied with the terms of the Act."

Lord Mortmain shook his head. "I am sorry to deny you at such a moment, Piers, but I think you knew very well that I would not be able to consent to such a—a reckless expenditure."

Mr. Perowne's expression was quite ugly for a moment. I turned from it to pick up the tray. Lord Mortmain prevented me, however, and in a stern voice desired me to step outside with him for a moment. I followed unwillingly, while Sir Hugo scampered upstairs.

"Well, my lord?" I said coldly, upon the bridge. "What do you want of me?"

He gazed at me enigmatically and I was obliged to drop my eyes. "Very little, at this moment," he assured me with a coldness to match mine. "Only to warn you to stay within the house tonight. You are aware of what date it is, I suppose?"

I stared at him in surprise. "The—thirtieth of April, sir, I think. Yes, it must be."

"Just so. May Eve, Miss Fabian, or Beltane, as some call it. Ah, here is Lady Martha's carriage. Excuse me, ma'am, while I have a word with her ladyship—and don't forget my warning."

Lady Perowne was so delighted by the news Sir Hugo brought her that his uncle and Miss Curtis did not hate each other any more that she raised no objection to my proposal of walking with her son to visit Granny Cumber in order to inform her of Old Nurse's progress, though I had expected her to do so, for the wind was cold.

We set out briskly, and Hugo informed me that his boat was finished.

"Perhaps I'll sail it in the quarry," he suggested as we passed it.

"It is far too cold today. Besides, those ponds are often deeper than they look. Oh, there is a man with a dog to drive the cows away. You will be glad of that, I suppose."

We paused on the safe side of the stile while I discovered from the cowherd that the beasts were in fact being taken to market and the field would be left empty until haymaking. We followed him at a discreet distance until Hugo suddenly seized my arm.

"What is it now, sir?"

"It's Smithers, Miss Fabian—Born-To-Be-Hanged!"

We watched the carter as he rode past us and neatly jumped the stile. Why was he going to Holy Mote, I wondered: to see his friend Piers Perowne, his brother the footman—or for some reason connected with Beltane?

"What is the matter, Hugo?" I said, seeing he was rather pale. "Does Smithers still frighten you?"

"I keep remembering that day I saw him in the attic," he burst out.

"Hugo—you mean—but you said that man had a foxy face?"

"Oh, not that man," he said scornfully. "This was years ago. It was in that room where you got trapped, where I saw the witches' knife. He—Smithers—had the horns, and you know that means he is the leader of the witches."

I took his hand and held it tightly. "The witchcraft is all over now," I reminded him. "You saw the bonfire."

Hugo gave a sudden grin. "Yes, that's true. It was beyond anything great! Come, let us hurry and see Granny Cumber."

"Don't tell her about it—"

"I should think not, indeed." He ran ahead to knock on the old woman's door and a moment later we entered the cottage.

As soon as my eyes adjusted to the dimmer light I saw that Granny Cumber seemed to be sitting very still. Her eyes were open and gleaming, however, and I realized that she owed her immobility to the fact that her black cat was lying on the top of her head like some monstrous fur turban.

She greeted us with her cackling laugh and the piously expressed hope that we had not come to tell her young Betty was dead. "Zounds," she cried, evidently enchanted by the thought, "but I little guessed I'd outlive my daughter, and she so strong and healthy all her life—though I did wonder when she dropped that cup t'other day and stood so still and strange a moment—but I should have thought I'd have known of it if she'd been dead."

"No, no, Mrs. Cumber—your daughter has to keep to her bed at present, but the doctor hopes that with care—"

"The doctor, eh?" Without moving her head, Granny Cumber contrived to spit into the fireplace. "Why do you plague her with doctors, miss? If anything is to finish her, 'twould be arguing with one of the likes of them. Doctors, pah! Did we have a doctor to Cumber, when the old Earl ran him down in his new carriage, may his lordship roast in Hell?"

"Old Lord Mortmain ran down your husband?" I queried, recalling that the wicked Earl had also accounted for Black Dick, the highwayman. He seemed to have had scant respect for human life; and it was now plain to see why Old Nurse hated the Quentins.

"Ay, curse him! But Cumber survived the accident without benefit of doctors—and dragged out his life for another twelvemonth. I've seen a mort of folk sent to their accounts by doctors, I can tell you, miss." She sighed with what seemed like satisfaction at the thought, while Hugo wriggled with delight. "I wish I could get to Betty now myself," she went on in her shrill voice. "At one time I brought about all manner of cures, here in Lindenfold—ay, and there was scarce a babe born in the village that I did not deliver—even my own Betty, when I thought she was dying that time—years ago, it was, I suppose, though that seems like yesterday—even then, did I fetch a doctor to her? No, miss, I did not, for that would have been to finish her, surely. I brought her back to the living with my Master's help—but it don't do to speak of him by daylight, eh. Tonight, ay, tonight will be the time for him . . . Beltane. Emmie!"

"Yes, ma'am? Just coming, ma'am."

"Well, hasten, girl—where is the tea? Do you want to keep miss waiting—and little Sir Hugo? Down, kitty, down," she muttered. "My old head is tiring now, too tired for the likes of you. . . ."

She nodded and the cat jumped neatly down to lick its fur by the empty fireplace, just as Emmie hurried in with the tea tray.

"Ay, Betty was took bad last Sunday week when she was here," the old woman went on. "My lady gave her leave to stay all evening but she was gone by dark saying she had to get back to rest . . . ay, they'll carry me to her funeral before long, I don't doubt it."

Her voice died to a murmur while Hugo and I sipped our tea. By the time we had emptied our cups the old woman was asleep and we were free to escape into the sunshine with, for my part, an unexpressed but fervent feeling of relief.

18

Mr. Shorncliffe and Mr. Pole-Carter joined us for a pleasant dinner that evening of fried eel and boiled beef, pease soup, pork, and suet pudding. Afterwards, Mr. Perowne excused himself early, confessing that his shoulder, which, as he explained somewhat unconvincingly, had been torn by a nail protruding from an old beam in the stables, had bled a good deal and still did not feel quite the thing. He wished, he said with a meaningful glance at his sister-in-law, to feel fit upon the morrow, when he intended to call in form on Mr. Curtis.

Lady Perowne detained him, however, as he was about to leave the room. "I think you had better see Cousin Mortmain first," she suggested in a low voice, "however unpleasant it may be for you—and I dare say it will be extremely awkward in the circumstances! But it is essential you should discuss with him what allowance you might expect from the estate as a married man."

"Oh, I tied all that up while you were out walking," he said easily. "Mortmain ain't wearing the willow for Thelma, I assure you. He cast no rub in the way, and I think the Curtises won't object to the terms of the settlement he has proposed."

"He has already determined how much the estate can afford?" she queried in surprise.

"Oh, ay. Had it all worked out to the last detail—said he had always known I was the kind of reckless noddle who would marry young—but by the same token perhaps marriage would steady me—patronizing, wasn't it? But I could hardly resent it, for some of the settlement is to come from his own estate by way of a wedding present."

"Well, that is certainly extremely obliging of Mortmain," Lady Perowne conceded doubtfully. "I wonder what he is thinking of?"

"Nothing that is not for the good of the Perownes, surely," Mr. Pole-Carter suggested, since her remark appeared to have been addressed to him. "I am aware that you think it a hardship to have had your fortune left in the Earl's keeping, ma'am, but he is a just steward, I am persuaded. Are you ready, Shorncliffe?" he added, rising as my godfather nodded. To Lady Perowne he explained rather whimsically that as it was May Eve, ordained clergymen had, he feared, no business to be abroad.

"Pray don't talk to me of such superstitious stuff," said Lady Perowne, shivering.

"But ma'am, your own Old Nurse—"

"And I will hear no word spoken against Mrs. Cumber," she cried. "Just because she brewed her own simples and understood disease, half the world believes she is a witch. If you must make such suggestions, make them elsewhere, sir."

He did not seem offended. "Very proper, ma'am. Your loyalty is to be commended. Pray forgive me."

After a moment she smiled and did so gracefully. "Good night to you both," she added. "And—a safe journey, since you think this night so dangerous."

Smiles and farewells were exchanged and, arguing gently as to the origins of the witches' festival of approaching summer, Mr. Pole-Carter and my godfather left.

I played Lady Perowne another song or two at her request and then she began to ply her fan and complain that she vowed she was exhausted by her walk. I helped her to her room and rang for Jackson.

"I was watching for you, miss," said Jackson, to my surprise. "Excuse me, my lady, for one moment." She whispered to me that Mrs. Cumber would be obliged if I would visit her before retiring.

I was puzzled and wary, but of course I could not refuse. What could the old woman do to me, I reasoned, bedridden as she was? I had only to remember that she was not really a witch and she would have no power over me. I took a candle and went to her room.

Again she did not answer to my knock, or so softly I could not hear her. I opened the door and looked in. She seemed to be sleeping, her face turned sideways on the pillow, but as the light fell on her cheek she started and her eyes opened.

"Hugo," she said, blinking. "Hugo," and with a noticeable effort she raised a finger to her lips.

I was mystified. "Hugo?" I repeated blankly. "Do you want me to bring him here?"

She frowned impatiently. "Gone," she said. "Don't tell. . . ."

"Sir Hugo has gone?"

This time she nodded. "Out . . . quarry. Bad . . . bad night for wandering."

"Don't worry about him. I will send Thomas or John to find him. Sir Hugo knows the quarry well. I am sure he will have come to no harm—" I broke off, reflecting that the quarry had an evil reputation. Old Nurse's features were contorted with pain and frustration as she tried to shake her head. "What is it, Mrs. Cumber?" I bit my lip. "This is so bad for you."

"Go . . . alone. Don't tell."

"Oh, I understand." I stood uncertainly, considering it. No doubt Mrs. Cumber had made Sir Hugo some sort of promise; perhaps she felt it might be the last thing she could do for her nurseling, to protect him from disgrace.

"Hurry," she muttered, her face purpling.

"Very well."

I shut the door and hastened to Sir Hugo's room. It was also possible that this was just a ruse to send me out into the night as some revenge for my having been the unwitting cause of the desecration of her sanctum, if one could use that term for a room so patently unholy. Perhaps I should find Sir Hugo asleep in his bed or sitting up putting the finishing touches to his model ship. But though his bed was dishevelled and warm, he was not in it, nor under it, nor in the closet beyond. I found his nightshirt crumpled on the floor and realized that he had taken it off and dressed again. He had really gone out, I thought with a chill of fear. He had gone out to try his boat, on May Eve, at an hour when honest folk should be abed and when witches were reputed to be abroad.

I did not wait to put on my cloak but ran straight down the stairs as I was. From the warmth of his discarded nightshirt I knew I could not be far behind him. The great front door was unbolted and unlocked, presumably by him, so I wasted no time opening it. Once outside I paused on the bridge to make certain that he was not crouched beside the moat, and then turned down the drive towards the meadow path and the old chalk-quarry.

I had forgotten to put down my candle. It blew out and I threw it away, expensive wax as it was, and hoped that I would be able to find it again. Without it, the night was dark. The stables were quiet. I could imagine the horses, already saddled and bridled, waiting patiently for their masked riders to lead them out to join the Wild Hunt, for I supposed that every stable in the district must contribute horses on such occasions, with or without the permission of their masters.

I thought I heard a twig snap underfoot. I strained my eyes and saw a movement in the wood ahead of me. The next instant I was certain it was Hugo, for he had paused beside the tree he never passed without touching.

Relief swept over me. I was in time, for though the moon had not yet risen I was certain to catch him before he reached the quarry. I did not want to call out while we were still so near the stables, so I picked up my skirts and ran after him.

There was one of those instants of unspecified warning which usually come too late to act on them. I had barely recognized it for what it was when something loomed over me, darker than the surrounding darkness—not a giant, as my horror-struck imagination suggested, but a horseman, leaning from the saddle. He seized me in strong arms and began to lift me, struggling, to his saddle bow. I tried to scream but his gloved hand crushed my lips. The horse, a great black beast, reared and pawed the air till I thought we would both fall. The rider, however, controlled him with the other hand and sent him forward, cantering dangerously through the trees. I thought of Hugo skipping ahead, intent on his boat, and struggled the more, but the hand tightened over my face so that I could hardly breathe and I had perforce to sit quietly. The hand relaxed a trifle and I drew in a breath tainted with the scent of Macouba.

Suddenly I saw Hugo. He looked round, and shrank back at the sight of us. I prayed that he would have the sense to run to the stables and tell them what he had seen but I could not help fearing that he might be too afraid of revealing his own escapade to risk discovery.

We cantered past the last of the slashing branches, from which the gloved hand at least protected my face, and out into the meadows, silvery in the starlight. The hand was taken from my mouth and gripped the reins, leaving me free to call out now that there was no one to hear me, and ensuring that I was securely trapped between his arms. We swept past the broken quarry fence and thundered down towards the stile. Terrified though I was, yet something deep within me responded to the wild ride. My fear was inextricably mingled with my old dream based on an enchanted memory, of Rowley setting me on his big black horse and riding off with me in just such a way long ago in the happy days when my kindly guardian lived, and everyone seemed gentle, and to be in one's own bed at night was to be in the safest place in the world.

The horse gathered itself to leap the stile. This was

a new alarm but I knew that at least I was in the hands of an expert rider. I gasped as we soared for an instant between earth and sky, and gasped again with the shock of landing. A few moments later we were cantering down the green ride by the church and I wondered if a scream would be heard at the vicarage above the pounding of the hooves, but as if he read my thoughts, the horseman lifted his hand from the reins and roughly covered my mouth again.

"Where are we going?" I cried, as soon as he released me. "What do you intend to do with me?"

I turned my head so that my hair blew all across my face and saw with a new shock of horror that he was masked—not with Herne's antlered mask this time but merely disguised with a small domino.

He did not immediately answer, and I wondered why I had been so foolish as to ask when I knew that he had ridden off with me to kill me, wherever he might intend the deed to take place.

He would never answer me, I thought. I would never know who he was—only whom I suspected him to be.

But then he spoke.

His voice was a low growl assumed like his mask, to disguise him, and his words—a shiver of surprise and a new terror shook me as I made out his words.

"We are riding in the Wild Hunt," he said.

I sat silent, my hands clutching the long black mane as it streamed in the wind, breathless and jolted by the movement of the horse, and deathly cold.

But what was to happen? What could he want with me in the Wild Hunt, unmasked, uncloaked, even, as I was? Did he believe it was enough for me to see it, that my death would be then ensured?

My surmises were abruptly driven from my head by a sudden screaming whinny from the horse that carried us, a whinny that was echoed by others as I made out the forms of riders in the lane, and more behind us, and still more coming trotting down a path to our left, all converging on the track that led to the Downs.

"Help!" I cried, and just as the horse's whinny had been echoed so was my cry taken up and drowned by the screams of those other riders, frightful cries that sounded as if they had been learned at the gates of Hell. They pressed close about us now, those riders with heads of dogs, cats, goats, cows, even horses, and I screamed again in vain, but as though my voice had been a signal the horsemen fanned out so that each man could make his own way up the hill.

The night was lighter now, the moon was near, and I could see one man out in front, leading the motley throng, and his mare could have been of any colour, but her white socks were unmistakable.

Surely Piers Perowne would save me, I thought; and I willed the man who held me to give his horse its head that we might draw closer to that antlered figure. All at once Herne reined back his horse and seemed to wait up there on the hill above us. The man who held me moved in the saddle and the great black leaped forward as if it were I, not they, who knew the dark secret ways of witchcraft.

The others, too, urged on their horses. Herne wheeled away before us but we were close enough for him to hear me.

"Piers!" I screamed. "Help me!"

He heard, I was sure of it, but he neither paused nor turned but galloped on towards the wood.

Holly Wood, I thought despairingly. Piers Perowne had not liked it when I had caught him there, kneeling before the great altar stone. Was it possible that I had been lured out tonight so that I should be brought to justice for that offense, shivering in my white virginal gown, on that same place of unholy sacrifice?

The hill was steep and the horses slowed but the wood came inexorably nearer, haloed now with an unearthly light. Quite suddenly the waning moon swam clear of the topmost branches brightening the silver landscape, deepening the shadows, turning the sea to milk and striking fire from bits and stirrups. A ragged howl rose all about us from the masked riders, a howl of savage anticipation. Then the thin branches were whipping my cheeks as we plunged into the darkness of the wood.

I remembered how I had hated Holly Wood the day I first ventured into it. I had shivered then, listening for the unheard, looking for the unseen. But now the evil, that then I had only sensed, was tangible. I could see and hear it all about me in the slashing whips, the panting riders, the wild brutal howling.

I tried to force it from my mind. I thought of Hugo, who might now be drowned in the quarry pool. Useless now to ask myself why I had allowed Old Nurse to persuade me to leave the house on a night when even my enemy Mortmain had warned me to stay inside. Why had he done so? But he had known he was safe, of course. He had known my simple foolish heart would lead me to disregard his warning. I remembered Born-To-Be-Hanged riding to the house this afternoon. That could have been when the message was passed to Old Nurse to tell her what to say to Hugo and what to me, that the victim might be led meekly to the slaughter. It was useless, too, to wish, as how bitterly I wished, that I had snatched up some weapon to bring with me, one of the guns or swords that hung in the Great Hall, instead of delivering myself unarmed into the hands of those for whom this night's work would no doubt serve to bring the prophecy true, that one of their leaders was born to be hanged.

But what had all this to do with my guardian's murder, years before? Was the man who abducted me the Earl of Mortmain, or had I been wrong from the very first?

The arms that held me tightened savagely as the horse plunged to a standstill and was backed between the bushes, off the path. We had halted on the edge of the moonlit glade and Herne had halted within it, bringing his horse to stand on that very sunken stone which I feared so much.

The other riders swept past, the night ringing with their wordless cries. They circled the stone and then rode on, to my amazement, into the farther side of the wood, dogs, goats, cats, horses, howling away into the dark trees across the glade.

The black horse stamped impatiently and quivered, and stepped sideways from time to time, but his rider kept him under iron control and did not allow him to follow the rest, seeming to be waiting for the moment when the whole Wild Hunt should have passed us by.

At last they had all gone. As the final straggler on his sweating horse jogged past him, Herne leaped forward on the chestnut mare, slashing the laggard rider mercilessly with his whip, driving the hunt before him through the wood.

Silence crept back.

The black horse shook himself suddenly, setting his bit jingling, the saddle leather flapping. My heart leapt with terror. I could hear the suddenly increased breathing of the man who held me as he took his right hand off the reins. I felt him fumble with something behind me. There was a curious whisper of sound and then a bright flash in the moonlight as his hand came forward and pointed a curving knife blade at my breast.

Time seemed to stand still while voices echoed in my stupefied mind. . . .

"Three great stones, the center one of which was surely a pagan altar, the scene, no doubt, of many a human sacrifice. . . ."

"Artemis for me. I fancy our Master likes them young. . . ."

"Others more powerful than I will deal with you. . . ."

"Alone I cannot do it . . . Herne must aid me . . . a sacrifice to Artemis. . . ."

"Is not the season past for sacrifice?"

"Ah, if it were but Candlemas!"

And lastly that fatal, decisive, "Even so!"

Now, I thought, now is the destined hour for the

sacrifice to Artemis. He will kill me here and now upon the altar and when my body is found all will know it was the work of witches and none will therefore dare to cast a stone . . . not even the noble magistrate, whether or no it is he who sits behind me here holding the witch-knife to my heart.

There was but one card left to play, a poor low card at that—but late in the game even a low card may sometimes take the trick.

I said in a voice that clung strangely to my throat, "Lord Mortmain—"

He started. He had not supposed I recognized him, then.

"You know me?" he whispered. And added chillingly, "That adds savour to the jest, indeed."

I tried to disregard the cruelty of his words.

"I am no danger to you, sir. I will even—marry you, if you wish—"

"Will you, by Herne!"

It was useless. Mary had warned me. I should have heeded her and never come to Lindenfold. Things were different here, where babies were no doubt dedicated to Herne in their cradles . . . where Piers Perowne led the Wild Hunt, an initiated witch . . . and where the Earl of Mortmain held ready the athame to plunge into my heart, to make a sacrifice to his dark master or to Artemis, goddess of the maiden moon.

Beneath me I felt the horse tremble as he received the office to advance. He took a step forward, only to be instantly reined back on his haunches. The knife was whipped away, the gloved hand covered my mouth. I watched incredulously as the wood seemed to come alive before me, shadows slipping in and out of bars of moonlight, but substantial shadows that snapped sticks beneath their feet.

The black horse was backed into hiding again and halted there with the glade still visible between the sheltering branches. The shadows became horsemen, masked, but not like animals, and strangely burdened. Their leader dismounted first. He was plainly dressed and held one arm stiffly to his side.

With a sudden leap of the heart I recognized his neat mare as Brown Bess, the man himself as Piers Perowne.

For the second time that night, I saw the flash of moonlight on steel. Mr. Perowne too had pulled out a knife and was dropping to his knees before the altar stone, just as he had been the day that I surprised him here. The arms that held me tightened and I did not even struggle, so intent was I on watching what was happening in the glade.

I do not know quite what I had expected, but it was nothing like what actually occurred. Mr. Perowne, after glancing at the blade of his knife, merely ran it horizontally along the length of the stone's edge. Four men then stepped forward and lifted the altar precisely as if it were a lid, revealing it to be a slab of stone rather than a deep sunken boulder. They stood at the four corners of it like mutes about a bier; and then Mr. Perowne threw up his arm and out of the wood rode the masked men, silent but for the creaking of their saddles and the faint jingling of their blackened bridles. Each man as he passed the stone leaned down and tossed something into the abyss it had protected. Two of the men had now provided themselves with long sticks which appeared to have been hidden in the bushes and with these they helped to dispose the offerings tidily, for they were of varied shapes, packages and parcels, and rolls of what looked like cloth—probably silk, I thought, suddenly realizing what this scene portended.

Piers Perowne was the leader of the freetraders, not of the Wild Hunt; and I remembered now a letter written in French that had not seemed like the love letter he had pretended it to be, stained with sea water as it had been. I remembered other things too: a woman rewarded by a half-anker of brandy for holding her tongue; a Revenue cutter patrolling the coast, close inshore; mention of an innkeeper who made a lucrative living—but not from hospitality—and a bullet wound in Perowne's shoulder that could have come from the gun of an exciseman. Indeed, there had been some topic of conversation at church the next morning—and Lady Perowne had called it disgraceful, while her brother-in-law had preferred to describe it as exciting . . . but witchcraft had come before freetrading, I thought, and still existed here. Piers Perowne had merely used it as a cloak for his own activities.

My thoughts were suddenly recalled to the present and to the danger I was in, deferred though it had been by the timely appearance of Mr. Perowne, by the tightening of the masked man's rough hand over my face. He had grown impatient, apparently, and began to back his horse through the bushes, slowly, carefully, one step at a time. Then, when he judged it safe, he wheeled the animal and cantered out of the wood.

Once out on the empty Downs, washed with moonlight, he mended his pace and we galloped recklessly down the winding zigzag path. He was angry, I sensed, angry that his plans had been frustrated. Perhaps his anger would overset his judgment and that might give me a chance of outwitting him. I wished I knew what was in his mind. Where was he taking me? Back to Lindenfold?

But he swung left at the bottom of the Downs, and spurred into the green church ride. A light burned in one of the vicarage bedrooms and I wondered briefly whose it was. The hooves made a different sound as we came out on to the lane and softened again as the horse was turned to the right up the path to Holy Mote and slowed to collect himself for the jump over the stile.

Were we going back to the house? And what would happen to me there?

The horse was tiring. He jumped the stile, but rapped it with his hind feet. He stumbled on landing and nearly threw me. We cantered more slowly on, across the great field where the cattle had lately grazed. We neared the quarry and the horse began to trot and then to walk. My heart plunged again. It was the quarry which was his objective. Holly Wood was barred to him. The deserted churchyard of All Saints would have served his purpose very well, but that it was not to be deserted on this night of Beltane. There remained the quarry, a hidden, secret place with a pool where a body might lie weighted . . . a place associated with witchcraft, where no one ever went —no one but Sir Hugo Perowne. If he were there—

But at that moment I saw one of the broken fence posts appear to move. A figure detached itself from the fence—too tall for Hugo and with a shambling gait familiar to me. His face washed to a semblance of mildness by the moonlight, my godfather stepped into our path.

The man behind me caught his breath. We halted as Mr. Shorncliffe put his thin hand on the reins and peered up at my captor. There was nothing mild about his voice as he said accusingly, "Mr. Exham, I think?"

There was a moment of stunned silence. Then the horse reared beneath us as the rider wrenched round its head. From some holster behind the saddle he dragged out a pistol and cocked it just as another horseman seemed to leap out of the trees towards us. I screamed and knocked the gun aside as a stab of flame came from it and a shattering explosion. Another report echoed the first and there was a grunt behind me of what sounded like surprise.

Mr. Shorncliffe gave a shout of protest. "Enough, Mortmain, you've winged him, I believe—"

I could see now that the other horseman was Lord Mortmain, grim-faced on his big bay, the gun smoking in his hand. He plunged to a halt beside us and the black horse shied away. The rider fell against me and his hand dropped from the reins. Then he toppled, limp as a man of straw, onto the broken railing, hung there a timeless moment and then dived through the little bushes and loose chalk stones, into the quarry.

Lord Mortmain thrust his pistol back into the holster

and dragged me from the saddle, onto his own. Immediately the black horse turned and began to gallop back towards the village, reins and stirrups flying.

"Miss Fabian!" cried Lord Mortmain, forcibly turning my face towards him. "Are you unhurt?"

I nodded, finding myself, for some ridiculous reason, unable to speak.

"By God," he exclaimed wrathfully, "you deserve not to be!"

Somehow that released my tongue. "I thought you were the murderer," I told him. "I thought you brought me here from school to—to silence me!"

He held me close. His face was taut with anger, his eyes burned with it and I was again aware of the extraordinary feeling of power which emanated from him.

"And why did you suppose I was thinking of marrying you?" he asked silkily.

"Well, of course—to prevent me from giving evidence against you—"

"Enough," he said sternly. "I must go and see what has become of Exham. Shorncliffe has already gone down, I think. Will you stay here and hold my horse, or would you prefer to make your way back to Holy Mote?"

Uncomfortable as was my position, bruised by his saddle-bow, held tightly in his arms, I found I did not want to alter it. Besides, I had a hundred questions to ask.

"Who is Exham?" I began, and yet even as I said it I felt I had heard the name before. "Why should he wish to kill me—as I assume he does," I added, beset by a sudden doubt. I had apparently been wrong about so much perhaps I had deluded myself about that also.

"Oh, he does," said Mortmain grimly. "He has been my enemy and therefore yours ever since Black Dick was hanged—"

"Black Dick!" I cried. "But that was years ago—and had nothing to do with me. I thought—I thought all this must have had some connection with my guardian's murder. I suspected you—"

"Ah! And now we come to it. What grounds did you have for suspecting me, may I ask?"

"Why—that I saw you coming from our house the very night Medlicott was killed."

"So when you said you remembered me, that was what you were thinking of?" He sounded somewhat abstracted.

"You—you don't deny it, then?" I said uncertainly.

"It would be useless, if you saw me. Listen!"

Thinly, Mr. Shorncliffe's voice exhorted the Earl to join him.

"I must go—"

"But Hugo!" I clutched his coat. "I had forgotten Hugo! He—"

"Hugo is asleep in bed by now, I trust. Tonight has made a man of him."

"Why, what do you mean?"

"When he saw you abducted he ran to bridle his pony and rode to find me—he has come to trust me lately, unlike yourself. As it happened, I was already setting out with the intention of watching over the house tonight—"

"You warned me," I cried. "How did you know I was in danger?"

"Your godfather and I had discussed everything with Pole-Carter—a strange tale it was we pieced together—and we concluded that, after what you had overheard with reference to sacrifice, it was likely to be tonight and probably here, for we felt Holly Wood and All Saints were likely to be overpopulated, though Pole-Carter did take the precaution of riding after you to the wood, just in case. But I agreed with Shorncliffe, and thank God we were right, and you are safe."

He leaned over and set me lightly on the ground, dismounting as soon as he saw that I stood steadily. The torn grass smelt fresh and green. The dew began to soak into my skirts and slippers. I shivered, but not, I think, from cold. I looked up at Holy Mote, crouched on its hill among the trees, its many windows staring in the moonlight, while behind me came a soft rattle of chalk stones as the Earl took a backward step onto the path.

I spun round and seized his hand. "Don't leave me, pray!"

He smiled ruefully. "I think I can't; but stay close by me." He slide his coat about my shoulders and then, his fingers tight on mine, began to lead me down into the pit.

It was a horrid place. Though the soft moonlight filled it and the water in the green pool lapped gently, there was a stagnant smell, an air of corruption to remind me that a sheep's carcase lay rotting there; and the brambles snatched at my skirt as if they wished to keep me prisoner.

Exham had fallen close against the side. For an instant as I looked at the broken body and the two men bending over it, I almost turned and ran. But then my enemy raised his head and stared directly at me, and it seemed cowardly to run from the gaze of a man who was probably dying.

"Ah," he said, and his voice was stronger than I had expected. "So you have followed me down here, have you, miss? That was obliging of you, sure. This time I'll finish you for certain."

His hand was inside his coat. He dragged it out and I heard Lord Mortmain catch his breath as the moonlight glanced off the blade of the long curved knife.

I gazed wonderingly at Exham. Was his venom prompted by madness, or something more? I thought I recognized his voice as the one I had heard in the chimney of my room, for his tone was still servile, regardless of the words he used, as if he had played a part so long he could not change his ways. His face, too, I knew, now that the mask hung torn from one ear, under the black wig, for once dishevelled. He was Holloway, the gentleman who owned the black stallion and lived at the Pines, who was cared for by old Mrs. Randall and had bowed to me from Mortmain's gate as I drove out of the deer park at Lady Martha's side.

Even as I frowned at him, bewildered, he swept off the wig.

"Do you recognize me now, Chantal Fabian?" he demanded. "I knew my disguise had taken you in but it was only a matter of time before you discovered my secret. Well, miss? I thought you had long hated me for laughing at your wretched attempt to save your beloved Rowley from drowning. I thought my features would have been graven on your memory. Do you not remember? You used to call me Foxy when I followed you through the woods."

19

Lord Mortmain made a sudden movement. Exham brandished the knife, reducing us to stillness as if it were a sorcerer's wand. I could see now that his was the foxy face that had haunted my nightmares—and I had seen it elsewhere, too.

"Back, dog of a Quentin," he cried, yet still with that incongruous air of politeness. "I have waited overlong to get my hands on this girl's throat, or a knife in her heart, so I can live in peace again after all these years—and you shall not stop me."

He coughed, and the knife flashed in his hand. He spat out blood and it lay in the white dust, a small dark pool.

Lord Mortmain drew in his breath. "It is really you, Dick Exham? I had suspected that you were the man Hugo saw in the attic at Holy Mote, that you had found sanctuary there—but never that you were also Holloway, that quiet respectable man."

"How should you, my lord, when I chose to deceive you? But I remember when we were boys together—ay, I remember you well enough."

"Go on," I whispered, a sudden excitement gripping me.

"I have forgot nothing, even that I used to call you Ivo once. I remember waiting in the charcoal burner's hut for my father to come home—and I remember that it was your father's men who came instead—do you think a child would forget such a thing as that? I remember well how they told me there had been an attempt at highway robbery in the forest and that your father had shot mine—and that mine would hang. Not the one who fired the shot, no, never that. Property is more sacred than life in England, to be sure, so they hanged Black Dick—even protected as he was . . . but perhaps he is closer to his Master now." He sighed and shifted a little to turn his pain-narrowed eyes towards the Earl. Yes, it was those eyes, I thought. I had seen them glinting through the slits of masks and now I saw them plain but I still could not recall of whom they reminded me.

"Your lady mother took me in, at Horsefield," Exham went on. "How I hated her charity . . . vowed myself to revenge . . . stole from you all." He turned his head to frown at my godfather. "You are the clergyman who caught me—not the last time when I broke into your church, but six years ago. You made me sign that cursed paper and lodged it with old Medlicott . . . I had to get it back. Night after night I lay outside his window waiting for him to open that confounded safe, until the night my enemy Ivo came to make his will before a duel. I had to listen to him talking of leaving everything to a foundling brat, a child who had no claim on him—"

I exclaimed, but Mortmain shot me a warning glance and I pressed my lips tightly together.

Exham moved restlessly. "At last Medlicott persuaded him his gesture was unnecessary, though he would allow him to be one of her trustees, if Ivo should survive him—and he unlocked the safe to get out his own will to amend it . . . Are you there, Mother? I thought I heard you call. Bring me water . . . water, in Herne's name."

Mr. Shorncliffe soaked his handkerchief in the evil-looking pond and moistened Exham's lips. "I can smell the Macouba now," he murmured to me. "Curious, as soon as you mentioned Macouba I thought of Exham, though it is such a common snuff."

I looked at the dying man. "So it was you who killed

my guardian! And to think I saw you leaving the house after—Lord Mortmain!"

Exham stared at me. "You saw Ivo Rolland that night—and did not tell the world! I could not point the finger at him myself, though I wrote anonymously to his father, but I counted on your knowing him."

"Ivo Rolland," I repeated in a voice of wonder. Rowley . . . of course, but I had not recognized him then, for it had been years since I had seen him, all the years he had been in Edinburgh.

"I never knew," I whispered, "that Rowley was Lord Rolland . . . before he became Lord Mortmain."

"You did not play the part for which I let you live," Exham muttered obliviously. "You saw me, too, and were a danger to me. I tested you, to see if you recognized me—looked into your face not four weeks ago at Horsefield, near to the very gibbet where my father hanged in chains—and waited for your Rowley to bring you here. I did not waste my time, did I, Mother?" Exham's voice trembled slightly. "Are you there? It grows dark. . . . I served my apprenticeship in witchcraft as I waited for Ivo to bring the girl to Lindendfold. I led two lives here, one as the rich and sober Holloway in the dark wig who keeps a respectable establishment at the Pines—and the other as Cumber's Familiar, sandy-haired and freckled, who haunts the attics of Holy Mote and rides Herne, turn and turn about with the one who is born to be hanged—ay," he added more strongly, "and hanging's too good for him, since he must have betrayed all the secrets of our coven just for gain. Smithers has been a smuggler for years, like his father before him—and a foolhardy one, which is how he came by his nickname—but I always believed until tonight that the Old Religion came first with him. It must have been Piers Perowne who corrupted him. . . . Perowne came back to Holy Mote with time on his hands, and I thought him converted to our ways when I heard of his excitements at Oxford, but all the time he must have looked on the Wild Hunt as nothing more than a convenient way to run a cargo past the excisemen. He must have persuaded Smithers to enroll him as a freetrader—and then to change horses with him while he led his own Wild Hunt—a Wild Hunt that owed nothing to our Master's will but was for gain alone—I'll set our Wild Hunt on their heels next time I lead it . . . to dare to desecrate the stone . . . he'll regret the day he crossed me, he and Smithers. . . ."

Exham's voice died away. "All in the end are pawns in my hands," he whispered. "Ivo, who brought the girl to Holy Mote just as I knew he would—her ladyship, who raised no objection to Chantal sleeping in the room I chose, that was best fitted for my purpose—even the little baronet who carved a doll in her likeness and helped us to strengthen it with the girl's hair and handkerchief so that we could stick in the magic pins to head and shoulder and foot—and, tonight, the last one in her heart—"

"But you failed, Exham," said Lord Mortmain in a hard voice. "You failed to kill her."

"I failed," he echoed weakly. "I thought I failed because I had wasted the chance of a human sacrifice, but then why did Baal spurn her, and now Ashtaroth, in whose name I would have knifed her on the stone if it had not been for Perowne and his men, and Nergal, lord of the flood, for whom I would have drowned her there in the pool—if I had only invoked his aid when I pushed her into the moat! But they have failed us!" He twisted about, staring at the blank walls of the quarry. "Mother, where are you? Have you lost your power?"

"Your mother—" I gasped, and he looked up at me, his light eyes glinting with malevolence. "Old Nurse—your eyes are hers!"

"Ay, it was long a secret . . . they could forgive her being a witch, no doubt, but they could never have forgiven her being a mother. . . . Mother, do you hear me now? Where are you? Have you come for me?"

He coughed and fell against my godfather's shoulder. Lord Mortmain murmured to Mr. Shorncliffe and then, putting his arm about my shoulders, began to lead me away.

I took a deep breath as we started to climb the path together. The nightmare, I told myself, was over. Foxy, whose face had watched me through the reeds, was the son of Old Nurse and Black Dick, the highwayman who had danced on the stars when he was hanged. Foxy was Exham and Holloway and sometimes Herne. He was the murderer of my guardian and my own enemy. But more important even than this was the realization that Mortmain, whom I had been thinking of as the wicked Earl well fitted to wear his father's shoes, was not only Rowley, hero of my childhood, but also my secret benefactor. . . .

The Earl helped me through the fence. I felt a slight ache in the ankle I had injured. Was there something in magic, after all? I had hurt my head, my shoulder, my ankle—and I had certainly felt some kind of dagger in my heart this night.

The bay horse raised his head from the dewy grass and whinnied softly. Lord Mortmain stood still and turned me round to face him. His eyes seemed to gleam in the moonlight with the strange power I had associated with him from the first.

"Chantal—"

I said quickly, defensively, "I thought Mr. Shorncliffe was my benefactor—he allowed me to think so, I suppose, lest I put too much dependence on the hope that it was my grandfather, who had been responsible for me. But now I must hasten to thank you, my lord, for your maintenance of me all these years. I do not know how to begin to thank you, indeed. The sense of obligation that I feel—"

"The Devil!" exclaimed the Earl, dropping his arms and beginning to pace about. "This is what Medlicott was afraid of—this is why he swore me to secrecy and made me promise never to remind you of the past unless you remembered it yourself, or fell in love with me—not that he knew quite how beholden to me circumstances would force you to feel, admittedly, but that he was afraid the past would fetter you. He did not want you to feel obliged to me, in a word; and God knows, I certainly do not. Let me inform you that your debt is paid. I have had my reward for anything I chose to do for you in those disapproving Reports of Mrs. Mason's which described the Chantal I had known as a child—though I was well aware that the child was growing fast to womanhood. Besides," he added, coming back to stand beside me again, "don't you know we belong together?"

I gazed up at him, my feelings no doubt plain to see in the moonlight.

"I suspected you!" I said in a low voice.

"I am not surprise at it. You saw me in a compromising situation that night six years ago. Good God, when I reflect how careful I was to keep that assignation with your guardian secret—for I had no desire for my father to get wind of what I was about—and that I even bribed the groom at the livery stables not to reveal that I was there that night, I consider that I am lucky not to have hanged for the crime. I probably would have been hanged if I had not been so badly wounded a few hours later that I lost all recollection of my meeting with Medlicott until my father questioned me about it." He frowned. "My startled and guilty response to his questions must have contributed largely to my father's death, I fear."

I put my hand on his arm. "Exham was to blame, not you."

He caught his breath. "Chantal, I love you! And have done so for years, though I have tried to tell myself a thousand times that my love for you had more to do with youth and summer days than reality. I have tried a thousand ways to forget you—and all the time I feared that when we did come face to face you would recoil from me, as indeed you did—"

"It was the memory—the association with my guardian's death—and not the scar—" I managed to say.

"I know that now," he assured me. "Don't you want to know when I finally accepted that the absurd fancy was fact? It was when I unpacked your miniature and looked upon your face again. Then you caught the measles and

I thought you would die before you could be mine. That was when I determined to wait no longer and devised my plot to bring you here. But it was a devilish task to keep my word to Medlicott and treat you coolly, as a stranger. I was half-afraid at first you might have had a hint of my identity from Mary Bowers and, knowing me, found you disliked the man who had sprung from the boy you were once not indifferent to." He smiled into my eyes. "I told my mother the whole tale this spring, when I was staying with her in the house at Horsefield where I passed all my boyhood; you met her once or twice, though you may not remember it, and though she is a frail invalid now she longs to meet you again—as my wife, Chantal."

"You do me too much honour, sir," I felt bound to protest, while my heart thundered in recognition of just what kind of power it was that this particular Quentin exercised over me so strongly.

"How dare you say that!" he cried. "The Fabians can trace back their ancestry farther than the Quentins—but you do not know perhaps that I have made it my business to become acquainted with your father's family—and a stiff-necked tribe they are, I assure you. But I can prophesy that your grandfather will attend our wedding and give us his blessing, if you want him to do so after his abominable neglect of you. Or if it is a lack of dowry that troubles you, I am sure there must be diamonds in Exham's house that will eventually be proved to be yours. Now, will your pride allow you to accept my hand and heart, or do I have to beat you into submission?"

I leaned my face against his shirt. "You may do so if you wish, my lord, but I assure you there is no need for it. I was only saying what Mrs. Mason instructed us was proper in such a case, but if you want me to forget all my expensive schooling, I suppose no one has a better right."

"Yes," he said in an odd tone, turning my face up with his hand and studying it intently. "I think you will do well to forget it."

He bent his head and, putting his lips to mine, proceeded to further my education along very different lines from those explored at Horsefield House.

It was some time later that, the Earl having given a gape-mouthed stable boy instructions to take a hurdle to the quarry, we approached the front of the house and saw the doctor's gig being driven to the stables by his groom, just as a small white-clad figure flung itself upon me.

"Oh, Miss Fabian, I am so glad you are all right—"

"I have you to thank for that, Sir Hugo, it seems."

"But Old Nurse, Miss Fabian—she's dead!"

"Dead?" I echoed. I held him closely, recalling Exham's last remarks. "I am sorry, Hugo. It is very distressing, but remember that she was old and ill—"

"You don't understand. It was all my fault. Dr. Burton said she would not be able to stand another shock—and I gave her one." I felt him shaking in my arms.

"You, Hugo? When you left the house, do you mean?"

"No," he said indignantly. "She told me to do that herself. She said I must watch for you to pass the end of the passage and then I could try my boat on the quarry pond. It was when I came back—Cousin Mortmain said I must tell her I was safe, but not what I had seen and done—but —but when I went to her room she did not hear the knock. I opened the door and she—oh, Miss Fabian!" He pressed his face against me.

"Go on, Hugo," the Earl insisted. "What happened then?"

"It was horrible. She was holding the figure she had made me carve of Miss Fabian. It was quite a good one, with hair and everything, dressed in one of her handkerchiefs. But Old Nurse—she had a long pin in her hand—she was holding it like a dagger—she pushed it into your chest, Miss Fabian—"

He twisted his head and stared up into my face as if

to assure himself that I had come to no harm. "I gave a shout," he confessed, "and she—she just fell down in the bed. It was horrible! I ran to Mama and she went in to her—but she came out after a while and told me Old Nurse was dead."

"Hugo," I said firmly, "Old Nurse may have had a fright at you catching her when she thought she was alone but I don't believe that was why she died. I think the time had come for her to die because she had just lost the only person in the world who really needed her. You must not blame yourself."

"No," said Lord Mortmain over my shoulder. "No, Hugo, there is no need for you to do that. Her race was run."

Hugo looked from one of us to the other and his eyes widened. Before he could voice whatever thought had entered his head I said quickly, "Now, sir, I have some news that will interest you. Do you remember that I told you of a boy who rode a black horse—a *blood* horse—and carved wonderful boats to sail on a green and golden lake? Your cousin Lord Mortmain was that very boy, Sir Hugo."

There was a moment's silence. "You said his name was Rowley," Hugo pointed out.

"Lord Rolland was his courtesy title—I must have got it from that."

Sir Hugo looked at his cousin, and decided to believe it. "Will you teach me how to make boats? I can do the hulls, but it is hard to get the rigging just as it should be—you will, won't you, sir?"

"With pleasure, Hugo, when I can find the time—"

I dared to interrupt him. "Why, sir, what else can occupy your time that is of greater importance, since you know what boat-making means to a boy? And I have a great debt to Hugo, who probably saved my life this evening, so I shall be happy to help your secretary, while you teach him."

The Earl turned and put the reins over his horse's

head. "We have both a great debt to Hugo, but there are one or two matters to settle first. I have to divorce my cousin Piers from his dangerous occupation, secure his betrothal to Miss Curtis, persuade Smithers the carter that Wild Hunts will be no longer tolerated at Lindenfold for whatever purpose, urge my decorators to complete their work—and bully my bride to agree to an immediate wedding."

"The latter task at least will not delay you long, I venture to predict. As for the others—"

The Earl swiftly turned me to face him. " 'Impulsive, quick-tempered, and impetuous!' " he exclaimed.

"You have it on the best authority," I admitted. "Also: 'of a somewhat passionate character,' " I reminded him softly.

I heard him catch his breath. Then his hands were firm about my waist as he lifted me on to the saddle-bow.

"What are you doing, Cousin?" Hugo cried, as the Earl swung himself into place behind me. "Where are you going?"

"We are going for a gallop, Hugo; while you, I fear, must go back to bed."

But Hugo was not to be distracted. "A gallop, sir?"

"Yes, my boy—such a gallop as was never seen. I am taking this young lady riding on a Wild Hunt of our own —the last of all the Wild Hunts at Lindenfold."

I leaned back a little in Rowley's arms and felt his warm breath on my hair, which I had never done in dreams. Then the air rushed past us as the great horse leaped forward along the drive under the dark trees, and turned down the chalky road that ran white in the moonlight, behind the village to the sea.